The Toxicologist's Pocket Handbook

Third Edition

T0141211

The Toxicologist's Pocket Handbook

Third Edition

Michael J. Derelanko

CRC Press

Taylor & Francis Group

Boca Raton London New York

CRC Press is an imprint of the
Taylor & Francis Group, an **informa** business

CRC Press
Taylor & Francis Group
6000 Broken Sound Parkway NW, Suite 300
Boca Raton, FL 33487-2742

© 2018 by Taylor & Francis Group, LLC
CRC Press is an imprint of Taylor & Francis Group, an Informa business

No claim to original U.S. Government works

Printed at CPI on sustainably sourced paper

International Standard Book Number-13: 978-1-138-62640-9 (Paperback)

Visit the Taylor & Francis Web site at
http://www.taylorandfrancis.com

and the CRC Press Web site at
http://www.crcpress.com

To all who use this book,
may they always find what they are looking for.

Contents

Preface

Toxicologists rely on a large information base to design, conduct and interpret toxicology studies, and to perform risk assessments. Reference books such as the *CRC Handbook of Toxicology* kept in the toxicologist's office supply ready access to this information. However, reference books of this nature tend to be quite large in size and are not easily carried in a briefcase. The goal in producing the first edition of *The Toxicologist's Pocket Handbook* was to provide a small, easily carried reference source of basic toxicological information for toxicologists and other health and safety professionals. This third edition of the *Toxicologist's Pocket Handbook* continues to meet the need of the toxicologist traveling outside the lab or office for toxicology reference information in a convenient pocket-size format.

As with previous editions, the third edition of *The Toxicologist's Pocket Handbook* contains selected tables and figures from the larger *CRC Handbook of Toxicology* of the most frequently used toxicology reference information. A glossary of commonly used toxicological terms is also included. As with the larger handbook, this book has been designed to allow basic reference information to be located quickly. Tables and figures have been placed in sections specific to various subspecialties of toxicology. The detailed table of contents contains a listing of all of the tables and figures contained in the book. This expanded edition features 35 tables not found in the second edition added to the sections on lab animals, general toxicology, dermal and ocular toxicology, genetic toxicology/carcinogenesis, neurotoxicology, immunotoxicology, reproductive/developmental toxicology, industrial chemical and pharmaceutical toxicology. New information is presented for additional laboratory animals such as swine and primates, infusion recommendations, newer methods such as the local lymph node assay, reference safety pharmacology values for standard species and more current classification of chemicals based on the globally harmonized system (GHS). Expanded information on typical genetic toxicology and immunotoxicology assays as well as *in vitro* assays for eye irritation has been added. Some tables from the second edition have been updated to include new information that has arisen since the earlier edition went to press. While some information from the second edition has been deleted such as regulatory requirements that are no longer applicable, other outdated material such as previously used chemical hazard classifications has been retained for historical purposes to provide the reader with a source of information that may be needed to understand and interpret older reports and publications and to allow comparison with current practices. As always, the reader is advised to verify that any information provided in this handbook is still relevant for his/her current situation and needs.

The reader may occasionally note that there may be slight disagreement between data presented in a table in one section with similar data presented in other sections. For example, a range given for a hematological parameter such as erythrocyte count may differ slightly in tables in different sections. Such information was obtained from different sources and is presented for different comparative purposes. The differences represent normal laboratory variation and the information is presented solely for the purpose of making relative comparisons. Therefore, as to not compromise the relative relationships being presented, no attempt was made to standardize the information between tables. As with previous editions, although the information presented was obtained from reliable and respected sources, I cannot attest to the accuracy and/or completeness of such information and cannot assume any liability of any kind resulting from the use of or reliance on the information provided. Mention of vendors, trade names, or commercial products does not constitute endorsement or recommendation for use.

Acknowledgments

The author recognizes the following individuals who contributed or provided critical review of the information included in this book or gave assistance in other ways toward the publication of *The Toxicologist's Pocket Handbook*:

M. B. Abou-Donia, C. S. Auletta, T. J. Baird, W. H. Baker, K. L. Bonnette, D. M. Brecha, W. J. Brock, L. A. Buckley, F. G. Burleson, G. R. Burleson, J. A. Dalton, J. Dolgin, D. A. Douds, B. J. Dunn, D. J. Ecobichon, H. C. Fogle, D. V. Gauvin, S. J. Hermansky, J. W. Harbell, R. M. Hoar, G. M. Hoffman, M. A. Hollinger, R. V. House, B. Jacob, V. J. Johnson, S. A. Kutz, H. Lameris, B. S. Levine, K. M. MacKenzie, R. C. Mandella, T. N. Merriman, M. A. Morse, P. E. Newton, J. Neis, R. M. Parker, J. C. Peckham, W. J. Powers, H. Raabe, R. E. Rush, G. M. Rutledge, A. M. Sargeant, G. E. Schulze, J. C. Siglin, S. A. Smith, G. P. Thomas, P. T. Thomas, S. Wilder, and R. R. Young.

Author

Michael J. Derelanko, PhD, DABT, Fellow ATS, has spent most of his career as a toxicologist in the pharmaceutical and chemical industries. Dr. Derelanko earned a BS at Saint Peter's University (formerly Saint Peter's College). He was a National Institutes of Health pre-doctoral trainee in the Albert S. Gordon Laboratory of Experimental Hematology at New York University, earning an MS and a PhD. He was a recipient of the New York University Gladys Mateyko Award for Excellence in Biology. Following a two-year postdoctoral fellowship in gastrointestinal pharmacology at Schering-Plough Corporation, he began his career in industrial toxicology as a research toxicologist in the laboratories of Allied Chemical Corporation in New Jersey which eventually became AlliedSignal and later Honeywell International. Over a 20-year period, Dr. Derelanko held various positions of increasing responsibility, becoming corporate manager of toxicology and risk assessment. More recently he was director of nonclinical drug safety assessment for Adolor Corporation in Pennsylvania where he contributed to the discovery and development of novel drugs for nearly 10 years. Dr. Derelanko is currently employed by Critical Path Services, a Knoell company, located in Garnet Valley, Pennsylvania, where he is a group leader consulting in toxicology and risk assessment.

Dr. Derelanko is a Diplomate of the American Board of Toxicology and a Fellow of the Academy of Toxicological Sciences. He is a member of the Society of Toxicology, the Society for Experimental Biology and Medicine, and the honorary research society Sigma Xi. He has served on the content advisory committee of the New Jersey Liberty Science Center, has chaired or been a member of industrial and government toxicology advisory committees, was actively involved with the Chemical Industry Institute of Toxicology, and has served on the speaker's bureau of the New Jersey Association for Biomedical Research. Dr. Derelanko is an adjunct professor in the Department of Pharmacology and Physiology of Drexel University College of Medicine.

Dr. Derelanko has authored numerous papers in experimental hematology, gastrointestinal pharmacology, and toxicology. He has been actively involved in educating the public about toxicology, particularly at the middle school level. He has delivered invited lectures on this subject at national meetings. In 2003, Dr. Derelanko received the Society of Toxicology award for Contributions to the Public Awareness of the Importance of Animals in Toxicology Research. He is the coeditor of all three editions of the *CRC Handbook of Toxicology* and author of both earlier editions of *The Toxicologist's Pocket Handbook*.

1

Lab Animals

TABLE 1

Guiding Principles in the Use of Animals in Toxicology

1. The use, care, and transportation of animals for training and for toxicological research and testing for the purpose of protecting human and animal health and the environment must comply with all applicable animal welfare laws.

2. When scientifically appropriate, alternative procedures that reduce the number of animals used, refine the use of whole animals, or replace whole animals (e.g., *in vitro* models, invertebrate organisms) should be considered.

3. For research requiring the use of animals, the species should be carefully selected and the number of animals kept to the minimum required to achieve scientifically valid results.

4. All reasonable steps should be taken to avoid or minimize discomfort, distress, or pain of animals.

5. Appropriate aseptic technique, anesthesia, and postoperative analgesia should be provided if a surgical procedure is required. Muscle relaxants or paralytics are not to be used in place of anesthetics.

6. Care and handling of all animals used for research purposes must be directed by veterinarians or other individuals trained and experienced in the proper care, handling, and use of the species being maintained or studied. Veterinary care is to be provided in a timely manner when needed.

7. Investigators and other personnel shall be qualified and trained appropriately for conducting procedures on living animals, including training in the proper and humane care and use of laboratory animals.

8. Protocols involving the use of animals are to be reviewed and approved by an IACUC before being initiated. The composition and function of the committee shall be in compliance with applicable animal welfare laws, regulations, guidelines, and policies.

9. Euthanasia shall be conducted according to the most current guidelines of the American Veterinary Medical Association (AVMA) Panel on Euthanasia or similar bodies in different countries.

Source: From Society of Toxicology (1999). With permission.

TABLE 2

Guiding Principles for Humane Treatment of Animals in Toxicology Studies

- Animals should be considered to experience comparable pain and distress in situations and procedures where pain and distress would occur in humans.
- Death, severe pain, and distress should be avoided where possible.
- Studies should be designed so that animals experience no more pain and distress than necessary to achieve the scientific objectives of the study.
- The duration of the study should be no longer than necessary to achieve its objectives.
- Specific endpoints for humanely terminating an animal should be established prior to the start of the study based on previous experience with the material being tested. Input should come from scientists, veterinarians, animal care personnel, and IACUC or similar ethical review board.
- Staff should be properly trained to recognize signs of pain and distress in the species being investigated.
- Clear roles, responsibilities, and authority should be established for making decisions on humane termination of an animal.
- Animals should be monitored with sufficient frequency to allow termination as soon as possible when humane endpoints have been reached.

Source:　　OECD (2000), Demers, G. et al. (2006).

TABLE 3

Signs Indicative of Pain, Suffering, and Distress in Animals[a]

Behavioral
- Vocalization in animals that do not normally vocalize on handling
- Writhing
- Tremors
- Lethargy, decreased activity, and reluctance to move
- Abnormal aggressive behavior, restlessness, agitation, and abnormal reaction to handling
- Wary or overly cautious behavior
- Licking, scratching, chewing on body part, or self-mutilation

Appearance
- Porphyrin staining around eyes and nares—rats

Movement
- Gait irregularities, non-weight-bearing movements

Physical Signs
- Rapid or labored respiration
- Excessive salivation
- Changes in posture/hunched posture
- Piloerection, unkempt appearance
- Reduced food/water consumption (loss of body weight or evidence of dehydration)

Source: Hawkins, P. (2002), OECD (2000).

[a] Individual signs may result from physiological/pharmacological mechanisms not associated with pain. The animal should be assessed in context with the study, design, and procedures performed on the animal.

TABLE 4

Signs of Moribundity[a] as Criteria for Humane Sacrifice[b]

- Prolonged impairment of locomotion, preventing access to food and water
- Sustained 10% decrease in body temperature
- Continued severe diarrhea
- Excessive weight loss/emaciation (>25% over 7 days or more)
- Eyes fixed or sunken
- Severe dehydration
- Significant blood loss from any orifice
- Evidence of irreversible major organ failure
- Prolonged absence of voluntary responses to external stimuli
- Prolonged inability to remain upright
- Persistent convulsions
- Continued difficult, labored breathing
- Self-mutilation

Source: Adapted from OECD (2000).

[a] State of dying or inability to survive.
[b] This list is not all-encompassing. A decision to sacrifice an animal may rely on the assessment of the general condition of the animal by a trained professional.

TABLE 5

General Information Sources for the Care and Use of Research Animals

1. *Public Health Service Policy on Humane Care and Use of Laboratory Animals.* PHS (Public Health Service), 1996, U.S. Department of Health and Human Services, Washington, DC 22 pp. [PL 99-158. Health Research Extension Act, 1985].

2. The Animal Welfare Act of 1966 (P.L. 89-544) as amended by the Animal Welfare Act of 1970 (P.L. 91-579); 1976 Amendments to the Animal Welfare Act (P.L. 94-279); the Food Security Act of 1985 (P.L. 99-198), Subtitle F (Animal Welfare File Name: PL99198); and the Food and Agriculture Conservation and Trade Act of 1990 (P.L. 101-624), Section 2503, Protection of Pets (File Name: PL 101624). Rules and regulations pertaining to implementation are published in the Code of Federal Regulations, Title 9 (Animals and Animal Products), Chapter 1, Subchapter A (Animal Welfare). Available from Regulatory Enforcement and Animal Care, APHIS, USDA, https://www.aphis.usda.gov/aphis/ourfocus/animalwelfare

3. *Guide for the Care and Use of Laboratory Animals.* Institute of Laboratory Animal Resources, Commission on Life Sciences, National Research Council, National Academy Press, Washington, DC, 1996 or succeeding revised editions. www.nap.edu/readingroom/books/labrats

4. *International Guiding Principles for Biomedical Research Involving Animals.* Council for International Organizations of Medical Sciences (CIOMS), Geneva, 1985.

5. *Interdisciplinary Principles and Guidelines for the Use of Animals in Research, Testing, and Education.* Ad Hoc Animal Research Committee, New York Academy of Sciences, 1988.

6. *Recognition and Alleviation of Pain and Distress in Laboratory Animals. A report of the Institute of Laboratory Animal Resources Committee on Pain and Distress in Laboratory Animals.* NCR (National Research Council). Washington, DC: National Academy Press, 1992.

7. *Education and Training in the Care and Use of Laboratory Animals: A Guide for Developing Institutional Programs.* AVMA (American Veterinary Medical Association). Report of the AVMA panel on euthanasia. *J. Am. Vet. Med. Assoc.* 218(5), 669–696, 2001.

8. *Guide to the Care and Use of Experimental Animals.* CCAC (Canadian Council on Animal Care), Vol. 1, 2nd ed., Edited by E. D. Olfert, B. M. Cross, and A. A. McWilliam. Ontario, Canada: Canadian Council on Animal Care, 1993. 211 pp.

9. *European Convention for the Protection of Vertebrate Animals Used for Experimental and Other Scientific Purposes.* Council of Europe. ETS No. 123, 1986—conventions.coe.int/treaty/en/treaties/html/123.htm

Source: Compiled by the Society of Toxicology.

TABLE 6

Approximate Daily Food and Water Requirements
for Various Species

Species	Daily Food Requirement	Daily Water Requirement
Mouse	3–6 g	3–7 mL
Rat	10–20 g	20–30 mL
Hamster	7–15 g	7–15 mL
Guinea pig	20–30 g[a]	12–15 mL/100 g
Rabbit	75–100 g	80–100 mL/kg
Cat	100–225 g	100–200 mL
Dog	250–1200 g	100–400 mL/day
Primate	40 g/kg[a]	350–1000 mL

[a] Like humans, guinea pigs and nonhuman primates require a continuous supply of vitamin C (ascorbic acid) in the diet.

TABLE 7

Common Stocks and Strains of Laboratory Mice

Strain	Description
CD-1 Mice	A multipurpose outbred albino strain descended from "Swiss" mice. Variants include the athymic CD-1 Nude Mouse often used in tumor xenograft research.
CF-1 Mice	Inbred from albino mice of likely wild origin for over 20 generations and then outbred to produce the current stock. A multipurpose model often used in infectious disease research.
Swiss–Webster Mice	A multipurpose outbred albino strain from selective inbreeding of Swiss mice by Dr. Leslie Webster.
SKH1 (Hairless) mice	Outbred strain that originated from an uncharacterized strain. Although the SKH-1 mouse resembles the nude mouse in appearance, this strain is euthymic and immunocompetent. Research uses include wound healing, dermatology, and UV-induced tumorigenesis.
BALB/c mice	An inbred albino strain developed originally by H.J. Bagg (Bagg albino). Frequently used for monoclonal antibody production, hybridoma development, and infectious disease research. Variants include the BALB/c Nude Mouse used for tumor xenografts and tumor biology.
C3H mice	An inbred agouti strain developed originally from the cross of a "Bagg albino" female and a DBA male. Used for oncology, neurologic research, and to study spontaneous retinal degeneration.
C57BL/6 mice	An inbred black strain developed originally by C.C. Little. A multipurpose model often used as the parental background strain for transgenic and knockout models. Variants include the Pound Mouse, isolated from a C57BL/6 colony at Charles River and used in research on diabetes and obesity.
DBA/2 mice	The oldest of all inbred mouse strains, the DBA/2 mouse is an inbred nonagouti, dilute brown strain developed originally by C.C. Little. A general model often used in immunology and experimental epilepsy.
FVB mice	An inbred albino strain derived originally from an outbred Swiss colony. Used primarily for transgenic and knockout model development.

(Continued)

TABLE 7 (*Continued*)

Common Stocks and Strains of Laboratory Mice

Strain	Description
AKR mice	An inbred albino strain originally developed by Furth as a high leukemia strain. Used for cancer, atherosclerosis, and metabolic research.
B6C3F1 mice	A hybrid agouti strain from female C57BL/6N × male C3H/He mice. Used by the National Toxicology Program as the mouse strain for carcinogenicity studies.
Notes:	The two transgenic lines that follow (i.e., the rasH2 and the p53 mouse) have been commonly used as alternatives to the standard 2-year mouse carcinogenicity study. Not listed below, the Tg.AC mouse (carrier of an activated mouse H-*ras* oncogene) was used as an alternative to the 2-year mouse carcinogenicity study for dermal products but is now seldom used due to a high incidence of false positives.
rasH2 mouse	The *ras* oncogenes are involved in approximately one-third of human cancers. The rasH2 mouse is produced on a CB6F1 mouse background and has the human c-Ha-*ras* oncogene in addition to its native murine Ha-ras proto-oncogene. Due to its sensitivity to both genotoxic and nongenotoxic compounds and its low spontaneous tumor incidence at 6 months, this transgenic model has become the primary replacement for the standard 2-year mouse carcinogenicity assay.
p53$^{+/-}$ mouse	The p53 tumor suppressor gene, often mutated or deleted in human and rodent tumors, is critical to cell cycle control and DNA repair The heterozygous version of the p53 model has one functional wild-type p53 allele and one inactivated allele, imparting sensitivity to the mutational and carcinogenic effects of genotoxic chemicals. The model tends to be rather insensitive, and most of the studies conducted in this model for regulatory purposes have yielded negative results.

TABLE 8

Common Stocks and Strains of Laboratory Rats

Strain	Description
Sprague–Dawley rats	Outbred albino strain originated by R.W. Dawley from a hybrid hooded male and female Wistar rat. This designation includes variants such as the CD®IGS rat. In general, the Sprague–Dawley rat and its variants constitute a hardy multipurpose animal model with a rapid growth rate.
Wistar Han rats	Outbred albino strain originally bred at the Hannover Institute as a distinct line divergent from the strain that originated at the Wistar Institute. This strain is preferred for carcinogenicity studies due to its smaller body size, lower spontaneous tumor incidence, and greater survival.
Long–Evans rats	Outbred white with black or occasional brown hood; originated by Drs. Long and Evans by cross of white Wistar females with wild gray male. A multipurpose model often used in obesity and behavioral research.
Zucker Diabetic Fatty rats	Inbred line derived from outbred obese Zucker rats used primarily for research on diabetes and obesity.
Fischer 344 (F-344) rats	Multipurpose inbred albino strain originated from mating #344 of rats obtained from local breeder (Fischer).
Lewis rats	Inbred albino strain originally developed by Dr. Lewis from Wistar stock. Principal research uses include induced anemia, induced arthritis/inflammation, induced type I diabetes, and transplantation research.
Wistar Kyoto (WKY) rats	Inbred albino strain originated from outbred Wistar stock from Kyoto School of Medicine. Used as a control for the SHR rat and ADHD model.
Spontaneously hypertensive (SHR) rats	Inbred albino strain developed from Wistar Kyoto rats with spontaneous hypertension. Used in hypertension research and as an animal model for ADHD. Variants include the Spontaneously Hypertensive Heart Failure (SHHF) Rat, the Spontaneously Hypertensive Obese (SHROB) Rat, and the Spontaneously Hypertensive Stroke-Prone (SHRSP) Rat.

TABLE 9

Physical and Physiological Parameters of Mice

Life span	1–2 years
Male adult weight	20–35 g
Female adult weight	20–35 g
Birth weight	1.0–1.5 g
Adult food consumption	3–6 g/day
Adult water consumption	3–7 mL/day
Male breeding age/weight	6–8 weeks/20–35 g
Female breeding age/weight	6–8 weeks/20–30 g
Placentation	Discoidal endotheliochorial
Estrous cycle	4–5 days (polyestrous)
Gestation period	19–21 days
Weaning age/weight	21 days/8–12 g
Average litter size	10–12 pups
Mating system(s)	1:1 or 1 male to multiple females
Adult blood volume	6%–7% of body weight
Maximum safe bleed	7–8 mL/kg
Red cell count	$7–12 \times 10^6/mm^3$
White cell count	$3–12 \times 10^3/mm^3$
Hemoglobin	13–17 g/dL
Hematocrit	40%–54%
Mean corpuscular volume	43–54
Mean corpuscular hemoglobin	13–18
Mean corpuscular hemoglobin concentration	31–34
Platelet count	$1000–1600 \times 10^3/mm^3$
Heart rate	300–600 beats/min
Respiration rate	90–180 breaths/min
Rectal temperature	37.5°C
Urine pH	6.0–7.5
Urine volume	1–3 mL/day
Chromosome number	$2n = 40$

Source: Evans, I.E. and Maltby, C.J. (1989), Williams, C.S.F. (1976), LAMA (1988).

TABLE 10

Physical and Physiological Parameters of Rats

Life span	2–3 years
Male adult weight	350–400 g
Female adult weight	180–200 g
Birth weight	5–6 g
Adult food consumption	10–20 g/day
Adult water consumption	20–30 mL/day
Male breeding age/weight	10–12 weeks/300–350 g
Female breeding age/weight	8–10 weeks/200–300 g
Placentation	Discoidal hemochorial
Estrous cycle	4–5 days (polyestrous)
Gestation	20–22 days
Weaning age/weight	21 days/35–45 g
Average litter size	10–12 pups
Mating system(s)	1:1 or 1 male to multiple females
Adult blood volume	6%–7% of body weight
Maximum safe bleed	5–6 mL/kg
Red cell count	$6-10 \times 10^6/\text{mm}^3$
White cell count	$7-14 \times 10^3/\text{mm}^3$
Hemoglobin	11–18 g/dL
Hematocrit	34%–48%
Mean corpuscular volume	50–65
Mean corpuscular hemoglobin	19–23
Mean corpuscular hemoglobin concentration	32–38
Platelet count	$800-1500 \times 10^3/\text{mm}^3$
Heart rate	250–500 beats/min
Respiration rate	80–150 breaths/min
Rectal temperature	37.5°C
Urine pH	6.0–7.5
Urine volume	10–15 mL/day
Chromosome number	$2n = 42$

Source: Evans, I.E. and Maltby, C.J. (1989), Williams, C.S.F. (1976), LAMA (1988).

TABLE 11

Physical and Physiological Parameters of Dogs

Life span	12–14 years
Male adult weight	6–25 kg
Female adult weight	6–25 kg
Birth weight	300–500 g
Adult food consumption	250–1200 g/day
Adult water consumption	100–400 mL/day
Breeding age (males)	9–12 months
Breeding age (females)	10–12 months
Estrous cycle	Biannual, monestrous
Gestation	56–58 days
Weaning age	6–8 weeks
Litter size	4–8
Mating	Pairs, 1 male to multiple females
Adult blood volume	8%–9%, 75–110 mL/kg
Maximum safe bleed	8–10 mL/kg
Red cell count	$5.5–8.5 \times 10^6/mm^3$
White cell count	$6–14 \times 10^3/mm^3$
Hemoglobin	13–18 g/dL
Hematocrit	38%–52%
Platelet count	$200–600 \times 10^3/mm^3$
Heart rate	80–140 beats/min
Respiration rate	10–30 breaths/min
Rectal temperature	38.5°C
Urine pH	7.0–7.8
Urine volume	25–45 mL/kg
Chromosome number	$2n = 78$

Source: Evans, I.E. and Maltby, C.J. (1989), LAMA (1988).

TABLE 12

Physical and Physiological Parameters of Rabbits

Life span	5–7 years
Male adult weight	4.0–5.5 kg
Female adult weight	4.5–5.5 kg
Birth weight	90–110 g
Adult food consumption	75–100 g
Adult water consumption	80–100 mL/kg body weight
Dietary peculiarities	Pelleted diet
Male breeding age/weight	6–7 months/3.5–4.0 kg
Female breeding age/weight	5–6 months/4.0–4.5 kg
Placentation	Discoidal hemoendothelial
Estrous cycle	Polyestrous, induced
Gestation	30–32 days
Weaning age/weight	6–7 weeks/1.0–1.5 kg
Litter size	4–12
Mating system(s)	1:1 or via artificial insemination
Adult blood volume	6% of body weight
Maximum safe bleed	6.5–7.5 mL/kg
Red cell count	$4.5–7.0 \times 10^6/mm^3$
White cell count	$5–12 \times 10^3/mm^3$
Hemoglobin	11–14 g/dL
Hematocrit	32%–48%
Mean corpuscular volume	58–72
Mean corpuscular hemoglobin	18–24
Mean corpuscular hemoglobin concentration	30–35
Platelet count	$250–750 \times 10^3/mm^3$
Heart rate	250–300 beats/min
Respiration rate	35–55 breaths/min
Rectal temperature	39.5°C
Urine pH	8.2
Urine volume	50–130 mL/kg
Chromosome number	$2n = 44$

Source: Evans, I.E. and Maltby, C.J. (1989), Williams, C.S.F. (1976), LAMA (1988).

TABLE 13

Physical and Physiological Parameters of Guinea Pigs

Life span	4–6 years
Male adult weight	1000–1200 g
Female adult weight	850–900 g
Birth weight	90–120 g
Adult food consumption	20–30 g/day
Adult water consumption	12–15 mL/100 g
Dietary peculiarities	Vitamin C required to avoid scurvy
Male breeding age/weight	11–12 weeks/600–700 g
Female breeding age/weight	7–8 weeks/350–450 g
Placentation	Discoidal hemochorial
Estrous cycle	16–18 days
Gestation	65–70 days
Weaning age/weight	7–14 days/150–200 g
Litter size	2–5
Mating	1M:1F or 1M:10F
Adult blood volume	6%–7% body weight
Maximum safe bleed	7–8 mL/kg
Red cell count	$4.5–7 \times 10^6/mm^3$
White cell count	$5–15 \times 10^3/mm^3$
Hemoglobin	11–17 g/dL
Hematocrit	39%–47%
Platelet count	$250–750 \times 10^3/mm^3$
Heart rate	230–300 beats/min
Respiration rate	60–110 breaths/min
Rectal temperature	39.5°C
Urine pH	8.0–9.0
Urine volume	15–75 mL/day
Chromosome number	$2n = 64$

Source: Evans, I.E. and Maltby, C.J. (1989), Williams, C.S.F. (1976), LAMA (1988).

TABLE 14

Physical and Physiological Parameters of
Miniature Swine[a]

Life span	10–15 years
Adult weight	12–55 kg
Birth weight	0.4–0.7 kg
Adult food consumption	1.0–1.5 kg/day
Adult water consumption	80–120 mL/kg/day
Minimum breeding age	5–6 months
Estrous cycle (days)	14
Gestation	114 days
Weaning age	4–5 weeks
Litter size	5–8
Red cell count	$5.3–9.25 \times 10^6/mm^3$
White cell count	$4.4–26.4 \times 10^3/mm^3$
Hemoglobin	9.0–15.8 g/dL
Hematocrit	32%–61%
Platelet count	$148–898 \times 10^3/mm^3$
Heart rate	83 ± 15 beats/min
Respiration rate	20 ± 9 breaths/min
Body temperature	37°C–38°C
Chromosome number	$2n = 38$

Source: Bollen, P.J.A. et al. (2000).
[a] For Gottingen and Sinclair minipigs and Yucatan micropigs.

TABLE 15

Physical and Physiological Parameters of Old World Monkeys

Parameter	*Macaca fascicularis*	*Macaca mulatta*
Adult male weight	4–8 kg	6–11 kg
Adult female weight	2–6 kg	4–9 kg
Respiratory rate	30–54 breaths/min	30–50 breaths/min
Heart rate	115–243 bpm	92–122 bpm
Rectal temperature	37°C–39°C (98.6°F–103.1°F)	37°C–39°C (98.6°F–103.1°F)
Daily food consumption	350–550 g	400–600 g
Daily water consumption	350–550 mL	400–600 mL
Urinary excretion	150–550 mL/day	Not listed
Blood volume	55–75 mL/kg	50–96 mL/kg
Red cell count	$5.3–6.3 \times 10^6$/mL	$5.1–5.6 \times 10^6$/mL
White cell count	$6.1–12.5 \times 10^3$/mL	$4.2–8.1 \times 10^3$/mL
Hemoglobin	11.0–12.4 g/dL	12.0–13.1 g/dL
Hematocrit	33%–75%	37%–40%
Platelet count	$300–512 \times 10^3$/mL	$260–361 \times 10^3$/mL
Chromosome number	$2n = 42$	$2n = 42$

Source: Fortman, J.D. et al. (2002).

TABLE 16

Physical and Physiological Parameters of New World Monkeys

Parameter	*Saimiri sciureus*	*Callithrix jacchus*
Adult male weight	0.9–1.1 kg	0.3–0.5 kg
Adult female weight	0.7 kg	0.3–0.4 kg
Respiratory rate	55–58 breaths/min	Not listed
Heart rate	215–263 bpm	194–242 bpm
Rectal temperature	39°C–40°C (101.7°F–103.6°F)	37°C–39°C (98.2°F–101.5°F)
Daily food consumption	45–60 g	20 g plus fruit
Red cell count	$6.3–7.1 \times 10^6$/mL	$4.6–6.6 \times 10^6$/mL
White cell count	$6.0–9.1 \times 10^3$/mL	$4.9–11.3 \times 10^3$/mL
Hemoglobin	12.2–13.6 g/dL	12.6–19.6 g/dL
Hematocrit	39%–44%	32%–54%
Platelet count	Not listed	$180–382 \times 10^3$/mL
Chromosome number	$2n = 44$	$2n = 46$

Source: Fortman, J.D. et al. (2002).

TABLE 17

Body Weight and Food Consumption—CD-1 Mice[a]

Week	Males		Females	
	Body Weight (g)	Food Consumption (g/kg/day)	Body Weight (g)	Food Consumption (g/kg/day)
−1	23.64	—	19.24	—
0	27.46	235.55	21.62	293.20
1	28.82	212.60	22.66	267.00
2	30.00	198.36	24.24	262.42
3	31.02	196.14	24.80	273.80
4	31.92	196.60	25.92	252.74
5	32.46	182.76	26.32	269.70
6	33.36	181.46	27.06	245.48
7	34.44	170.20	27.80	247.56
8	34.40	180.72	27.96	240.86
9	34.88	169.70	28.50	222.38
10	35.08	166.00	28.90	216.24
11	35.36	166.90	29.26	211.50
12	36.14	153.94	29.34	209.74
13–14	36.30	166.34	29.80	218.14
17–18	37.62	146.16	30.98	209.00
21–22	38.00	152.86	31.54	204.54
25–26	38.24	153.40	32.28	190.88
27–30	38.40	149.40	32.30	190.95
31–34	39.40	134.50	33.12	177.30
35–38	39.66	135.48	33.48	166.90
39–42	40.26	136.42	33.28	180.42
43–46	40.03	131.80	34.10	165.70

(Continued)

TABLE 17 (*Continued*)

Body Weight and Food Consumption—CD-1 Mice[a]

	Males		Females	
Week	Body Weight (g)	Food Consumption (g/kg/day)	Body Weight (g)	Food Consumption (g/kg/day)
47–50	40.40	126.46	34.34	156.52
51–54	39.90	128.72	35.10	149.16
55–58	40.16	135.80	34.46	173.96
59–62	39.80	137.88	35.20	158.84
63–66	40.46	133.70	34.76	172.72
67–70	40.24	133.18	34.78	161.12
71–74	39.88	142.22	35.14	167.62
75–79	40.56	142.66	35.48	171.02

[a] Typical chronic study, age at –1 week is approximately 5 weeks.

TABLE 18

Body Weight and Food Consumption—Sprague-Dawley Rats[a]

Week	Males Body Weight (g)	Males Food Consumption (g/kg/day)	Females Body Weight (g)	Females Food Consumption (g/kg/day)
−1	132.46	—	103.82	—
0	188.64	146.00	139.14	122.30
1	236.12	115.98	161.56	122.58
2	286.94	101.12	182.80	111.00
3	327.78	90.28	203.46	104.30
4	362.78	82.02	222.10	94.04
5	393.70	75.12	232.70	87.44
6	417.26	68.98	243.04	84.82
7	432.70	63.32	250.68	77.12
8	446.16	62.08	255.40	76.44
9	457.80	58.44	261.36	73.40
10	475.74	60.96	268.62	75.84
11	481.42	57.14	274.52	73.34
12	497.12	59.78	278.24	73.26
13	508.12	56.14	282.46	70.68
17–18	549.84	50.30	299.70	67.42
21–22	577.62	47.62	311.82	63.30
25–26	603.22	45.04	327.22	62.18
29–30	616.76	44.36	334.22	61.84
33–34	635.46	43.94	348.96	61.74
37–38	651.96	42.60	363.38	58.44
41–42	660.34	41.12	374.96	56.40
45–46	678.16	40.12	389.65	52.63
49–50	695.72	39.24	409.40	52.34
53–54	703.72	39.20	412.76	51.28

(Continued)

TABLE 18 (*Continued*)

Body Weight and Food Consumption—Sprague-Dawley Rats[a]

| Week | Males | | Females | |
	Body Weight (g)	Food Consumption (g/kg/day)	Body Weight (g)	Food Consumption (g/kg/day)
57–58	720.88	37.70	430.90	49.16
61–62	728.74	37.56	441.18	48.82
65–66	735.76	38.00	455.06	48.26
69–70	735.04	38.46	460.24	48.24
72–74	737.54	37.80	465.70	47.78
77–78	736.58	39.34	467.70	46.78
81–82	738.04	38.48	471.72	47.74
85–86	733.70	39.22	473.98	47.38
89–90	725.80	38.52	479.42	47.82
93–94	723.62	37.92	490.06	45.68
97–98	721.50	37.48	494.10	45.04
101–102	703.84	35.60	498.56	44.10

[a]　Typical chronic study, age at –1 week is approximately 5 weeks.

TABLE 19

Body Weight and Food Consumption—Fischer 344 Rats[a]

	Males		Females	
Week	Body Weight (g)	Food Consumption (g/kg/day)	Body Weight (g)	Food Consumption (g/kg/day)
−1	86.03	—	68.43	—
0	118.27	137.73	89.90	146.93
1	151.50	112.73	108.07	124.63
2	183.60	99.33	123.70	107.00
3	211.07	85.73	135.70	96.43
4	228.17	77.73	142.80	90.93
5	244.10	75.07	150.60	89.03
6	256.27	67.13	155.77	80.47
7	263.53	63.07	162.07	72.73
8	275.23	60.40	164.40	69.67
9	281.90	56.17	166.47	68.20
10	290.47	54.20	170.20	66.27
11	296.60	55.63	172.23	67.40
12	300.23	54.40	173.07	69.10
13	301.10	54.25	174.75	64.10
16	319.07	51.10	181.53	65.63
20	331.67	48.17	186.87	62.63
24	343.40	47.97	193.67	62.07
28–30	355.70	46.67	201.90	59.87
32–34	366.23	45.33	206.70	60.00
36–38	374.67	44.23	211.37	61.10
40–42	382.90	44.07	215.40	58.27
44–46	384.13	42.63	218.90	56.00
48–50	385.90	38.17	223.53	57.07
52–54	396.23	44.17	228.47	57.37

(Continued)

TABLE 19 (*Continued*)

Body Weight and Food Consumption—Fischer 344 Rats[a]

	Males		Females	
Week	Body Weight (g)	Food Consumption (g/kg/day)	Body Weight (g)	Food Consumption (g/kg/day)
56–59	400.87	43.57	236.17	55.23
60–62	401.90	40.97	241.47	50.63
64–66	404.87	43.77	246.37	53.43
68–70	406.57	44.37	253.80	54.60
72–74	410.20	43.67	259.53	54.57
76–78	404.93	46.00	265.07	54.77
80–82	394.77	43.60	262.80	51.50
84–86	393.60	43.37	264.47	50.37
88–90	397.07	43.73	270.70	51.50
92–94	391.43	42.93	275.13	50.83
96–98	388.77	42.47	277.17	49.63
100–102	388.03	43.33	277.70	51.57

[a] Typical chronic study, age at −1 week is approximately 5 weeks.

TABLE 20

Typical Routes and Dosages of Several Sedative, Analgesic, and Anesthetic Agents

Agents	Dosage and Route in Species						
	Mouse	Rat	Hamster	Guinea Pig	Rabbit	Dog	Primate
Chlorpromazine (mg/kg)	3–35 (IM) 6 (IP)	1–20 (IM) 4–8 (IP)	0.05 (IM)	5–10 (IM)	10–25 (IM)	1–6 (IM) 0.5–8 (PO)	1–6 (IM)
Promazine (mg/kg)	0.5 (IM)	0.5–1 (IM)	0.5–1 (IM)	0.5–1 (IM)	1–2 (IM)	2–4 (IM)	2–4 (IM)
Acepromazine (mg/kg)	—	—	—	—	1 (IM)	0.5–1 (IM) 1–3 (PO)	0.5–1 (IM)
Meperidine (mg/kg)	60 (IM) 40 (IP)	44 (IM) 50 (IP) 25 (IV)	2 (IM)	1 (IP) 2 (IM)	10 (IV)	0.4–10 (IM)	3–11 (IM)
Innovar-Vet (mL/kg)	0.05 (IM)	0.13–0.16 (IM)	—	0.08–0.66 (IM)	0.2–0.3 (IM)	0.13–0.15 (IM)	0.05 (IM)
Ketamine (mg/kg)	25 (IV) 25–50 (IP) 22 (IM)	25 (IV) 50 (IP) 22 (IM)	40 (IM) 100 (IP)	22–64 (IM)	22–44 (IM)	—	5–15 (IM)
Pentobarbital (mg/kg)	35 (IV) 40–70 (IP)	25 (IV) 40–50 (IP)	50–90 (IP)	24 (IV) 30 (IP)	25 (IV) 40 (IP)	30 (IV)	25–35 (IV)
Thiopental (mg/kg)	25–50 (IV)	40 (IM) 25–48 (IP)	—	55 (IM) 20 (IP)	25–50 (IV)	16 (IV)	25 (IV)

Note: Drugs and dosages presented are to serve only as guidelines. Selection and administration of specific agents and dosages should be supervised by a qualified veterinarian. See Chapter 1, *CRC Handbook of Toxicology*, 3rd edition, Derelanko, M.J. and Auletta, C.S., Eds., CRC Press, Boca Raton, 2014, for additional information on anesthetics.

TABLE 21

Guiding Principles for Animal Euthanasia

- Whenever an animal's life is to be taken, it should be treated with the highest respect.
- Euthanasia should place emphasis on making the animal's death painless and distress free.
- Euthanasia techniques should result in rapid loss of consciousness, followed by cardiac or respiratory arrest and ultimate loss of brain function.
- Techniques should require minimum restraint of the animal and should minimize distress and anxiety experienced by the animal before loss of consciousness.
- Techniques should be appropriate for the species, age, and health of the animal.
- Death must be verified following euthanasia and before disposal of the animal.
- Personnel responsible for carrying out the euthanasia techniques should be trained (1) to carry out euthanasia in the most effective and humane manner; (2) to recognize signs of pain, fear, and distress in relevant species; and (3) to recognize and confirm death in relevant species.
- Human psychological responses to euthanasia should be taken into account when selecting the method of euthanasia, but should not take precedence over animal welfare considerations.
- Ethics committees should be responsible for approval of the method of euthanasia (in line with any relevant legislation). This should include euthanasia as part of the experimental protocol, as well as euthanasia for animals experiencing unanticipated pain and distress.
- A veterinarian experienced with the species in question should be consulted when selecting the method of euthanasia, particularly when little species-specific euthanasia research has been done.

Source: Demers, G. et al. (2006). With permission.

TABLE 22

Acceptable and Conditionally Acceptable Agents and Methods of Euthanasia

Method	Rodents	Rabbits	Dogs	Swine	Primates
CO₂ (bottled gas only)	A	A	A	A	C
CO (bottled gas only)	A	A	A	C	C
N₂ or Argon (only if <2% O₂ is achieved rapidly and animals are heavily sedated or anesthetized)	C	C	C		C
Inhalant anesthetics (halothane preferred)	A	A	A	C	C
Methoxyflurane (slow action, may be accompanied by agitation)	C				
Ether (irritating to eyes and nose, may cause stress, hazardous)	C				
Barbiturates (IV; IP acceptable only for small animals and in special situations) (preferred method for dogs)	A	A	A	A	A
Potassium chloride (in conjunction with general anesthesia)	A	A	A	A	
Chloral hydrate (IV after sedation)				C	
Cervical dislocation (requires training; rats < 200 g and rabbits <1 kg)	C	C			
Decapitation (with trained operator, sharp, well-maintained device) (only for small rabbits).	C	C			

(*Continued*)

TABLE 22 (*Continued*)

Acceptable and Conditionally Acceptable Agents and Methods of Euthanasia

Method	Rodents	Rabbits	Dogs	Swine	Primates
Blow to the head (requires training, swine < 3 wk of age)				C	
Penetrating captive bolt (not preferred method for dogs)		C	C	A	
Gunshot				C	
Microwave irradiation (brain) (rats and mice only. Requires specialized, high-power euthanasia microwave device)	A				
Electrocution (only with animal rendered unconscious)			C	C	
Exsanguination (only with highly sedated or anesthetized animals).					

Source: Adapted from AVMA (2007).

A = Acceptable method. Consistently produces death when used as the sole means of euthanasia.

C = Conditionally acceptable method. By nature of the technique or because of greater potential for operator error, might not consistently produce humane death; are safety hazards;or are methods not well documented in the scientific literature.

TABLE 23

Summary of the Characteristics of Several Euthanasia Methods

Euthanasia Method	Mechanism of Action	Effectiveness	Personnel Safety
Inhalant anesthetics	Hypoxia due to depression of vital centers	Moderately rapid onset of anesthesia; initial excitation may occur	Minimize exposure to personnel by scavenging or venting
Carbon dioxide	Hypoxia due to depression of vital centers	Effective in adult animals; may be prolonged in immature and neonatal animals	Minimal hazard
Carbon monoxide	Hypoxia due to inhibition of O_2-carrying capacity of hemoglobin	Effective and acceptable with proper equipment and operation	Extremely hazardous; difficult to detect
Barbiturates	Hypoxia due to depression of vital centers	Highly effective when administered appropriately	Safe, except human abuse potential of controlled substances(s)
Inert gasses (Ni, Ar)	Hypoxic hypoxemia	Effective, but other methods are preferable; acceptable only if animal is heavily sedated or anesthetized	Safely used in ventilated area
Cervical dislocation	Hypoxia due to disruption of vital centers, direct depression of brain	Effective and irreversible; requires training, skill, and IACUC approval; aesthetically displeasing	Safe
Decapitation	Hypoxia due to disruption of vital centers, direct depression of brain	Effective and irreversible; requires training, skill, and IACUC approval; aesthetically displeasing	Potential injury due to guillotine

References

AVMA (2007), *AVMA Guidelines on Euthanasia*, American Veterinary Medical Association, Schaumburg, IL, 2007.

Bollen, P.J.A., Hansen, A.K., and Rasmussen, H.J. (2000), *The Laboratory Swine*, CRC Press, Taylor & Francis Group, Boca Raton, FL.

Demers, G., Griffin, G., De Vroey, G., Haywood, J.R., Zurlo, J., and Bedard, M. (2006), Harmonization of Animal Care and Use Guidance, *Science* 312, 700–701, Policy Forum-Animal Research. www.sciencemag.org.

Evans, I.E. and Maltby, C.J. (1989), *Technical Laboratory Animal Management*, MYM Associates, Virginia.

Fortman, J.D., Hewett, T.A., and Bennett, B.T. (2002), *The Laboratory Nonhuman Primate*, CRC Press, Boca Raton, FL.

Hawkins, P. (2002), Recognizing and assessing pain, suffering and distress in laboratory animals: A survey of current practices in the UK with recommendations, *Lab. Anim.* 36, 378–395.

LAMA (1988), *LAMA Lines*, Newsletter of the Laboratory Animal Management Association, 4, September/October.

OECD (2000), Guidance Document on the Recognition, Assessment and Use of Clinical Signs as Humane Endpoints for Experimental Animals Used in Safety Evaluation, OECD Environmental Health and Safety Publications, Series on Testing and Assessment, No. 19, November, Paris.

Williams, C.S.F. (1976), *Practical Guide to Laboratory Animal Care*, C.V. Mosby, St. Louis, MO.

2

General Toxicology

TABLE 24

Minimum Requirements for an Acceptable Toxicology Study

1. The study should be conducted at a laboratory recognized by accreditation and/ or reputation as having the scientific capability, expertise, and experience to conduct the study of interest.

2. The study should be conducted according to Good Laboratory Practices (GLP).

3. The objectives and design of the study should be specified in a study-specific protocol approved by the study director and the sponsor (if applicable).

4. The chemical nature of the tested material should be precisely defined and documented including chemical identity, stability, and degree of purity with any impurities clearly defined.

5. The specificity of any methodology used should be adequate for the degree of detection of the end points to be evaluated. Such methods must be validated. Positive controls and standards should be used as necessary.

6. The number of test and control animals should be sufficient to allow the detection of biological variability in response to exposure, to allow trends to be appreciated, and to be sufficient for statistical analyses.

7. Ideally, doses or exposure levels should be sufficient to detect toxicity, define thresholds, and establish no-effect levels.

8. Statistical procedures used should be appropriate for the type of data analyzed.

9. The study should be reported in a clear and unambiguous manner with all necessary detail to allow the reader to understand the study design, interpret the results, and draw conclusions. All deviations from the protocol that have occurred should be clearly stated and the potential impact on the study assessed.

10. The report should be signed by the study director to indicate agreement with the results and conclusions. In addition, the study director should sign the GLP compliance statement that states whether the study was conducted in full GLP compliance and if not, the areas of noncompliance. The report should also contain the signed quality assurance statement providing dates of inspection and reporting to management.

TABLE 25

Typical Contents of a GLP Protocol

Minimum Required

- A descriptive title and statement of the purpose of the study
- Identification of the test article by name and/or code number
- The name of the sponsor of the study
- The name and address of the testing facility
- The number, body weight range, sex, source of supply, species, strain, substrain, and age of the test system (animals)
- The procedure to be used for identification of the test system
- A description and/or identification of the diet used in the study to include a statement of contaminants
- A description of the experimental design, including the methods for the control of bias
- A description and/or identification of solvents, emulsifiers, and/or other materials used to solubilize or suspend the test article
- Dose levels in appropriate units, and method and frequency of administration
- The type and frequency of tests, analyses, and measurements to be conducted
- A statement of the proposed statistical methods to be used
- The records to be maintained
- Date of approval by the sponsor and dated signature of the study director

Additional Information Frequently Provided

- Contact information for study director, principal investigators, and other key personnel
- Proposed experimental start and termination dates
- Statement of regulatory guideline compliance
- Storage conditions for test and control articles
- Reserve archive samples to be taken
- Justification for selection of the test system
- Animal care and use statement
- Animal husbandry practices
- Justification of dose and route of administration
- Test article disposition
- Computer systems utilized
- Proposed content of study report

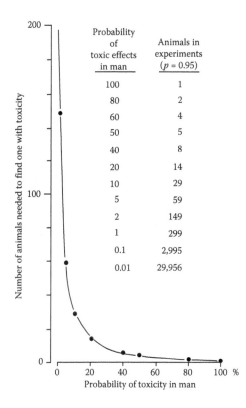

FIGURE 1
Animal number and predictive probability. Data derived from Zbinden, G. (1973) and Ecobichon, D.J. (1992). With permission.

TABLE 26

Suggested Dose Volumes (mL/kg) for Test Material Administration

	Route													
	Gavage		Dermal		IV		IP		SC[a]		IM[a]		NASAL[b]	
Species	Ideal	Limit	Ideal	Limit	Ideal	Limit	Ideal	Limit	Ideal	Limit	Ideal	Limit	Ideal	Limit
Mouse	10	20–50	—	—	5	15–25	5–10	25	1–5	10–20	0.1	0.5–1	—	—
Rat	10	20–50	2	—	1–5	10–20	5–10	25	1	10–20	0.1	1	0.1	0.2
Rabbit	10	10–20	2	—	1–3	5–10	—	—	1–2.5	5–10	0.1	1	0.2	1
Dog	10	10–20	—	—	1	5–10	1–3	5	0.5	1–2	0.1	1	0.2	2
Monkey	10	10	—	—	1	5–10	1–3	5	0.5	1–2	0.1	1	0.2	1
Minipig	10	15	0.5	2	1	5–10	1	20	1	2	0.1	0.5	—	—

Source: Adapted from *SYNAPSE* (1991). Some adaptations have been made based on experience in the laboratory. Note that there are many guidelines for this, both published and internal (laboratory-specific) and there is no universal agreement on specific numbers. Recommended dose volumes vary from publication to publication and laboratory to laboratory.

[a] Dose per site.

[b] Nasal doses (mL/animal, 1/2 dose per nostril) based on experience in the laboratory.

TABLE 27

Suggested Dosing Apparatus/Needle Sizes (Gauge) for Test Material Administration[a]

Species	Gavage Recommended	IV Ideal	IV Range	IP Ideal	IP Range	SC Ideal	SC Range	IM Ideal	IM Range
Mouse	Premature infant feeding tube cut to 70 mm, marked at 38 mm; French rubber catheter—flexible, 18–20 gauge, 30 mm, bulb-shaped tip	25 or 27	25–30	25 or 27	22–30	25 or 27	22–30	25 or 27	22–30
Rat	3-in. ball-tipped intubation needle; French rubber catheter, size 8 or 10	25	25–30	25	22–30	25	25–30	25	22–30
Rabbit	No. 18 French catheter, cut to 15 in., marked at 12 in.	21	21–22	21	18–23	25	22–25	25	22–30
Dog	Kaslow stomach tube 12 Fr ≥ 24 in., Davol 32 Fr intubation tube; 24–32 French rectal tube—~20 in. long	21	21–22	—	—	22	20–23	21 or 25	20–25
Monkey	No. 8 French tube (nasogastric gavage); 14–18 French rectal tube	25	25–30	—	—	22	22–25	25	22–25

[a] Recommended gavage equipment and ideal needle sizes are those used in our laboratory. Suggested ranges of needle sizes are from: *Laboratory Manual for Basic Biomethodology of Laboratory Animals*, MTM Associates, Inc.

TABLE 28

Guidelines for Dose Administration for Intravenous Infusion Studies

		Species			
		Dog	Primate	Rat	Rabbit
Recommended IV volume (taken from EFPIA guidelines)	IV Bolus (mL/kg)	2.5	2	5	2
	IV Slow injection (mL/kg—max vol)	5	Similar to dog	20	10
	Continuous IV infusion (mL/kg/h)	1	2.5	2.5	1
Surgery not required bolus or intermittent IV infusion (<4 h)		28 or fewer doses (regardless of 1×/day, 2×/day, 1×/week)	14 or fewer doses (regardless of 1×/day, 2×/day, 1×/week)	14 or fewer doses AND fewer than 50 animals	8 or fewer doses
Surgery required	Duration of dose	Over 4 h to continuous	Over 4 h to continuous	Over 4 h to continuous	Over 4 h to continuous
	Number of doses	More than 28 doses	More than 14 doses	More than 14 doses OR over 50 animals	More than 8 doses
VAP/rodent intermittent infusion model		Repeat intermittent infusion dosing for 24 h or less	Repeat intermittent infusion dosing for 24 h or less	2 or fewer doses per week	Repeat intermittent infusion dosing for over 24 h or less
Permanent exteriorized catheter		Continuous dosing for over 24 h	Continuous dosing for over 24 h	More than 2 doses per week	Continuous dosing for over 24 h

Note: EFPIA, European Federation of Pharmaceutical Institutes and Associations.

TABLE 29

Body Weight: Surface Area Conversion

Species	Representative Body Weight to Surface Area[a]		
	Body Weight (kg)	Surface Area (m²)	Conversion Factor (km)
Mouse	0.02	0.0066	3
Rat	0.15	0.025	5.9
Monkey	3	0.24	12
Dog	8	0.4	20
Human			
Child	20	0.8	25
Adult	60	1.6	37

Source: Adapted from Freireich, E.J. et al. (1966).

[a] Example: To express a mg/kg dose in any given species as the equivalent mg/m² dose, multiply the dose by the appropriate km. In human adults, 100 mg/kg is equivalent to $100 \text{ mg/kg} \times 37 \text{ kg/m}^2 = 3700 \text{ mg/m}^2$.

TABLE 30

Equivalent Surface Area Dosage Conversion Factors[a]

From	To				
	Mouse (20 g)	Rat (150 g)	Monkey (3 kg)	Dog (8 kg)	Human (60 kg)
Mouse	1	1/2	1/4	1/6	1/12
Rat	2	1	1/2	1/3	1/6
Monkey	4	2	1	3/5	1/3
Dog	6	4	3/2	1	1/2
Man	12	7	3	2	1

Source: Adapted from Freireich, E.J. et al. (1966).

Example: To convert a dose of 50 mg/kg in the mouse to an equivalent dose in the monkey, assuming equivalency on the basis of mg/m², multiply 50 mg/kg × 1/4 = 13 mg/kg.

[a] This table gives approximate factors for converting doses expressed in terms of mg/kg from one species to an equivalent surface area dose expressed as mg/kg in the other species tabulated.

TABLE 31

Comparison of Dosage by Weight and Surface Area

Species	Weight (g)	Dosage (mg/kg)	Dose (mg/Animal)	Surface Area (cm²)	Dosage (mg/cm²)
Mouse	20	100	2	46	0.043
Rat	200	100	20	325	0.061
Guinea pig	400	100	40	565	0.071
Rabbit	1,500	100	150	1,270	0.118
Cat	2,000	100	200	1,380	0.145
Monkey	4,000	100	400	2,980	0.134
Dog	12,000	100	1,200	5,770	0.207
Human	70,000	100	7,000	18,000	0.388

Source: From Amdur, M.O. et al., Eds. (1991), *Casarett and Doull's Toxicology*, 4th ed., Pergamon Press, New York, 1991. With permission.

TABLE 32

Approximate Diet Conversion Factors (ppm to mg/kg)

Species	Age	Conversion Factor (Divide ppm by)
Mice	Young (1–12 weeks of study)	5
	Older (13–78 weeks of study)	6–7
Rats	Young (1–12 weeks of study)	10
	Older (13–104 weeks of study)	20
Dogs		40

Note: To estimate the approximate test material intake of rats receiving a 1000 ppm dietary concentration during a 4-week study: 1000 ppm ÷ 10 = 100 mg/kg b.w./day.

TABLE 33

Clinical Signs of Toxicity

Clinical Observation	Observed Signs	Organs, Tissues, or Systems Most Likely to Be Involved
I. Respiratory: blockage in the nostrils, changes in rate and depth of breathing, changes in color of body surfaces	A. Dyspnea: difficult or labored breathing; essentially gasping for air; respiration rate usually slow	
	1. Abdominal breathing; breathing by diaphragm, greater deflection of abdomen upon inspiration	CNS respiratory center, paralysis of costal muscles, cholinergic inhibition
	2. Gasping; deep labored inspiration, accompanied by a wheezing sound	CNS respiratory center, pulmonary edema, secretion accumulation in airways (increase cholinergic)
	B. Apnea: a transient cessation of breathing following a forced respiration	CNS respiratory center, pulmonary cardiac insufficiency
	C. Cyanosis: bluish appearance of tail, mouth, foot pads	Pulmonary–cardiac insufficiency, pulmonary edema
	D. Tachypnea: quick and usually shallow respiration	Stimulation of respiratory center, pulmonary–cardiac insufficiency
	E. Nostril discharges: red or colorless	Pulmonary edema, hemorrhage

(Continued)

TABLE 33 (*Continued*)
Clinical Signs of Toxicity

Clinical Observation	Observed Signs	Organs, Tissues, or Systems Most Likely to Be Involved
II. Motor activities: changes in frequency and nature of movements	A. Decrease or increase in spontaneous motor activities, curiosity, preening, or locomotions	Somatomotor, CNS
	B. Somnolence: animal appears drowsy, but can be aroused by prodding and resumes normal activities	CNS sleep center
	C. Loss of righting reflex: loss of reflex to maintain normal upright posture when placed on the back	CNS, sensory, neuromuscular
	D. Anesthesia: loss of righting reflex and pain response (animal will not respond to tail and toe pinch)	CNS, sensory
	E. Catalepsy: animal tends to remain in any position in which it is placed	CNS, sensory, neuromuscular, autonomic
	F. Ataxia: inability to control and coordinate movement while animal is walking with no spasticity, epraxia, paresis, or rigidity	CNS, sensory, autonomic
	G. Unusual locomotion: spastic, toe walking, pedaling, hopping, and low body posture	CNS, sensory, neuromuscular
	H. Prostration: immobile and rests on belly	CNS, sensory, neuromuscular
	I. Tremors: involving trembling and quivering of the limbs or entire body	Neuromuscular, CNS
	J. Fasciculation: involving movements of muscles, seen on the back, shoulders, hind limbs, and digits of the paws	Neuromuscular, CNS, autonomic

(Continued)

TABLE 33 (*Continued*)

Clinical Signs of Toxicity

Clinical Observation	Observed Signs	Organs, Tissues, or Systems Most Likely to Be Involved
III. Convulsion (seizure): marked involuntary contraction or seizures of contraction of voluntary muscle	A. Clonic convulsion: convulsive alternating contraction and relaxation of muscles	CNS, respiratory failure, neuromuscular, autonomic
	B. Tonic convulsion: persistent contraction of muscles, attended by rigid extension of hind limbs	
	C. Tonic–clonic convulsion: both types may appear consecutively	
	D. Asphyxial convulsion: usually of clonic type, but accompanied by gasping and cyanosis	
	E. Opisthotonos: tetanic spasm in which the back is arched and the head is pulled toward the dorsal position	
IV. Reflexes	A. Corneal (eyelid closure): touching of the cornea causes eyelids to close	Sensory, neuromuscular
	B. Pinnal: twitch of external ear elicited by light stroking of inside surface of ear	Sensory, neuromuscular, autonomic
	C. Righting: ability of animal to recover when placed dorsal side down	CNS, sensory, neuromuscular
	D. Myotact: ability of animal to retract its hind limb when limb is pulled down over the edge of a surface	Sensory, neuromuscular
	E. Light (pupillary): constriction of pupil in the presence of light	Sensory, neuromuscular, autonomic
	F. Startle reflex: response to external stimuli such as touch, noise	Sensory, neuromuscular

(*Continued*)

TABLE 33 (*Continued*)
Clinical Signs of Toxicity

Clinical Observation	Observed Signs	Organs, Tissues, or Systems Most Likely to Be Involved
V. Ocular signs	A. Lacrimation: excessive tearing, clear or colored	Autonomic
	B. Miosis: constriction of pupil regardless of the presence or absence of light	Autonomic
	C. Mydriasis: dilation of pupils regardless of the presence or absence of light	Autonomic
	D. Exophthalmos: abnormal protrusion of eye from orbit	Autonomic
	E. Ptosis: dropping of upper eyelids, not reversed by prodding animal	Autonomic
	F. Chromodacryorrhea (red lacrimation)	Autonomic, hemorrhage, infection
	G. Relaxation of nictitating membrane	Autonomic
	H. Corneal opacity, iritis, conjunctivitis	Irritation of the eye
VI. Cardiovascular signs	A. Bradycardia: decreased heart rate	Autonomic, pulmonary–cardiac insufficiency
	B. Tachycardia: increased heart rate	Autonomic, pulmonary–cardiac insufficiency

(*Continued*)

TABLE 33 (*Continued*)
Clinical Signs of Toxicity

Clinical Observation	Observed Signs	Organs, Tissues, or Systems Most Likely to Be Involved
	C. Vasodilation: redness of skin, tall, tongue, ear, foot pad, conjunctivae, and warm body	Autonomic, CNS, increased cardiac output, hot environment
	D. Vasoconstriction: blanching or whitening of skin, cold body	Autonomic, CNS, cold environment, cardiac output decrease
	E. Arrhythmia: abnormal cardiac rhythm	CNS, autonomic, cardiacpulmonary insufficiency, myocardial infarction
VII. Salivation	A. Excessive secretion of saliva: hair around mouth becomes wet	Autonomic
VIII. Piloerection	A. Contraction of erectile tissue of hair follicles resulting in rough hair	Autonomic
IX. Analgesia	A. Decrease in reaction to induced pain (e.g., hot plate)	Sensory, CNS
X. Muscle tone	A. Hypotonia: generalized decrease in muscle tone	Autonomic
	B. Hypertonia: generalized increase in muscle tension	Autonomic

(Continued)

TABLE 33 (*Continued*)

Clinical Signs of Toxicity

Clinical Observation	Observed Signs	Organs, Tissues, or Systems Most Likely to Be Involved
XI. Gastrointestinal signs: droppings (feces)	A. Solid, dried, and scant	Autonomic, constipation, GI motility
	B. Loss of fluid, watery stool	Autonomic, diarrhea, GI motility
Emesis	A. Vomiting and retching	Sensory, CNS, autonomic (in rat, emesis is absent)
Diuresis	A. Red urine (hematuria)	Damage in kidney
	B. Involuntary urination	Autonomic, sensory
XII. Skin	A. Edema: swelling of tissue filled with fluid	Irritation, renal failure, tissue damage, long term immobility
	B. Erythema: redness of skin	Irritation, inflammation, sensitization

Source: From Chan, P.K. and Hayes, A.W. (1989). With permission.

TABLE 34

Autonomic Signs

Sympathomimetic	Piloerection
	Partial mydriasis
Sympathetic block	Ptosis
	Diagnostic if associated with sedation
Parasympathomimetic	Salivation (examined by holding blotting paper)
	Miosis
	Diarrhea
	Chromodacryorrhea in rats
Parasympathomimetic block	Mydriasis (maximal)
	Excessive dryness of mouth (detect with blotting paper)

Source: From Chan, P.K. and Hayes, A.W. (1989). With permission.

TABLE 35

Clinical Chemistry Parameters of Subchronic and Chronic Studies—Standard Study Guidelines

Parameter	OECD	JMHW	RDBK	OCSPP	JMAFF	METI
	\-	\-	Guidelines[a]	\-	\-	\-
Alkaline phosphatase		×		×	×	×
Alanine aminotransferase (ALT)	×	×	×	×	×	×
Aspartate aminotransferase (AST)	×	×	×	×	×	×
γ-glutamyltransferase (GGT)	×		×	×		×
Glucose	×	×	×	×	×	×
Bilirubin: total	×	×	×	×[c]		
Creatinine	×	×	×	×		×
Urea (BUN)	×	×	×	×	×	×
Total protein	×	×	×	×	×	×
Albumin	×	×	×	×	×	
Albumin/globulin ratio		×				
Electrolytes (Na, K, Cl, Ca, P)	×	×	×	×	×	×
Cholesterol		×		×		×
Triglycerides		×		×		×
Ornithine decarboxylase (ODC)	b	b				b
Protein-electrophoretogram		×				
Lactate dehydrogenase						c
Creatine kinase						c
Phospholipids						c
Uric acid						c

[a] OECD = Organization for Economic Cooperation and Development; JMHW = Japanese Ministry of Health and Welfare; RDBK = FDA Redbook; OCSPP = Office of Chemical Safety and Pollution Prevention (Environmental Protection Agency); JMAFF = Japanese Ministry of Agriculture, Forestry and Fisheries; METI: Ministry of Economy, Trade and Industry (Japan).

[b] Ornithine decarboxylase is a tissue enzyme; no acceptable analytical procedure for blood exists.

[c] Chronic studies only.

TABLE 36

Hematology Parameters of Subchronic and Chronic Studies—
Standard Study Guidelines[a]

Hematology

All Guidelines

Erythrocyte count

Hematocrit

Hemoglobin concentration

Leukocyte count (total and differential)

Some measure of clotting function

    Suggestions: Clotting time

                Platelet count

                Prothrombin time (PT)

                Activated partial thromboplastin time (APTT)

Exceptions/Additions

OCSPP: MCV, MCH, MCHC

MHW: in addition, reticulocyte count, PT, APTT

JMAFF: specifies platelet count

METI chronic studies: in addition, reticulocyte count; specifies platelet count

EC: Guidelines do not specify parameters

[a] OCSPP = Office of Chemical Safety and Pollution Prevention (EPA); JMHW: Japanese Ministry of Health and Welfare; JMAFF = Japanese Ministry of Agriculture, Forestry and Fisheries; METI: Ministry of Economy, Trade and Industry (Japan); EC = European Community.

TABLE 37

Urinalysis Parameters of Subchronic and Chronic Studies—Standard
Study Guidelines

	Guideline[a]				
Parameter	OECD	JMHW	OCSPP	JMAFF	METI (Chronic)
Appearance	×		×	×	
Volume	×	×	×	×	×
Specific gravity	×	×	×	×	
Protein	×	×	×	×	×
Glucose	×	×	×	×	×
Ketones		×		×	×
Occult blood[b]	×	×	×	×	×
Sediment microscopy[b]		×	×	×	c
pH	×	×	×		×
Bilirubin		×	×		×
Urobilinogen		×			×
Electrolytes (Na, K, etc.)		×			

[a] OECD = Organization for Economic Cooperation and Development; JMHW = Japanese Ministry of Health and Welfare; RDBK = FDA Redbook; OCSPP = Office of Chemical Safety and Pollution Prevention (Environmental Protection Agency); JMAFF = Japanese Ministry of Agriculture, Forestry and Fisheries; METI: Ministry of Economy, Trade and Industry (Japan).

[b] Semiquantitative evaluation.

[c] When necessary.

TABLE 38

Organ Weight Requirements—Standard Study Guidelines

Organ to be Weighed	Guideline[a]					
	OECD	JMHW	RDBK	OCSPP	JMAFF	METI
Adrenal glands	×	×	×	×	×	×
Kidneys	×	×	×	×	×	×
Liver	×	×	×	×	×	×
Testes	×	×	×	×	×	×
Epididymides	×					
Ovaries	×	×		×		×
Thyroid/parathyroids	NR	[b]	NR	NR	NR	
Brain	×	×		×	Chronic	Chronic
Heart	×	×		×		Chronic
Lungs		[b]				Chronic
Spleen	×	×		×		Chronic
Pituitary		×				Chronic
Salivary gland		[b]				
Seminal vesicles		[b]				
Thymus	×	[b]		×		
Uterus	×	[b]		×		

[a] OECD = Organization for Economic Cooperation and Development; JMHW = Japanese Ministry of Health and Welfare; RDBK = FDA Redbook; OCSPP = Office of Chemical Safety and Pollution Prevention (Environmental Protection Agency); JMAFF = Japanese Ministry of Agriculture, Forestry and Fisheries; METI: Ministry of Economy, Trade and Industry (Japan); NR = Nonrodent.

[b] Guidelines state that these organs are "often weighed."

TABLE 39

Microscopic Pathology Requirements—Standard Study Guidelines—Tissues Most Often Recommended for Chronic Studies[a]

Tissues	OECD	EC	JMHW	RDBK	OCSPP	JMAFF	METI
Adrenal glands	×	×	×	×	×	×	×
Bone (sternum/femur/vertebrae)	S, F	S, E, or V	S, F	S	×	S, F	S, E, or V
Bone marrow (sternum/femur/vertebrae)	S	S, E, or V	S, F	S	×	S, F	S
Brain (medulla/pons, cerebrum, cerebellum)	×	×	×	×	×	×	×
Esophagus	×	×	×	×	×	×	×
Heart	×	×	×	×	×	×	×
Kidney	×	×	×	×	×	×	×
Large intestine (cecum, colon, rectum)	×	Colon	×	×	×	×	×
Liver	×	×	×	×	×	×	×
Lung (with mainstem bronchi)	×	×	×	×	×	×	×
Lymph node (representative)	×	×	×	×	×	×	×
Mammary gland	♀	×	×	×	♀	♀	♀
Ovaries	×	×	×	×		×	×
Pancreas	×	×	×	×	×	×	×
Pituitary	×	×	×	×	×	×	×
Prostate	×	×	×	×	×	×	×
Salivary glands	×	×	×	×	×	×	×

(Continued)

TABLE 39 (*Continued*)

Microscopic Pathology Requirements—Standard Study Guidelines—Tissues Most Often Recommended for Chronic Studies[a]

Tissues	OECD	EC	JMHW	RDBK	OCSPP	JMAFF	METI
Small intestine (duodenum, ileum, jejunum)	x	x	x	x	x	x	x
Spleen	x	x	x	x	x	x	x
Stomach	x	x	x	x	x	x	x
Testes (with epididymides)	x	x	x	x	x	x	x
Thymus	x	x	x	x	x	x	x
Thyroid (with parathyroids)	x	x	x	x	x	x	x
Trachea	x	x	x	x	x	x	x
Urinary bladder	x	x	x	x	x	x	x
Uterus	x	x	x	x	x	x	x
Gross lesions/masses/ target organs	x	x	x	x	x	x	x

[a] OECD = Organization for Economic Cooperation and Development; EC = European Community; JMHW = Japanese Ministry of Health and Welfare; RDBK = FDA Redbook; OCSPP = Office of Chemical Safety and Pollution Prevention (Environmental Protection Agency); JMAFF = Japanese Ministry of Agriculture, Forestry and Fisheries; METI: Ministry of Economy, Trade and Industry (Japan).

TABLE 40

Microscopic Pathology Requirements—Tissues Occasionally Recommended for Chronic Studies—Standard Study Guidelines[a]

Tissue	OECD	EC	JMHW	RDBK	JMAFF	OCSPP	METI
Aorta	×			×	×	×	
Eyes	×	×	×	×	×	×	×
Gallbladder (not present in rats)		×	×	×	×	×	
Lacrimal gland (rodent only)			×				
Larynx					×		×
Fallopian tubes					×		
Muscle (skeletal, usually biceps femoris)	×			×	×	×	
Nerve (peripheral/sciatic)	×			×	×	×	×
Nose					×		
Pharynx					×		
Seminal vesicles (not present in dogs)	×		×	×	×	×	×
Skin	×		×	×	×	×	×
Smooth muscle			×				
Spinal cord (number of sections; total number indicated)	3	×	×	2	×	3	×
Tongue			×				
Vagina			×				×

Note: Additional tissues sometimes taken—Harderian gland, clitoral gland, preputial gland, zymbal gland, and nasal turbinates.

[a] OECD = Organization for Economic Cooperation and Development; EC = European Community; JMHW = Japanese Ministry of Health and Welfare; RDBK = FDA Redbook; OCSPP = Office of Chemical Safety and Pollution Prevention (Environmental Protection Agency); JMAFF = Japanese Ministry of Agriculture, Forestry and Fisheries; METI: Ministry of Economy, Trade and Industry (Japan).

TABLE 41

Effect of Decreased Body Weights on Relative Organ
Weights[a] of Rats[b]

Decrease	No Change	Increase
Liver (?)	Heart	Adrenal glands (?)
	Kidneys	Brain
	Prostate	Epididymides
	Spleen	Pituitary
	Ovaries	Testes
		Thyroid (?)
		Uterus

Note: ?, Differences slight or inconsistent.

[a] Relative weights: organ/body weight ratios.

[b] For absolute weights, all except thyroids decrease. Summary of
results reported in Schwartz, E.R. et al. (1973) and Scharer, K.
(1977).

TABLE 42

Common Abbreviations and Codes Used in Histopathology

Code	Finding or Observation
+ (1)	Minimal grade lesion
++ (2)	Mild or slight grade lesion
+++ (3)	Moderate grade lesion
++++ (4)	Marked or severe grade lesion
+++++ (5)	Very severe or massive grade lesion
(No Entry)	Lesion not present or organ/tissue not examined
+	Tissue examined microscopically
–	Organ/tissue present, no lesion in section
A	Autolysis precludes examination
B	Primary benign tumor
I	Incomplete section of organ/tissue or insufficient tissue for evaluation
M	Primary malignant tumor
M	Organ/tissue missing, not present in section
N	No section of organ/tissue
N	Normal, organ/tissue within normal limits
NCL	No corresponding lesion for gross finding
NE	Organ/tissue not examined
NRL	No remarkable lesion, organ/tissue within normal limits
NSL	No significant lesion, organ/tissue within normal limits
P	Lesion present, not graded (e.g., cyst, anomaly)
R	Recut of section with organ/tissue
U	Unremarkable organ/tissue within normal limits
WNL	Organ/tissues within normal limits
X	Not remarkable organ/tissue, normal
X	Incidence of listed morphology, lesion present

TABLE 43

Examples of Frequently Used Grading Schemes[a] for Histopathology Findings

A1	A2	B	C	D	E
(1) Minimal <1%–25%	(1) Minimal <1%–15%	(1) Minimal <1%	(1) Slight 1%–25%	(1) Minimal <1%	1 = 1–4 foci
(2) Mild 26%–50%	(2) Mild 16%–35%	(2) Slight 1%–25%	(2) Mild 26%–50%	(2) Mild 1%–30%	2 = 5–8 foci
(3) Moderate 51%–75%	(3) Moderate 36%–60%	(3) Moderate 26%–50%	(3) Moderate 51%–75%	(3) Moderate 31%–60%	3 = 9–12 foci
(4) Marked 76%–100%	(4) Marked 61%–100%	(4) Moderately severe 51%–75%	(4) Severe 76%–100%	(4) Severe 61%–90%	4 = >12 foci
		(5) Severe 76%–100%		(5) Very severe or massive 91%–100%	

Source: Adapted by Dr. John Peckham from Hardisty, J.F. and Eustis, S.L. (1990), World Health Organization (1978).

Note: The relative proportion of an affected organ associated with specific severity term: *Minimal* is a very small amount, *Slight* is a very small to small amount, *Mild* is a small amount, *Moderate* is a middle or median amount, *Marked* is a large amount, *Moderately severe* is also a large amount, *Severe* is a very large amount, and *Very severe or massive* is also a very large amount.

[a] A1 and A2 are examples of four-severity grade schemes commonly used by pathologists in the National Toxicology Program; B and D are examples of five-severity grade schemes that have been used by other researchers, often for pharmaceutical companies; C is another four-severity grade scheme similar to A1 using different terminology; and E is an example for grading of a quantifiable lesion.

References

Amdur, M.O., Doull, J., and Klaassen, C.D., Eds. (1991), *Casarett and Doull's Toxicology*, 4th ed., Pergamon Press, New York.

Chan, P.K. and Hayes, A.W. (1989), Principles and methods for acute toxicity and eye irritancy, *in Principles and Methods of Toxicology*, 2nd ed., A.W. Hayes, Ed., Raven Press, New York.

Ecobichon, D.J. (1992), *The Basis of Toxicity Testing*, CRC Press, Boca Raton, FL, chap. 2.

Freireich, E.J. et al. (1966), Quantitative comparison of toxicity of anti-cancer agents in mouse, rat, dog and monkey and man, *Cancer Chemother. Rep.* 50, 219.

Hardisty, J.F., and Eustis, S.L. (1990), Toxicological pathology: A critical stage in study interpretation, in *Progress in Predictive Toxicology*, Clayson, D.B., Munro, I.C., Shubik, P., and Swenberg, J.A., Eds., Elsevier, New York.

Laboratory Manual for Basic Biomethodology of Laboratory Animals, MTM Associates, Inc.

Scharer, K. (1977), The effect of underfeeding on organ weights of rats: How to interpret organ weight changes in cases of marked growth retardation in toxicity tests. *Toxicology* 7, 45.

Schwartz, E., Tomaben, J.A., and Boxill, G.C. (1973), The effects of food restriction on hematology, clinical chemistry and pathology in the albino rat, *Toxicol. Appl. Pharmacol.* 25, 515.

SYNAPSE (1991), American Society of Laboratory Animal Practitioners, Vol 24, March.

World Health Organization (1978), *Principles and Methods for Evaluating the Toxicity of Chemicals. Part 1, Environmental Health Criteria 6*, World Health Organization, Geneva.

Zbinden, G. (1973), *Progress in Toxicology*, Vol. 1, Springer-Verlag, New York.

3

Inhalation Toxicology

TABLE 44

Body Weight and Lung Volumes in Fischer-344 Rats at Various Ages[a]

Parameter	3 Months	18 Months	27 Months
Body weight (g)	222 ± 61	334 ± 106	332 ± 71
Total lung capacity (TLC) (mL)	11.9 ± 1.7	13.9 ± 2.2	14.4 ± 1.9
TLC/body weight (mL/kg)	56 ± 8	42 ± 7	43 ± 6
Vital capacity (mL)	11.0 ± 1.8	13.4 ± 2.3	13.4 ± 1.7
Functional residual capacity (mL)	2.1 ± 0.3	1.7 ± 0.3	2.7 ± 0.4
Residual volume (RV) (mL)	1.0 ± 0.3	0.6 ± 0.2	1.1 ± 0.5
RV/TLC, (mL/mL)	0.08 ± 0.03	0.04 ± 0.01	0.07 ± 0.03

Source: Adapted from Mauderly, J.L. (1982). From Sahebjami, H. (1992).

[a] Values are means ± SD.

TABLE 45

Body Weight and Lung Volumes in Adult and Older Hamsters[a]

Parameter	15 Weeks	65 Weeks	p Value
Body weight (g)	126 ± 12	125 ± 7	>0.20
Total lung capacity (mL)	9.6 ± 1.3	11.1 ± 1.0	<0.02
Vital capacity (mL)	6.9 ± 1.0	7.8 ± 0.9	<0.10
Functional residual capacity (mL)	3.5 ± 0.5	4.3 ± 0.3	<0.05
Residual volume (RV) (mL)	2.7 ± 0.60	3.3 ± 0.3	<0.05
RV/TLC (%)	28 ± 5	30 ± 5	>0.20

Source: Adapted from Mauderly, J.L. (1979). From Sahebjami, H. (1992).
[a] Values are means ± SD.

TABLE 46

Ventilatory Parameters in Fischer-344 Rats at Various Ages[a]

Parameter	3 Months	18 Months	27 Months
Respiratory frequency (breath/min)	48 ± 6	54 ± 7	54 ± 6
Tidal volume (mL)	1.1 ± 0.3	1.5 ± 0.3	1.5 ± 0.3
Minute ventilation (mL/min)	54 ± 14	82 ± 23	82 ± 18
Minute ventilation (mL/min/kg body weight)	254 ± 48	251 ± 45	252 ± 52

Source: Adapted from Mauderly, J.L. (1982). From Sahebjami, H. (1992).
[a] Values are means ± SD.

TABLE 47

Ventilatory Parameters in Hamsters at Various Ages[a]

Parameter	15 Weeks	65 Weeks
Respiratory frequency (breath/min)	24 ± 2.7	25 ± 3.9
Tidal volume (mL)	1.2 ± 0.2	1.1 ± 0.2
Minute volume (mL/min)	27.8 ± 3.3	28.1 ± 4.0

Source: Adapted from Mauderly, J.L. (1979). From Sahebjami, H. (1992).
[a] Values are means ± SD.

TABLE 48

Morphometric Values in Sprague-Dawley Rats at Various Ages[a]

Parameter	4 Months	8 Months	18 Months
V_L body weight (mL/kg)	21.7 ± 1.0	30.9 ± 1.5	38.4 ± 2.8
Lm (μm)	54 ± 2	71 ± 2	87 ± 7
ISA (cm²)	5.571 ± 445	7.979 ± 318	8.733 ± 721

Source: Adapted from Johanson, W.G., Jr. and Pierce, A.K. (1973); Sahebjami, H. (1992).
[a] Values are means ± SEM. V_L postfixation lung volume; Lm, mean chord length; ISA, internal surface area.

TABLE 49

Normal Cytology of BALF (Percentage of Total Cells)

Animal	Macrophages	Neutrol	EOS	Lymph
Rat, mouse, rabbit, Syrian hamster	95	<1	<1	<1
Guinea pig	90	—	10	—
Rabbit	95	<1	<1	4
Dog	85	5	5	5
Sheep	70	5	5	15
Horse	83	5	<1	10
Monkey	89	—	—	10
Human (nonsmoker)	88	<1	<1	10

Source: From Henderson, R.F. (1989). With permission.
Note: BALF, bronchoalveolar lavage fluid; Neutro, neutrophil; EOS, eosinophils; Lymph, lymphocytes.

TABLE 50

Normal Biochemical Content of BALF, $\overline{X}$ (SE)[a]

Animal	n	LDH (mIU/mL)	Alkaline Phosphatase (mIU/mL)	Acid Phosphatase (mIU/mL)	β-Glucuronidase (mIU/mL)	Protein (mg/mL)
Rat	240–280	109 (2)	53 (1)	2.4 (0.1)	0.34 (0.02)	0.39 (0.02)
Mouse	45–95	233 (13)	2.5 (0.2)	7.5 (0.8)	0.53 (0.08)	0.82 (0.07)
Guinea pig	6	69 (26)	5.7 (1.6)	2.5 (0.2)	0.65 (0.12)	0.13 (0.03)
Syrian hamster	6	72 (7)	3.6 (1.0)	2.0 (0.1)	0.57 (0.09)	0.37 (0.03)
Rabbit	6	27 (6)	8.5 (4.4)	5.3 (0.5)	0.37 (0.02)	0.44 (0.10)
Dog	4–12	134 (25)	22 (5)	1.4 (0.1)	0.30 (0.04)	0.35 (0.18)
Chimpanzee	5	51 (12)	53 (3)	—	—	0.01 (9.01)

Source: From Henderson, R.F. (1989). With permission.

[a] Values are normalized per milliliter of lung volume washed.

TABLE 51

Tracheal Mucociliary Clearance

Species	Mucous Velocity[a] (mm/min)
Mouse	+
Rat	1.9 ± 0.7
	5.1 ± 3.0
	5.9 ± 2.5
Ferret	+
	18.2 ± 5.1
	10.7 ± 3.7
Guinea pig	2.7 ± 1.4
Rabbit	3.2 ± 1.1
	+
Chicken	*
Cat	2.5 ± 0.8
Dog	21.6 ± 5.0
	9.8 ± 2.1
	19.2 ± 1.6
	7.5 ± 3.7
	14.5 ± 6.3
Baboon	+
Sheep	17.3 ± 6.2
	10.5 ± 2.9
Pig	*
Cow	*
Donkey	14.7 ± 3.8
Horse	16.6 ± 2.4
	17.8 ± 5.1
Human	3.6 ± 1.5
	5.5 ± 0.4
	5.1 ± 2.9

(*Continued*)

TABLE 51 (*Continued*)

Tracheal Mucociliary Clearance

Species	Mucous Velocity[a] (mm/min)
	11.5 ± 4.7
	10.1 ± 3.5
	21.5 ± 5.5
	15.5 ± 1.7

Source: From Wolff, R.K. (1992). With permission.

Note: *, Transport studied but no velocity given; +, Inhalation study, clearance measured but no tracheal velocities given.

[a] Mean ± SD.

TABLE 52

Nasal Mucociliary Clearance

Species	Velocity[a] (mm/min)
Rat	2.3 ± 0.8
Dog	3.7 ± 0.9
Man	5.2 ± 2.3
	5.5 ± 3.2
	5.3 (0.5–23.6)
	8.4 ± 4.8
	6.8 ± 5.1
	7 ± 4

Source: From Wolff, R.K. (1992). With permission.

[a] Mean ± SD.

TABLE 53

Ammonia Concentrations in an Inhalation Chamber

Animal Loading (%)	Chamber Air Flow (L/min)	No. of Air Changes per Hour	Hour of Sample (ppm $NH_3 \pm SE$)		
			2	4	6
1	13	8	0.38 ± 0.08	0.48 ± 0.07	0.46 ± 0.13
1	26	16	0.20 ± 0.01	0.24 ± 0.02	0.45 ± 0.06
1	40	24	0.19 ± 0.04	0.24 ± 0.05	0.22 ± 0.03
3.1	13	8	0.84 ± 0.14	1.13 ± 0.14	1.11 ± 0.27
3.1	26	16	0.60 ± 0.09	1.04 ± 0.23	1.60 ± 0.22
3.1	40	24	0.19 ± 0.02	0.33 ± 0.05	0.39 ± 0.05
5.1	13	8	1.23 ± 0.18	1.51 ± 0.16	2.42 ± 0.38
5.2	26	16	0.66 ± 0.06	1.23 ± 0.20	2.05 ± 0.41
5.2	40	24	0.46 ± 0.08	1.02 ± 0.11	1.30 ± 0.27

Source: From Phalen, R.F. (1984). With permission.

TABLE 54

Conversion Table for Gases and Vapors (1 mg/L → ppm, and 1 ppm → mg/L, 25°C and 760 mmHg)[a]

Molecular Weight	1 mg/L ppm	1 ppm mg/L	Molecular Weight	1 mg/L ppm	1 ppm mg/L	Molecular Weight	1 mg/L ppm	1 ppm mg/L
1	24,450	0.0000409	25	987	0.001022	49	499	0.002004
2	12,230	0.0000818	26	940	0.001063	50	489	0.002045
3	8,150	0.0001227	27	906	0.001104	51	479	0.002086
4	6,113	0.0001636	28	873	0.001145	52	470	0.002127
5	4,890	0.0002045	29	843	0.001186	53	461	0.002168
6	4,075	0.0002454	30	815	0.001227	54	453	0.002209
7	3,493	0.0002863	31	789	0.001268	55	445	0.002250
8	3,056	0.000327	32	764	0.001309	56	437	0.002290
9	2,717	0.000368	33	741	0.001350	57	429	0.002331
10	2,445	0.000409	34	719	0.001391	58	422	0.002372
11	2,223	0.000450	35	699	0.001432	59	414	0.002413
12	2,038	0.000491	36	679	0.001472	60	408	0.002454
13	1,881	0.000532	37	661	0.001513	61	401	0.002495
14	1,746	0.000573	38	643	0.001554	62	394	0.00254
15	1,630	0.000614	39	627	0.001595	63	388	0.00258
16	1,528	0.000654	40	611	0.001636	64	382	0.00262
17	1,438	0.000695	41	596	0.001677	65	376	0.00266

(Continued)

TABLE 54 (*Continued*)

Conversion Table for Gases and Vapors (1 mg/L $\to$ ppm, and 1 ppm $\to$ mg/L)[a]

Molecular Weight	1 mg/L ppm	1 ppm mg/L	Molecular Weight	1 mg/L ppm	1 ppm mg/L	Molecular Weight	1 mg/L ppm	1 ppm mg/L
18	1,358	0.000736	42	582	0.001718	66	370	0.00270
19	1,287	0.000777	43	569	0.001759	67	365	0.00274
20	1,223	0.000818	44	556	0.001800	68	360	0.00278
21	1,164	0.000859	45	543	0.001840	69	354	0.00282
22	1,111	0.000900	46	532	0.001881	70	349	0.00286
23	1,063	0.000941	47	520	0.001922	71	344	0.00290
24	1,019	0.000982	48	509	0.001963	72	340	0.00294
73	335	0.00299	124	197.2	0.00507	175	139.7	0.00716
74	330	0.00303	125	195.6	0.00511	176	138.9	0.00720
75	326	0.00307	126	194.0	0.00515	177	138.1	0.00724
76	322	0.00311	127	192.5	0.00519	178	137.4	0.00728
77	318	0.00315	128	191.0	0.00524	179	136.6	0.00732
78	313	0.00319	129	189.5	0.00528	180	135.8	0.00736
79	309	0.00323	130	188.1	0.00532	181	135.1	0.00740
80	306	0.00327	131	186.6	0.00536	182	134.3	0.00744
81	302	0.00331	132	185.2	0.00540	183	133.6	0.00748
82	298	0.00335	133	183.8	0.00544	184	132.9	0.00753
83	295	0.00339	134	182.5	0.00548	185	132.2	0.00757

(*Continued*)

TABLE 54 (Continued)

Conversion Table for Gases and Vapors (1 mg/L → ppm, and 1 ppm → mg/L)[a]

Molecular Weight	1 mg/L ppm	1 ppm mg/L	Molecular Weight	1 mg/L ppm	1 ppm mg/L	Molecular Weight	1 mg/L ppm	1 ppm mg/L
84	291	0.00344	135	181.1	0.00552	186	131.5	0.00761
85	288	0.00348	136	179.8	0.00556	187	130.7	0.00765
86	284	0.00352	137	178.5	0.00560	188	130.1	0.00769
87	281	0.00356	138	177.2	0.00564	189	129.4	0.00773
88	278	0.00360	139	175.9	0.00569	190	128.7	0.00777
89	275	0.00364	140	174.6	0.00573	191	128.0	0.00781
90	272	0.00368	141	173.4	0.00577	192	127.3	0.00785
91	269	0.00372	142	172.2	0.00581	193	126.7	0.00789
92	266	0.00376	143	171.0	0.00585	194	126.0	0.00793
93	263	0.00380	144	169.8	0.00589	195	125.4	0.00798
94	260	0.00384	145	168.6	0.00593	196	124.7	0.00802
95	257	0.00389	146	167.5	0.00597	197	124.1	0.00806
96	255	0.00393	147	166.3	0.00601	198	123.5	0.00810
97	252	0.00397	148	165.2	0.00605	199	122.9	0.00814
98	249.5	0.00401	149	164.1	0.00609	200	122.3	0.00818
99	247.0	0.00405	150	163.0	0.00613	201	121.6	0.00822
100	244.5	0.00409	151	161.9	0.00618	202	121.0	0.00826
101	242.1	0.00413	152	160.9	0.00622	203	120.4	0.00830

(Continued)

TABLE 54 (Continued)

Conversion Table for Gases and Vapors (1 mg/L → ppm, and 1 ppm → mg/L)[a]

Molecular Weight	1 mg/L ppm	1 ppm mg/L	Molecular Weight	1 mg/L ppm	1 ppm mg/L	Molecular Weight	1 mg/L ppm	1 ppm mg/L
102	239.7	0.00417	153	159.8	0.00626	204	119.9	0.00834
103	237.4	0.00421	154	158.8	0.00630	205	119.3	0.00838
104	235.1	0.00425	155	157.7	0.00634	206	118.7	0.00843
105	232.9	0.00429	156	156.7	0.00638	207	118.1	0.00847
106	230.7	0.00434	157	155.7	0.00642	208	117.5	0.00851
107	228.5	0.00438	158	154.7	0.00646	209	117.0	0.00855
108	226.4	0.00442	159	153.7	0.00650	210	116.4	0.00859
109	224.3	0.00446	160	152.8	0.00654	211	115.9	0.00863
110	222.3	0.00450	161	151.9	0.00658	212	115.3	0.00867
111	220.3	0.00454	162	150.9	0.00663	213	114.8	0.00871
112	218.3	0.00458	163	150.0	0.00667	214	114.3	0.00875
113	216.4	0.00462	164	149.1	0.00671	215	113.7	0.00879
114	214.5	0.00466	165	148.2	0.00675	216	113.2	0.00883
115	212.6	0.00470	166	147.3	0.00679	217	112.7	0.00888
116	210.8	0.00474	167	146.4	0.00683	218	112.2	0.00892
117	209.0	0.00479	168	145.5	0.00687	219	111.6	0.00896
118	207.2	0.00483	169	144.7	0.00691	220	111.1	0.00900
119	205.5	0.00487	170	143.8	0.00695	221	110.6	0.00904

(Continued)

TABLE 54 (Continued)

Conversion Table for Gases and Vapors (1 mg/L → ppm, and 1 ppm → mg/L)[a]

Molecular Weight	1 mg/L ppm	1 ppm mg/L	Molecular Weight	1 mg/L ppm	1 ppm mg/L	Molecular Weight	1 mg/L ppm	1 ppm mg/L
120	203.8	0.00491	171	143.0	0.00699	222	110.1	0.00908
121	202.1	0.00495	172	142.2	0.00703	223	109.6	0.00912
122	200.4	0.00499	173	141.3	0.00708	224	109.2	0.00916
123	198.8	0.00503	174	140.5	0.00712	225	108.7	0.00920
226	108.2	0.00924	251	97.4	0.01027	276	88.6	0.01129
227	107.7	0.00928	252	97.0	0.01031	277	88.3	0.01133
228	107.2	0.00933	253	96.6	0.01035	278	87.9	0.01137
229	106.8	0.00937	254	96.3	0.01039	279	87.6	0.01141
230	106.3	0.00941	255	95.9	0.01043	280	87.3	0.01145
231	105.8	0.00945	256	95.5	0.01047	281	87.0	0.01149
232	105.4	0.00949	257	95.1	0.01051	282	86.7	0.01153
233	104.9	0.00953	258	94.8	0.01055	283	86.4	0.01157
234	104.5	0.00957	259	94.4	0.01059	284	86.1	0.01162
235	104.0	0.00961	260	94.0	0.01063	285	85.8	0.01166
236	103.6	0.00965	261	93.7	0.01067	286	85.5	0.01170
237	103.2	0.00969	262	93.3	0.01072	287	85.2	0.01174
238	102.7	0.00973	263	93.0	0.01076	288	84.9	0.01178

(Continued)

TABLE 54 (Continued)

Conversion Table for Gases and Vapors (1 mg/L → ppm, and 1 ppm → mg/L)[a]

Molecular Weight	1 mg/L ppm	1 ppm mg/L	Molecular Weight	1 mg/L ppm	1 ppm mg/L	Molecular Weight	1 mg/L ppm	1 ppm mg/L
239	102.3	0.00978	264	92.6	0.01080	289	84.6	0.01182
240	101.9	0.00982	265	92.3	0.01084	290	84.3	0.01186
241	101.5	0.00986	266	91.9	0.01088	291	84.0	0.01190
242	101.0	0.00990	267	91.6	0.01092	292	83.7	0.01194
243	100.6	0.00994	268	91.2	0.01096	293	83.4	0.01198
244	100.2	0.00998	269	90.9	0.01100	294	83.2	0.01202
245	99.8	0.01002	270	90.6	0.01104	295	82.9	0.01207
246	99.4	0.01006	271	90.2	0.01108	296	82.6	0.01211
247	99.0	0.01010	272	89.9	0.01112	297	82.3	0.01215
248	98.6	0.01014	273	89.6	0.01117	298	82.0	0.01219
249	98.2	0.01018	274	89.2	0.01121	299	81.8	0.01223
250	97.8	0.01022	275	88.9	0.01125	300	81.5	0.01227

Source: From Fieldner, A.C. et al. (1921); Clayton, G.D. and Clayton, F.E., Eds. (1991). With permission.

[a] For example - for a gas of molecular weight of 100, 1 mg/L = 244.5 ppm and 1 ppm = 0.00409 mg/L.

TABLE 55

Calculations Used in Inhalation Toxicology

- *Chamber air change time*

$$\text{Air change (min)} = V/F$$

where V = volume of the chamber (L)
F = flow rate through the chamber (L/min)

- *Time to chamber equilibration (T_x)*

$$T_x \text{ (min)} = K \times V/F$$

where V = volume of chamber (L)
F = flow rate through the chamber (L/min)
K = exponential constant = 4.6 (99% equilibration)
= 2.3 (90% equilibration)

- *Minimum flow rate for nose only chamber (Q)*

$$Q(L/min) = \text{number of animals} \times \text{minute volume}$$

- *Volume-to-volume concentration of gas or vapor in air*

$$\text{Concentration (ppm)} = \text{volume of vapor or gas (μL)/volume of air (L)}$$

- *Weight-to-volume/ppm conversion*

$$\text{Concentration (mg/m}^3) = \text{concentration (ppm)} \times MW/24.5$$

where MW = molecular weight
24.5 = universal gas constant at 25°C and 760 mmHg

- *Concentration of pure gas metered into a chamber*

$$\text{Concentration (ppm)} = \text{flow rate of gas(L/min)/flow rate of chamber(L/min)} \times 10^6 \text{ μL/L}$$

- *Maximum attainable concentration in air for a volatile liquid*

$$\text{Concentration (ppm)} = \text{vapor pressure (mmHg)/atmospheric pressure(mmHg)} \times 10^6 \text{ μL/L}$$

where atmospheric pressure = 760 mm/Hg at sea level

(Continued)

TABLE 55 (*Continued*)

Calculations Used in Inhalation Toxicology

- *Haber's rule*

$$\text{Response} = C \times T$$

where C = exposure concentration
T = time of exposure

- *Nominal concentration*

$$\text{Nominal concentration (mg/m}^3) = W/V \times 1000 \text{ L/m}^3$$

where W = quantity (mg) of test material consumed during the exposure
V = volume of air (L) through the chamber during the exposure

- *Theoretical resting ventilation rate for mammals (V_m)*

$$V_m \text{ (mL/min)} = 2.18 \text{ M}^{3/4}$$

where M = mass of the animal (g)

- *Theoretical dose from an inhalation exposure*

$$\text{Dose (mg/kg)} = C \times MV \times T \times D/BW$$

where C = concentration of test material in air (mg/L)
MV = minute volume of animal (L/min)
T = time of exposure duration (min)
D = deposition fraction into the respiratory tract
BW = body weight of animal (kg)

References

Clayton, G.D. and Clayton, F.E., Eds. (1991), *Patty's Industrial Hygiene and Toxicology*, 4th ed., John Wiley & Sons, New York.

Fieldner, A.C., Katz, S.H., and Kinney, S.P. (1921), Gas Masks for Gases Met in Fighting Fires, U.S. Bureau of Mines, Technical Paper No. 248.

Henderson, R.F. (1989), Bronchoalveolar lavage: A tool for assessing the health status of the lung. In *Concepts in Inhalation Toxicology*, McClellan, R.O., and Henderson, R.F., Eds., Hemisphere Publishing, New York, chap. 15.

Johanson, W.G., Jr. and Pierce, A.K. (1973), Lung structure and function with age in normal rats and rats with papain emphysema, *J. Clin. Invest.* 52, 2921.

Mauderly, J.L. (1982), The effect of age on respiratory function of Fischer-344 rats, *Exp. Aging Res.* 8, 31.

Mauderly, J.L. (1979), Ventilation, lung volumes and lung mechanics of young adult and old Syrian hamsters, *Exp. Aging Res.* 5, 497.

Phalen, R.F. (1984), *Inhalation Studies: Foundations and Techniques*, CRC Press, Boca Raton, FL.

Sahebjami, H. (1992), Aging of the normal lung, in *Treatise on Pulmonary Toxicology, Vol. 1: Comparative Biology of the Normal Lung*, Parent, R.A., Ed., CRC Press, Boca Raton, FL, chap. 21.

Wolff, R.K. (1992), Mucocilliary function, in *Treatise on Pulmonary Toxicology, Vol. 1: Comparative Biology of the Normal Lung*, Parent, R.A., Ed., CRC Press, Boca Raton, FL, chap. 35.

4

Dermal Toxicology

TABLE 56

Relative Ranking of Skin Permeability in Different Animal Species

Mouse > Guinea Pig > Rabbit > Dog > Monkey > Pig > Human > Chimpanzee
Most permeable ⟶ Least permeable

Source: Leung, H-W. and Paustenbach, D.J. (1999).

TABLE 57

Common Materials Used as Positive Controls

Material	CAS No.	Suggested Concentrations	Category
Sodium lauryl sulfate (SLS)	151-21-3	1.0%	Irritant
Hexylcinnamaldehyde (HCA)	101-86-0	25% in acetone:olive oil (4:1, v/v) (OECD 429/ OECD 442A/OECD 442B)	Mild-to-moderate sensitizer
Mercaptobenzothiazole	149-30-4	5% in N,N-dimethylformamide (OECD 429)	Mild-to-moderate sensitizer
Benzocaine	94-09-7	—	Mild-to-moderate sensitizer
p-Phenylenediamine	106-50-3	—	Sensitizer
2,4-Dinitrochlorobenzene (DNCB)	97-00-7	Induction: 0.1%–0.5%, 0.25% w/v in ethanol/acetone Challenge: 0.1%–0.3%, w/v in ethanol/acetone	Sensitizer

(Continued)

TABLE 57 (*Continued*)

Common Materials Used as Positive Controls

Material	CAS No.	Suggested Concentrations	Category
Eugenol	97-53-0	25% in acetone:olive oil (4:1, v/v) (OECD 442A/ OECD 442B)	Sensitizer
Potassium dichromate	7778-50-9	—	Sensitizer
Neomycin sulfate	1405-10-3	—	Sensitizer
Nickel sulfate	7786-81-4	—	Sensitizer
8-Methoxypsoralen (Oxsoralen Lotion®)	298-81-7	1.0%	Photoirritant
5-Methoxypsoralen (Bergapten)	298-81-7	1.0%	Photoirritant
2,4-Dinitro,3-methyl,6-tertiarybutyl- anisole (musk ambrette)	83-66-9	Induction: 10.0% w/v in ethanol/acetone Challenge: 0.5% w/v in ethanol/acetone	Photosensitizer
2-Chloro 10[3-dimethylaminopropyl] phenothiazine hydrochloride (chloropromazine)	50-53-3	Induction: 1.0% w/v in methanol Challenge: 0.1% w/v in methanol	Photosensitizer
3,3,4,5-Tetrachlorosalicylandide (TCSA)	1154-59-2	Induction: 1.0% w/v in acetone Challenge: 1.0% w/v in acetone	Photosensitizer (in mice and guinea pigs), possible sensitizer in guinea pigs

Source: From Organization for Economic Cooperation and Development (1992), The Commission of the European Communities (1992b), Springborn Laboratories, Inc. (1994a, 1994b, 1994c, 1994d), Hawkins, R.E. et al. (1961), Siglin, J.C. et al. (1991), Ichikawa, H. et al. (1981).

TABLE 58

Draize Dermal Irritation Scoring System

Erythema and Eschar Formation	Value	Edema Formation	Value
No erythema	0	No edema	0
Very slight erythema (barely perceptible)	1	Very slight edema (barely perceptible)	1
Well-defined erythema	2	Slight edema (edges of area well-defined by definite raising)	2
Moderate-to-severe erythema	3	Moderate edema (raised approximately 1 mm)	3
Severe erythema (beet redness) to slight, eschar formation (injuries in depth)	4	Severe edema (raised more than 1 mm and extending beyond the area of exposure)	4

Source: From Draize, J.H. (1959).

TABLE 59

Human Patch Test Dermal Irritation Scoring System

Skin Reaction	Value
No sign of inflammation; normal skin	0
Glazed appearance of the sites, or barely perceptible erythema	±(0.5)
Slight erythema	1
Moderate erythema, possible with barely perceptible edema at the margin; papules may be present	2
Moderate erythema, with generalized edema	3
Severe erythema with severe edema, with or without vesicles	4
Severe reaction spread beyond the area of the patch	5

Source: From Patrick, E. and Maibach, H.I. (1989).

TABLE 60

Chamber Scarification Dermal Irritation Scoring System

Skin Reaction	Value
Scratch marks barely visible	0
Erythema confined to scratches perceptible erythema	1
Broader bands of increased erythema, with or without rows of vesicles, pustules, or erosions	2
Severe erythema with partial confluency, with or without other lesions	3
Confluent, severe erythema sometimes associated with edema, necrosis, or bullae	4

Source: From Patrick, E. and Maibach, H.I. (1989).

TABLE 61

Magnusson Sensitization Scoring System

Skin Reaction	Value
No reaction	0
Scattered reaction	1
Moderate and diffuse reaction	2
Intense reddening and swelling	3

Source: From Magnusson, B. and Kligman, A. (1970).

TABLE 62

Split Adjuvant Sensitization Scoring System

Skin Reaction	Value
Normal skin	0
Very faint, nonconfluent pink	±
Faint pink	+
Pale pink to pink, slight edema	++
Pink, moderate edema	+++
Pink and thickened	++++
Bright pink, markedly thickened	+++++

Source: From Klecak, G. (1983).

TABLE 63

Buehler Sensitization Scoring System

Skin Reaction	Value
No reaction	0
Very faint erythema, usually confluent	±(0.5)
Faint erythema, usually confluent	1
Moderate erythema	2
Strong erythema, with or without edema	3

Source: From Buehler, E.V. and Griffin, F. (1975).

TABLE 64

Contact Photosensitization Scoring System

Skin Reaction	Value
No erythema	0
Minimal but definite erythema confluent	1
Moderate erythema	2
Considerable erythema	3
Maximal erythema	4

Source: From Harber, L.C. et al. (1993).

TABLE 65

Local Lymph Node Ear Scoring System

Skin Reaction	Value
No erythema	0
Very slight erythema (barely perceptible)	1
Well-defined erythema	2
Moderate-to-severe erythema	3
Severe erythema (beet redness) to eschar formation preventing grading of erythema	4

Source: Organization for Economic Cooperation and Development (2010a).

TABLE 66

Human Patch Test Sensitization Scoring System

Skin Reaction	Value
Doubtful reaction; faint erythema only	? or + ?
Weak positive reaction; erythema, infiltration, discrete papules	+
Strong positive reaction: erythema, infiltration, papules, vesicles	++
Extreme positive reaction; intense erythema, infiltration, and coalescing vesicles	+++
Negative reaction	—
Irritant reaction of different types	IR
Not tested	NT

Source: From Fischer, T. and Maibach, H.I. (1991).

TABLE 67

Environmental Protection Agency (EPA) Method of Calculating the Primary Irritation Index (PII) for Dermal Irritation Studies

Option 1

Separately add up each animal's erythema and edema scores for the 1, 24, 48, and 72 h scoring intervals. Add all six values together and divide by the (number of test sites × 4 scoring intervals).

Option 2

Add the 1, 24, 48, and 72 h erythema and edema scores for all animals and divide by the (number of test sites × 4 scoring intervals).

Source: From U.S. Environmental Protection Agency (1984a) and (1992).

TABLE 68

Federal Hazardous Substances Act (CPSC-FHSA) Method of Calculating the Primary Irritation Index (PII) for Dermal Irritation Studies

Option 1

Separately add up each animal's intact and abraded erythema and edema scores for the 25 and 72 h scoring intervals. Add all six values together and divide by the (number of test sites × 2 scoring intervals).

Option 2

Add the 25 and 72 h erythema and edema scores for all animals (intact and abraded sites) and divide by the (number of test sites × 2 scoring intervals).

Source: From U.S. Consumer Products Safety Commission (1993).

TABLE 69

European Economic Community's (EEC) Method of Calculating the Primary Irritation Index (PII) for Dermal Irritation Studies

For Six Animals

1. *Erythema*: Add all 24, 48, and 72 h erythema scores for each animal together and divide by the (number of test sites × 3 scoring intervals).
2. *Edema*: Add all 24, 48, and 72 h edema scores for each animal together and divide by the (number of test sites × 3 scoring intervals).

For Three Animals

1. *Erythema*: Add all 24, 48, and 72 h erythema scores of each animal individually and divide by the number of scoring intervals.
2. *Edema*: Add all 24, 48, and 72 h edema scores of each animal individually and divide by the number of scoring intervals.

Source: From the Commission of the European Communities (1992a).

TABLE 70

Environmental Protection Agency (EPA)
Dermal Classification System

Primary Irritation Index	Irritation Rating
0.00	Nonirritant
0.01–1.99	Slight irritant
2.00–5.00	Moderate irritant
5.01–8.00	Severe irritant

Source: From U.S. Environmental Protection Agency (1988).

TABLE 71

Environmental Protection Agency (EPA) Standard Evaluation
Procedure Dermal Classification System

Mean Score (Primary Irritation Index)	Response Category
0–0.4	Negligible
0.5–1.9	Slight
2–4.9	Moderate
5–8.0	Strong (primary irritant)

Source: From U.S. Environmental Protection Agency (1984b).

TABLE 72

Federal Fungicide, Insecticide, and Rodenticide Act (EPA-FIFRA)
Dermal Classification System

Toxicity Category	Warning Label
I	Corrosive. Causes eye and skin damage (or irritation). Do not get in eyes, on skin, or on clothing. Wear goggles or face shield and gloves when handling. Harmful or fatal if swallowed. (Appropriate first aid statement required.)
II	Severe irritation at 72 h. Causes eye (and skin) irritation. Do not get on skin or on clothing. Harmful if swallowed. (Appropriate first aid statement required.)
III	Moderate irritation at 72 h. Avoid contact with skin, eyes, or clothing. In case of contact, immediately flush eyes or skin with plenty of water. Get medical attention if irritation persists.
IV	Mild or slight irritation at 72 h. (No precautionary statements required.)

Source: From U.S. Environmental Protection Agency (1993).

TABLE 73

European Economic Community (EEC) Dermal
Classification System

Mean Erythema Score	Irritation Rating
0.00–1.99	Nonirritant
≥2.00	Irritant
Mean Edema Score	**Irritation Rating**
0.00–1.99	Nonirritant
≥2.00	Irritant

Source: From The Commission of the European Communities (1983).

TABLE 74

Federal Hazardous Substances Act (CPSC-FHSA)
Dermal Classification System

Primary Irritation Score	Irritation Rating
0.00–4.99	Nonirritant
≥5.00	Irritant

Source: From U.S. Consumer Products Safety Commission (1993).

TABLE 75

Draize Dermal Classification System

Primary Irritation Index	Irritation Rating
<2	Mildly irritating
2–5	Moderately irritating
>5	Severely irritating

Source: From Patrick, E. and Maibach, H.I. (1989).

TABLE 76

Environmental Protection Agency (EPA) Design
for the Environment Dermal Classification System
Based on Skin Irritation/Corrosivity

Classification	Criteria
Very high	Corrosive
High	Severe irritation at 72 hours
Moderate	Moderate irritation at 72 hours
Low	Mild or slight irritation at 72 hours
Very low	Not irritating

Source: U.S. Environmental Protection Agency (2011).

TABLE 77

United Nations Globally Harmonized System (GHS) for Dermal Classification Based on Skin Irritation/Corrosivity

Category	Classification	Criteria
1A	Corrosive	Corrosive in ≥1 of 3 animals ≤3 min exposure ≤1 hour observation
1B	Corrosive	>3 min ≤1 hour exposure ≤14 days observation
1C	Corrosive	>1 hour ≤4 hour exposure ≤14 days observation
2	Irritant	Mean value of ≥2.3 ≤4.0 for erythema/eschar or for edema in at least 2 of 3 tested animals from gradings at 24, 48, and 72 hours after patch removal or, if reactions are delayed, from grades on 3 consecutive days after the onset of skin reactions (or) Inflammation that persists to the end of the observation period normally 14 days in at least 2 animals, particularly taking into account alopecia (limited area), hyperkeratosis, hyperplasia, and scaling (or) In some cases where there is pronounced variability of response among animals, with very definitive positive effects related to chemical exposure in a single animal but less than the criteria above
3	Mild irritant	Mean value of ≥1.5 <2.3 for erythema/eschar or for edema in at least 2 of 3 tested animals from grades at 24, 48, and 72 hours or, if reactions are delayed, from grades on 3 consecutive days after the onset of skin reactions (when not included in the irritant category above)

Source: United Nations (2011).

TABLE 78

Department of Transportation (DOT), Occupational Safety and Health Administration (OSHA), and International Maritime Organization (IMO) Packing Group Classification System

Corrosive Subcategories (OHSA)/Packing Categories (DOT)	Definition
1A/I	Materials that cause full-thickness destruction of intact skin tissue within an observation period of up to 60 min starting after the exposure time of 3 min or less.
1B/II	Materials other than those meeting Packing Group I criteria that cause full-thickness destruction of intact skin tissue within an observation period of up to 14 days starting after the exposure time of more than 3 min but not more than 1 h.
1C/III	Materials, other than those meeting Packing Group I or II criteria 1. That cause full-thickness destruction of intact skin tissue within an observation period of up to 14 days starting after the exposure time of more than 1 h but not more than 4 h; or 2. That do not cause full-thickness tissue destruction of intact skin tissue but exhibit a corrosion rate on steel or aluminum surfaces exceeding 6.25 mm (0.25 in.)/year at a test temperature of 55°C (130°F).

Source: DOT (2012); International Maritime Dangerous Goods Code (1994); U.S. Occupational Safety and Health Administration (2012).

Note: Several of these agencies accept human and *in vitro* testing based on OECD guidelines in some cases.

TABLE 79

Maximization Sensitization Classification System

Sensitization Rate, %	Grade	Classification
0	—	Nonsensitizer
>0–8	I	Weak sensitizer
9–28	II	Mild sensitizer
29–64	III	Moderate sensitizer
65–80	IV	Strong sensitizer
81–100	V	Extreme sensitizer

Source: From Magnusson, B. and Kligman, A. (1970).

TABLE 80

Optimization Sensitization Classification System

Intradermal Positive Animals, %	Epidermal Positive Animals, %	Classification
s, >75	And/or s, >50	Strong sensitizer
s, 50–75	And/or s, 30–50	Moderate sensitizer
s, 30–50	n.s., 0–30	Weak sensitizer
n.s., 0–30	n.s., 0	No sensitizer

Source: From Patrick, E. and Maibach, H.I. (1989).
Note: s, significant; n.s., not significant (using Fisher's Exact Test).

TABLE 81

OECD and EPA Sensitization Classification Systems

Modified Buehler and Standard Buehler tests (BTs)	≥15% = a mild-to-moderate sensitizer (e.g., HCA) or ≥30% = a moderate-to-strong sensitizer (e.g., DNCB)
GPMT	≥30% = a mild-to-moderate sensitizer (e.g., HCA)

Source: U.S. Environmental Protection Agency (2003); Organization for Economic Cooperation and Development (1992).
Note: Dermal scores of 1 in both the test and control animals are generally considered equivocal unless a higher dermal response is noted in the test animals.

TABLE 82

Local Lymph Node Classification Systems

EPA 2600 and OECD 429 (Radiolabeled Assay)	SI > 3 (SI > 20 is considered excessive)
OECD 442A (DA Method)	SI > 1.8 (1.8–2.5 considered borderline; SI > 10 is considered excessive)
OECD 442B (BrdU ELISA Method)	SI > 1.6 (1–1.9 considered borderline; SI > 14 is considered excessive)

Source: U.S. Environmental Protection Agency (2003); Organization for Economic Cooperation and Development (2010a); Organization for Economic Cooperation and Development (2010b); Organization for Economic Cooperation and Development (2010c).

Note: SI, mean result from test group/mean result from vehicle control group. For borderline results, consider dose–response relationship, statistical differences, and consistency of vehicle control responses.

TABLE 83

Environmental Protection Agency (EPA) Design for the Environment Dermal Classification System Based on Skin Sensitization

Classification	Criteria
High	High frequency of sensitization in humans and/or high potency in animals[a]
Moderate	Low-to-moderate frequency of sensitization in humans and/or low-to-moderate potency in animals[b]
Low	Adequate data available but not high or moderate category

Source: U.S. Environmental Protection Agency (2011).

[a] Human: Positive response ≤500 μg/cm^2 human patch test induction threshold; Animal: LLNA EC3 value ≤2%; GPMT ≥30% responding at ≤0.1% intradermal induction dose or ≥60% responding at >0.1% to ≤1% intradermal induction dose; Buehler assay ≥15% responding at ≤0.2% topical induction dose or ≥60% responding at >0.2% to ≤20% topical induction dose.

[b] Human: Positive response >500 μg/cm^2 human patch test induction threshold; Animal: LLNA EC3 value >2%; GPMT ≥30% to <60% responding at >0.1% to <1% intradermal induction dose or ≥30% responding at >1% intradermal induction dose; Buehler assay ≥15% to <60% responding at >0.2% to ≤20% topical induction dose or ≥15% responding at >20% topical induction dose.

TABLE 84

United Nations Globally Harmonized System (GHS) for Dermal Classification Based on Skin Sensitization

Category	Classification	Criteria
1A	Skin sensitizer	Substances showing a high frequency of occurrence in humans and/or a high potency in animals can be presumed to have the potential to produce significant sensitization in humans. Severity of reaction may also be considered[a]
1B	Skin sensitizer	Substances showing a low-to-moderate occurrence in humans and/or a low-to-moderate potency in animals can be presumed to have the potential to produce sensitization in humans. Severity of reaction may also be considered[b]

Source: United Nations (2011).

[a] Human: Positive response ≤500 µg/cm^2 human patch test induction threshold; Animal: LLNA EC3 value ≤2%; GPMT ≥30% responding at ≤0.1% intradermal induction dose or ≥60% responding at >0.1% to ≤1% intradermal induction dose; Buehler assay ≥15% responding at ≤0.2% topical induction dose or ≥60% responding at >0.2% to ≤20% topical induction dose.

[b] Human: Positive response >500 µg/cm^2 human patch test induction threshold; Animal: LLNA EC3 value >2%; GPMT ≥30% to <60% responding at >0.1% to <1% intradermal induction dose or ≥30% responding at >1% intradermal induction dose; Buehler assay ≥15% to <60% responding at >0.2% to ≤20% topical induction dose or ≥15% responding at >20% topical induction dose.

References

Buehler, E.V. and Griffin, F. (1975), Experimental skin sensitization in the guinea pig and man, *Animal Models Dermatol.* 55.

DOT (2012), Research and Special Programs Administration, U.S. Department of Transportation, 49 CFR, Part 173. 136 and 137, September.

Draize, J.H. (1959), *Appraisal of the Safety of Chemicals in Foods, Drugs and Cosmetics*, The Association of Food and Drug Officials of the United States, p. 49.

Fischer, T. and Maibach, H.I. (1991), Patch testing in allergic contact dermatitis, in *Exogenous Dermatoses: Environmental Dermatitis*, Menne, T. and Maibach, H.I., Eds., CRC Press, Boca Raton, FL, chap. 7.

Hakim, R.E., Freeman, R.G., Griffin, A.C., and Knox, J.M. (1961), Experimental toxicologic studies on 8-methoxypsoralen in animals exposed to the long ultraviolet, *J. Pharmacol. Exp. Ther.* 131, 394.

Harber, L.C., Shalita, A.R., and Armstrong, R.B. (1993), Immunologically mediated contact photosensitivity in guinea pigs, in *Dermatotoxicology*, 2nd ed., Marzulli, F.N. and Maibach, H.E., Eds., Hemisphere Publishing Corporation, Washington, DC, chap. 16.

Ichikawa, H., Armstrong, R.B., and Harber, L.C. (1981), Photoallergic contact dermatitis in guinea pigs; Improved induction technique using Freund's complete adjuvant. *J. Invest. Dermatol.* 76, 498.

International Maritime Dangerous Goods Code (1994), Class 8 Corrosives, International Maritime Organization, London, England.

Klecak, G. (1983), Identification of contact allergies: Predictive tests in animals, in *Dermatotoxicology*, 2nd ed., Marzulli, F.N. and Maibach, H.I., Eds., Hemisphere Publishing Corporation, Washington, DC, chap. 9.

Leung, H-W., and Paustenbach, D.J. (1999), Percutaneous toxicity, in *General and Applied Toxicology*, Ballantyne, B., Marrs, T.C., and Syversen, T., Eds., Groves Dictionaries, New York, pp. 577–586, chap 29.

Magnusson, B. and Kligman, A. (1970), *Allergic Contact Dermatitis in Guinea Pigs*, C.C. Thomas, Springfield, IL.

Organization for Economic Co-operation and Development (1992), OECD Guidelines for Testing of Chemicals, Section 4: Health Effects, Subsection 406: Skin Sensitization, 1.

Organization for Economic Cooperation and Development (2010a), *OECD Guidelines for Testing of Chemicals, Section 4: Health Effects, Subsection 429: Skin Sensitization: Local Lymph Node Assay.*

Organization for Economic Cooperation and Development (2010b), *OECD Guidelines for Testing of Chemicals Section 4: Health Effects, Subsection 442A: Skin Sensitization: Local Lymph Node Assay: DA.*

Organization for Economic Cooperation and Development (2010c), *OECD Guidelines for Testing of Chemicals, Section 4: Health Effects, Subsection 442B: Skin Sensitization: Local Lymph Node Assay: BrdU-ELISA.*

Patrick, E. and Maibach, H.I. (1989), Dermatotoxicology, in *Principles and Methods of Toxicology*, 2nd ed., Hayes, A.W., Ed., Raven Press, Ltd., New York, chap. 32.

Siglin, J.C., Jenkins, P.K., Smith, P.S., Ryan, C.A., and Gerberick, G.F. (1991), Evaluation of a New Murine Model for the Predictive Assessment of Contact Photoallergy (CPA), *American College of Toxicology Annual Meeting, Savannah, GA.*

Springborn Laboratories, Inc. (1994a), Protocol for a Photoallergy Study in Mice, FDA/PHS-2-2/94, Spencerville, OH.

Springborn Laboratories, Inc. (1994b), Protocol for a Photoirritation Study in Rabbits, FDA/PHS-1-2/94, Spencerville, OH.

Springborn Laboratories, Inc. (1994c), Protocol for a Photosensitization Study in Guinea Pigs, FDA/PHS-1-2/94, Spencerville, OH.

Springborn Laboratories, Inc. (1994d), Protocol for a Primary Irritation Study in Rabbits, EPA/PSI-1-2/94, Spencerville, OH.

The Commission of the European Communities (1992a), *Official Journal of the European Communities, Part B: Methods for the Determination of Toxicity*, No. L 383 A/124, B.4. Acute Toxicity (Skin Irritation).

The Commission of the European Communities (1992b), *Official Journal of the European Communities, Part B: Methods for the Determination of Toxicity*, No. L 383 A/131, B.6: Skin Sensitization.

The Commission of the European Communities (1983), *Official Journal of the European Communities, Annex VI, General Classification and Labeling Requirements for Dangerous Substances*, No. L257/11.

United Nations (2011), *Globally Harmonized System of Classification and Labelling of Chemicals (GHS)*, fourth revision edition, New York and Geneva.

U.S. Consumer Products Safety Commission (1993), 16 CFR Chapter II, Subchapter C: Federal Hazardous Substances Act Regulation, Part 1500, Subsection 1500.3: Definitions, 381.

U.S. Environmental Protection Agency (1984a), Federal Insecticide, Fungicide, Rodenticide Act, Pesticide Assessment Guidelines, Subdivision F, Hazard Evaluation: Human and Domestic Animals, Series 81-5 Dermal Irritation, 55e.

U.S. Environmental Protection Agency (1984b), Federal Insecticide, Fungicide, Rodenticide Act, Pesticide Assessment Guidelines, Hazard Evaluation Division, Standard Evaluation Procedure, Guidance for Evaluation of Dermal Irritation Testing, 1.

U.S. Environmental Protection Agency (1988), Federal Insecticide, Fungicide, Rodenticide Act, Pesticide Assessment Guidelines, Subdivision F: Hazard Evaluation: Humans and Domestic Animals—Addendum 3 on Data Reporting.

U.S. Environmental Protection Agency (1992), Toxic Substances Control Act, Test Guidelines, 40 CFR Part 798, Subpart E—Specific Organ/Tissue Toxicity, Section 798.4470 Primary Dermal Irritation, 491.

U.S. Environmental Protection Agency (1993), Toxic Substances Control Act, Test Guidelines, 40 CFR chap.1 (7-1-93), Part 156: Labeling Requirements for Pesticides and Devices, Section 156.10, 75.

U.S. Environmental Protection Agency (2003), *Health Effects Guidelines, OPPTS 870.2600: Skin Sensitization Study.*

U.S. Environmental Protection Agency (2011), *Design for the Environment Program, Alternatives Assessment Criteria for Hazard Evaluation,* Version 2.0, Office of Pollution Prevention and Toxics, August.

U.S. Occupational Safety and Health Administration (1991), Labor, 29 CFR Chapter XVII, Part 1910, Appendix A to Section 1900.1200—Health Hazard Definitions (Mandatory), 364.

U.S. Occupational Safety and Health Administration (2012), Labor, 29 CFR Chapter XVII, Part 1910, Appendix A to Section 1900.1200—Health Hazard Definitions (Mandatory).

5

Ocular Toxicology

TABLE 85

Scale of Weighted Scores for Grading the Severity of Ocular Lesions
(developed by Draize et al.)

In 1944, Draize et al. described an eye irritancy grading system for evaluating drugs
and other materials intended for use in or around the eye. Numerical scores were
assigned for reactions of cornea, iris, and conjunctivae. The total ocular irritation
score was calculated by a formula that gave the greatest weight to corneal changes
(total maximum = 80). A total maximum score = 10 for the iris, and 20 for the
conjunctiva.

I. Cornea

A. Opacity Degree of Density (area which is most dense is taken for reading)

Scattered or diffuse area—details of iris clearly visible 1

Easily discernible translucent areas—details of iris clearly visible 2

Opalescent areas—no details of iris visible; size of pupil barely discernible 3

Opaque—iris invisible 4

B. Area of Cornea Involved

One quarter (or less) but not zero 1

Greater than one quarter—less than one half 2

Greater than one half—less than three quarters 3

Greater than three quarters—up to whole area 4

Score equals A × B × 5   Total maximum = 80

II. Iris

A. Values

Folds above normal, congestion, swelling, circumcorneal injection
(any one or all of these or combination of any thereof), iris still reacting
to light (sluggish reaction is positive) 1

No reaction to light, hemorrhage; gross destruction (any one or all of these) 2

Score equals A × 5   Total possible maximum = 10

(Continued)

TABLE 85 (*Continued*)

Scale of Weighted Scores for Grading the Severity of Ocular Lesions
Developed by Draize et al.

III. Conjunctivae

    A. Redness (refers to palpebral conjunctivae only)

        Vessels definitely injected above normal 1

        More diffuse, deeper crimson red, individual vessels not easily discernible...... 2

        Diffuse beefy red 3

    B. Chemosis

        Any swelling above normal (includes nictitating membrane).............................. 1

        Obvious swelling with partial eversion of the lids 2

        Swelling with lids about half closed... 3

        Swelling with lids about half closed to completely closed 4

    C. Discharge

        Any amount different from normal (does not include small amounts
observed in inner canthus of normal animals) 1

        Discharge with moistening of the lids and hairs just adjacent to the lids 2

        Discharge with moistening of the lids and considerable area around the eye ... 3

        Score $(A + B + C) \times 2$ Total maximum = 20

Note: The maximum total score is the sum of all scores obtained for the cornea, iris, and conjunctivae.

TABLE 86

Grades for Ocular Lesions

The following standardized grading system is used in testing guidelines of several US federal agencies (Consumer Product Safety Commission, Occupational Safety and Health Administration, FDA, Environmental Protection Agency, and Food Safety and Quality Service of the Department of Agriculture) and the Organization for Economic Cooperation and Development (OECD) member countries.

Cornea

Opacity: degree of density (area most dense taken for reading)

No ulceration or opacity 0

Scattered or diffuse areas of opacity (other than slight dulling of normal luster, details of iris clearly visible) .... 1[a]

Easily discernible translucent areas, details of iris slightly obscured 2

Nacreous areas, no details of iris visible, size of pupil barely discernible 3

Opaque cornea, iris not discernible through the opacity 4

Iris

Normal 0

Markedly deepened rugae, congestion, swelling, moderate circumcorneal hyperemia, or injection—any of these or any combination thereof, iris still reacting to light (sluggish reaction is positive) .... 1[a]

No reaction to light, hemorrhage, gross destruction (any or all of these) 2

Conjunctivae

Redness (refers to palpebral and bulbar conjunctivae excluding cornea and iris)

Blood vessels normal 0

Some blood vessels definitely hyperemic (injected) 1

Diffuse, crimson color, individual vessels not easily discernible .... 2[a]

Diffuse beefy red 3

Chemosis: lids and/or nictitating membranes

No swelling 0

Any swelling above normal (includes nictitating membranes) 1

Obvious swelling with partial eversion of lids .... 2[a]

Swelling with lids about half closed 3

Swelling with lids more than half closed 4

[a] Readings at these numerical values or greater indicate positive responses.

TABLE 87

Classification of Compounds Based on Eye Irritation Properties

This classification scheme developed by Kay and Calandra (1962) utilizes the Draize
scoring system to rate the irritating potential of substances.

Step 1

Using the Draize eye irritation scoring system, find the maximum mean total score for
all three tissues (cornea, iris, and conjunctivae) occurring within the first 96 h after
instillation for which the incidence of this score plus or minus 5 points is at least 40%.

Step 2

Choose an initial or "tentative rating" on the basis of the score found in Step 1 as
follows:

Score from Step 1	Tentative Eye Irritation Rating	Symbol
0.0–0.5 points	Nonirritating	N
0.5–2.5 points	Practically nonirritating	PN
2.5–15 points	Minimally irritating	M_1
15–25 points	Mildly irritating	M_2
25–50 points	Moderately irritating	M_3
50–80 points	Severely irritating	S
80–100 points	Extremely irritating	E
100–110 points	Maximally irritating	M_x
For borderline scores, choose the higher rating		

Step 3

Tentative Rating	Requirement for Maintenance
N	$MTS_{24} = 0$; for $MTS_{24} > 0$, raise one level
PN	As for N
M_1	$MTS_{48} = 0$; for $MTS_{48} > 0$, raise one level
M_2	$MTS_{96} = 0$; for $MTS_{96} > 0$, raise one level
M_3	1. $MTS \leq 20$; for $MTS_f > 20$, raise one level
	2. $ITS_f = 10$ (60%); if not true then no rabbit may show ITS_f 30; otherwise raise one level

(Continued)

TABLE 87 (*Continued*)

Classification of Compounds Based on Eye Irritation Properties

Tentative Rating	Requirement for Maintenance
S	1. As for M_3, except use $MTS_f \leq 40$
	2. As for M_3, except use $ITS_f \leq 30$ (60%) and 60 for high
E	1. As for M_3, except use $MTS_f \leq 80$
	2. As for M_3, except use $ITS_f \leq 60$ (60%) and 100 for high
M_x	1. $MTS_f > 80$ (60%); for $MTS_f \leq 80$, lower one level
	2. $ITS_f > 60$ (60%); otherwise lower one level

Note 1: Symbols: MTS, mean total score; ITS, individual rabbit total score; Subscripts denote scoring interval: 24, 48, or 96 h; f, final score (7 days).

Note 2: Two requirements must be met before a tentative rating may become final. First, the mean total score for the 7-day scoring interval may not exceed 20 points if the rating is to be maintained. Second, individual total scores for at least 60% of the rabbits should be 10 points or less and in no case may any individual rabbit's total score exceed 30. If either or both of these requirements are not met, then the "tentative rating" must be raised one level, and the higher level becomes the "final rating."

TABLE 88

NAS Classification Method Based on Severity and Persistence

This descriptive scale, adapted from work conducted by Green et al., (1978) attaches significance to the persistence and reversibility of responses. It is based on the most severe response observed in a group of animals rather than the average response.

1. Inconsequential or Complete Lack of Irritation

Exposure of the eye to a material under the specified conditions causes no significant ocular changes. No staining with fluorescein can be observed. Any changes that occur clear within 24 h and are no greater than those caused by isotonic saline under the same conditions.

2. Moderate Irritation

Exposure of the eye to the material under the specified conditions causes minor, superficial, and transient changes of the cornea, iris, or conjunctiva as determined by external or slit lamp examination with fluorescein staining. The appearance at the 24 h or subsequent grading interval of any of the following changes is sufficient to characterize a response as moderate irritation: opacity of the cornea (other than a slight dulling of the normal luster), hyperemia of the iris, or swelling of the conjunctiva. All observations resolve within 7 days.

3. Substantial Irritation

Exposure of the eye to the material under the specified conditions causes significant injury to the eye, such as loss of the corneal epithelium, corneal opacity, iritis (other than a slight injection), conjunctivitis, pannus, or bullae. The effects resolve within 21 days.

4. Severe Irritation or Corrosion

Exposure of the eye to the material under the specified conditions results in the same types of injury as in the previous category and in significant necrosis or other injuries that adversely affect the visual process. Injuries persist for 21 days or more.

Source: From National Academy of Sciences (1977).

TABLE 89

Modified NAS Classification Method Developed by Brendan J. Dunn, Department of Toxicology and Risk Assessment, Honeywell International Inc. (unpublished)

This classification scheme helps distinguish mildly irritating substances from moderately irritating substances, as well as identifying strongly and severely irritating substances. It is based on the most severe ocular response observed in a group of animals, rather than the average response, and on the persistence of the response.

1. Nonirritation

Exposure of the eye to the material under the specified conditions causes no ocular changes. No tissue staining with fluorescein is observed. Slight conjunctival injection (grade 1; some vessels definitely injected) that does not clear within 24 h is not considered a significant change. This level of change is inconsequential as far as representing physical damage to the eye and can be seen to occur naturally for unexplained reasons in otherwise normal rabbits.

2. Mild Irritation

Exposure of the eye to the material under the specified conditions causes minor and/or transient changes as determined by external or slit lamp examination or fluorescein staining. No opacity, ulceration, or fluorescein staining of the cornea (except for staining that is characteristic of normal epithelial desquamation) are observed at any grading interval. The appearance of any of the following changes is sufficient to characterize a response as mild irritation:

- Grade 1 hyperemia of the iris that is observed at 1 h, but resolves by 24 h
- Grade 2 conjunctival hyperemia (redness) that is observed at 1, 24, and/or 48 h, but resolves by 72 h
- Grade 2 conjunctival chemosis (swelling) that is observed at 1 h, but diminishes to grade 1 or 0 by 24 h; or Grade 1 conjunctival chemosis that is observed at 1 and/or 24 and/or 48 h, but resolves by 72 h

3. Moderate Irritation

Exposure of the eye to the material under the specified conditions causes major ocular changes as determined by external or slit lamp examination or fluorescein staining. The appearance of any of the following changes is sufficient to characterize a response as moderate irritation:

- Opacity of the cornea (other than slight dulling of the normal luster) is observed at any observation period, but resolves by day 7.
- Ulceration of the cornea (absence of a confluent patch of corneal epithelium) is observed at any observation period, but resolves by day 7.

(Continued)

TABLE 89 (*Continued*)

Modified NAS Classification Method Developed by Brendan J. Dunn, Department of Toxicology and Risk Assessment, Honeywell International Inc. (unpublished)

- Fluorescein staining of the cornea (greater than that which is characteristic of normal epithelial desquamation) is observed at 1, 2, 3, and/or 4 days, but no staining is found by day 7.
- Grade 1 or 2 hyperemia of the iris (circumcorneal injection, congestion) is observed and persists to 24 h or longer, but resolves by day 7.
- Grade 2 conjunctival hyperemia is observed and persists to at least 72 h, but resolves by day 7; or Grade 3 conjunctival hyperemia is observed at any observation period, but resolves by day 7.
- Grade 1 or greater conjunctival chemosis is observed and persists to 72 h or longer, but resolves by day 7.

4. Strong Irritation (Clearing within 21 Days)

Exposure of the eye to the material under the specified conditions results in the type of injury described in the former category, but the effects (possibly including pannus or bullae) heal or clear within 21 days.

5. Severe Irritation (Persisting for 21 Days) or Corrosion

Exposure of the eye to the material under the specified conditions results in the type of injury described in the two former categories, but causes significant tissue destruction (necrosis) or injuries that probably adversely affect the visual process. The effects of the injuries persist for at least 21 days.

TABLE 90

Environmental Protection Agency (EPA) Design for the Environment Classification System Based on Eye Irritation/Corrosivity

Classification	Criteria
Very high	Irritation persists for >21 days or corrosive
High	Clearing in 8–21 days, severely irritating
Moderate	Clearing in 7 days or less, moderately irritating
Low	Clearing in less than 24 hours, mildly irritating
Very low	Not irritating

Source: U.S. Environmental Protection Agency (2011).

TABLE 91

United Nations Globally Harmonized System (GHS) for Classification Based on Eye Irritation/Corrosivity

Category	Classification	Criteria
1	Irreversible effects on the eye	At least in one animal effects on the cornea, iris, or conjunctiva that are not expected to reverse or have not fully reversed within an observation period of normally 21 days and/or at least in 2 of 3 animals, a positive response of corneal opacity ≥3 and/or iritis ≥1.5 calculated as the mean scores following grading at 24, 48, and 72 hours after installation of the test material. Category 1 includes other severe reactions (e.g., destruction of cornea)
2A	Irritating to eyes	At least in 2 of 3 animals a positive response of corneal opacity ≥1 and/or conjunctival redness ≥2 and/or conjunctival edema (chemosis) ≥2 calculated as the mean scores following grading at 24, 48, and 72 hours after installation of the test material which fully reverses within an observation period of normally 21 days
2B	Irritating to eyes	As for category 2A when the listed effects are fully reversible within 7 days of observation

Source: United Nations (2011).

TABLE 92

Categorization of Substances Using the Slit Lamp Biomicroscope and Fluorescein

Site	"Accept"	"Accept with Caution"	"Probably Injurious to Human Eyes"
Conjunctiva	Hyperemia without chemosis	Chemosis, less than 1 mm at the limbus	Chemosis, greater than 1 mm at the limbus
Cornea	Staining, corneal stippling[a] without confluence at 24 h	Confluence[b] of staining at 24–48 h	Staining with infiltration or edema
Anterior chamber	0	0	Flare[c] (visibility of slit beam; rubeosis of iris)

Source: From Beckley, J.H. et al. (1969) and U.S. Environmental Protection Agency (1988).

[a] Corneal stippling: multiple discrete punctate irregularities in the corneal epithelial layer, which retain fluorescein.

[b] Confluence: uniform zones for fluorescein retention larger than 1 mm in diameter.

[c] Flare: Tyndall effect in a beam traversing the aqueous humor.

TABLE 93

Categorization and Labeling of Pesticides (Label Statements Regarding Eye Irritation Hazards due to Pesticides)

Toxicity Category	Signal Word	Skull and Crossbones and "Poison" Required	Precautionary Statement	Practical Treatment
I. Corrosive (irreversible destruction of ocular tissue), corneal involvement, or irritation persisting for more than 21 days.	Danger	No	Corrosive.[a] Causes irreversible eye damage. Harmful if swallowed. Do not get in eyes or on clothing. Wear goggles, face shield, or safety glasses.[b] Wash thoroughly with soap and water after handling. Remove contaminated clothing and wash before reuse.	*If in eyes*: Flush with plenty of water. Get medical attention. *If swallowed*: Drink promptly a large quantity of milk, egg whites, gelatin solution, or, if these are not available, drink large quantities of water. Avoid alcohol. *Note to physician*: Probable mucosal damage may contraindicate the use of gastric lavage.

(Continued)

TABLE 93 (Continued)

Categorization and Labeling of Pesticides (Label Statements Regarding Eye Irritation Hazards due to Pesticides)

Toxicity Category	Signal Word	Skull and Crossbones and "Poison" Required	Precautionary Statement	Practical Treatment
II. Corneal involvement or irritation clearing in 21 days or less.	Warning	No	Causes substantial but temporary eye injury. Do not get into eyes or on clothing. Wear goggles, face shield, or safety glasses.[b] Harmful if swallowed. Wash thoroughly with soap and water after handling. Remove contaminated clothing and wash before reuse.	*If in eyes:* Flush with plenty of water. Get medical attention. *If swallowed:* Drink promptly a large quantity of milk, egg whites, gelatin solution, or, if these are not available, drink large quantities of water. Avoid alcohol.
III. Corneal involvement or irritation clearing in 7 days or less.	Caution	No	Causes (moderate) eye injury (irritation). Avoid contact with eyes or clothing. Wash thoroughly with soap and water after handling.	*If in eyes:* Flush with plenty of water. Get medical attention if irritation persists.
IV. Minimal effects clearing in less than 24 h.	Caution	No	None required.	None required.

Source: From Camp, D.D (1984).

[a] The term "corrosive" may be omitted if the product is not actually corrosive.

[b] Choose appropriate form of eye protection. Recommendation for goggles or face shield is more appropriate for industrial, commercial, or nondomestic uses. Safety glasses may be recommended for domestic or residential use.

TABLE 94

Considerations in Selecting *In Vitro* Assays to Support Product Development Programs

Factors to Be Considered	Impact of Those Factors
Chemical Class of the Test Materials	
Alcohols Organics Preservatives Surfactants Acids/alkalis	Not all *in vitro* systems have been characterized for their performance with a variety of chemical classes. Carefully investigate individual *in vitro* systems to determine with which chemical classes they are compatible.
Physical Characteristics of the Test Material	
Solid/liquid Water-soluble/insoluble Extremes of pH Highly reactive amount of material available	Solid or water insoluble materials should generally not be tested with a monolayer cell culture system since the test materials may not reach the target cells. Topical application assays are preferred for such materials. Highly reactive materials may bind to the constituents of the tissue culture medium and thus be unavailable to the target tissue. Dilution into buffered culture medium will reduce extremes of pH. Although *in vitro* tests generally require far less test material than do animal tests, there is still a considerable range of requirement among the *in vitro* systems. Some have been designed to use microquantities of materials.

(Continued)

TABLE 94 (Continued)

Considerations in Selecting *In Vitro* Assays to Support Product Development Programs

Factors to Be Considered	Impact of Those Factors
Stage in Product Development	
Single (perhaps active) ingredient	Biological activity of single ingredients may be assessed in a variety of systems including *ex vivo* tissues, tissue constructs, and monolayer culture systems. Water solubility and the end point(s) of interest will guide selection.
Mixtures	Depending on the target tissue, testing of mixtures and final formulations may require
Final formulation	that the test system be able to model the expected exposure kinetics of the formulation as a whole. For example, dermal irritation studies may require a test system with a functional stratum corneum such as a tissue construct. In contrast, ocular irritation studies of surfactant formulations might well use *ex vivo* tissues, tissue constructs, and monolayer culture systems.
Expected Level of Toxicity	
Low	The dynamic range of response of the *in vitro* system should match the expected level of toxicity of the test material. Some *in vitro* systems are designed to differentiate between weakly reactive materials; more robust systems may be useful for highly toxic materials.
Medium	
High	
Expected Exposure to the Tissue of Interest	
Incidental/accidental	The exposure kinetics of the *in vitro* system can often be varied, and should closely match the expected *in vitro* exposures if an accurate estimate of toxicity is to be obtained.
Short-term or infrequent	
Leave-on vs. wash-off application	
Chronic application	
Use population (infants, adult, aged)	

(Continued)

TABLE 94 (*Continued*)

Considerations in Selecting *In Vitro* Assays to Support Product Development Programs

Factors to Be Considered	Impact of Those Factors
Resolution Required of the Test	
Differentiate among similar test formulations	It may require a more sophisticated *in vitro* system to differentiate closely related materials than it would to place test materials into general classifications.
Separating highly toxic from non-toxic materials	
Intended Use of the Data	
Safety/efficacy screen	The purpose of the testing should be matched to the test system; example, a simple, inexpensive *in vitro* model may be sufficient for use in screening, whereas a more sophisticated model might be necessary to characterize the effects of minor formulation changes.
Product development	
Formula optimization	
Claims support	
Resources Available	
Funding	Many *in vitro* tests can seem expensive if applied to a single test article, but are designed for easy batching of materials, which results in significant cost benefits.
Time	
Number of materials to be tested	

TABLE 95

Advantages and Disadvantages of Dilution-Based Assays[a] for Ocular Irritation

Advantages	Disadvantages
Rapid to execute using multiwell plate formats.	Cannot be used easily with water-insoluble materials.
Most are machine scored based on dye incorporation/reduction.	Dilution effects may mask toxicity of neat material (e.g., alcohols).
Generally very cost-effective—Multiple materials may be tested concurrently.	Change in the physical form, example, solids to solutions, may impact exposure kinetics.
Seem to work well with surfactants and surfactant-based formulations.	Buffering effects of the medium may affect toxicity significantly
Often differentiate well between very mild materials.	Possible reaction of the test material with the solvent/medium components.

[a] Assays in which serial dilutions of the test material are applied to the test system, and the end point is the concentration of test material that causes a selected response.

TABLE 96

Advantages and Disadvantages of Topical Application Assays[a] for Ocular Irritation

Advantages	Disadvantages
• Material is tested in its "native" form, that is, in the same form as an *in vivo* exposure	• Test substrate can often be expensive
• Exposure of the target tissue can be assured	• Exposure times may be inconveniently long, requiring work past the normal workday
• In some models, exposure time can be selected to match expected *in vivo* exposure	• Solid materials may require special handling to apply to the test system
• Exposure measured through the depth of the tissue	• Availability of tissue in some markets
	• Limited shelf life

[a] Assays in which only the neat or end-use concentration of test material is applied to the test system (i.e., *ex vivo* or reconstructed tissue), and the end point(s) depend on the dynamic range of the test system and/or exposure time vs. end point activity.

References

Beckley, J.H., Russell, T.J., and Rubin, L.F. (1969). Use of the Rhesus monkey for predicting human response to eye irritants, *Toxicol. Appl. Pharmacol.*, 15, 1.

Camp, D.D. (1984). Labeling requirements for pesticides and devices, *Federal Register*, 49, 188.

Draize, J.H., Woodard, G., and Calvery, H.O. (1944). Methods for the study of irritation and toxicity of substances applied topically to the skin and mucous membranes, *J. Pharmacol. Exp. Ther.*, 82, 377.

Environmental Protection Agency (1988). Guidance for evaluation of eye irritation testing, Hazard Evaluation Division Standard Evaluation Procedures, EPA-540/09-88-105, Washington, DC.

Green, W.R. et al. (1978). *A Systematic Comparison of Chemically Induced Eye Injury in the Albino Rabbit and Rhesus Monkey*, The Soap and Detergent Association, New York, 407.

Kay, J.H. and Calandra, J.C. (1962). Interpretation of eye irritation tests, *J. Soc. Cosmet. Chem.*, 13, 281.

National Academy of Sciences (1977), Committee for the Revision of NAS Publication 1138, Principles and Procedures for Evaluating the Toxicity of Household Substances, Washington, D.C.

United Nations (2011), *Globally Harmonized System of Classification and Labelling of Chemicals (GHS)*, fourth revision edition, New York and Geneva.

U.S. Environmental Protection Agency (2011), *Design for the Environment Program, Alternatives Assessment Criteria for Hazard Evaluation*, Version 2.0, Office of Pollution Prevention and Toxics, August.

6

Genetic Toxicology/Carcinogenesis

TABLE 97

Mutagenicity Assay Bacteria Strains[a] Overview

TA98 TA1537 TA97 TA97a	Reverted by frame-shift mutagens
TA100 TA1535 TA102[b] (detects oxidative mutagens and cross-linking agents) WP2 *uvrA*[b] WP2 *uvrA* (pKM101)[b] (detects cross-linking agents)	Reverted by base substitution mutagens

[a] A standard battery of strains for a full mutagenicity assay usually consists of five strains consisting of TA98, TA100, TA1535, TA1537 (or TA97 or TA97a), and TA102 (or WP2 uvrA or WP2 uvrA (pKM101)). Strains with TA designations are *S. typhimurium* and WP2 designations are *E. coli*.

[b] Strain TA102 and the *E. coli* strains possess A-T base pairs at the site of mutation. Other strains possess G-C base pairs at their mutation sites.

TABLE 98

The Genetic Code

Codon	Amino Acid	Codon	Amino Acid
UUU or UUC	Phenylalanine	UAA or UAG	Nonsense (ochre)
			Nonsense (amber)
UUA or UUG	Leucine	CAU or CAC	Histidine
CUU, CUC, CUA, or CUG	Leucine	CAA or CAG	Glutamine
AUU, AUC, or AUA	Isoleucine	AAU or AAC	Asparagine
AUG	Methionine	AAA or AAG	Lysine
GUU, GUC, GUA, or GUG	Valine	GAU or GAC	Aspartic acid
UCU or UCC	Serine	GAA or GAG	Glutamic acid
UCA or UCG	Serine	UGU or UGC	Cysteine
CCU, CCC, CCA, or CCG	Proline	UGA	Nonsense (umber)
ACU, ACC, ACA, or ACG	Threonine	CGU, CGC, CGA, or CGG	Arginine
GCU, GCC, GCA, or GCG	Alanine	AGU or AGC	Serine
UAU or UAC	Tyrosine	AGA or AGG	Arginine
		GGU, GGC, GGA, or GGG	Glycine

TABLE 99

Genetic Toxicology Assays

Assay	Evaluates	Comments
Bacterial mutagenesis (Ames)	Bacterial gene mutation	This most widely used *in vitro* assay for detecting gene mutations utilizes tester strains of *Salmonella typhimurium* (frequently used are TA98, TA100, TA102, TA1535, TA1537) and *Escherichia coli* (WP2 *uvrA* and WP2 *uvrA* (pKM101)). These bacteria have defects in one of the genes involved in histidine or tryptophan biosynthesis, respectively. As auxotrophic mutants they require the amino acid for growth, but cannot produce it. The method tests the capability of the test substance in creating mutations that result in a return of the bacteria to a prototrophic/wild state, so that they can grow on a histidine- or tryptophan-free medium. These mutations, which lead to a regaining of normal activity or function, are called *back* or *reverse* mutations and the process is referred to as *reversion*. Strains TA98, TA1537 are reverted by frameshift mutagens. Strains TA100, TA1535, TA102, WP2 *uvrA*, and WP2 *uvrA* (pKM101) are reverted by base substitution mutagens.
Mouse lymphoma	Mammalian cell gene mutation	A common *in vitro* assay for gene mutation in mammalian cells that utilizes cultured mouse lymphoma L5178Y cells. The assay, designed to detect forward mutations at a specific loci, is designated as the thymidine kinase +/− (TK+/−) mouse lymphoma mutation assay according to the *tk* target gene. These cells contain functional TK enzyme, which is involved in a salvage pathway for use/re-use of thymidine in the cell through phosphorylation of thymidine. Trifluorothymidine (TFT), the selective agent used in this assay, can also be phosphorylated by the TK enzyme, but is lethal to the cell. Forward mutation at the single functional TK gene forces the cell to switch to *de novo* thymidine synthesis. The loss of TK activity in mutant cells (the cells are tk−/−) make the mutant cells resistant to TFT and able to grow in the presence of TFT.

(Continued)

TABLE 99 (*Continued*)

Genetic Toxicology Assays

Assay	Evaluates	Comments
CHO/HPRT	Mammalian cell gene mutation	This *in vitro* assay for gene mutation in mammalian cells utilizes cultured Chinese hamster ovary (CHO) cells. The assay, designed to detect forward mutations at a specific loci, is designated as the CHO/hypoxanthine–guanine phosphoribosyl transferase (HPRT or HGPRT) mutation assay according to the *hprt* target gene. HPRT catalyzes phosphorylation of purines in the purine salvage pathway. The selective agent used in this assay, 6-thioguanine (6-TG), is also a substrate for this enzyme and cells that retain functional HPRT are killed by 6-TG. Forward mutations that result in the loss of the functional *hprt* render cells resistant to 6-TG.
Sister chromatid exchange (SCE)	Mutagenic potential via assessment of recombination process	The SCE assay detects the ability of a chemical to enhance the exchange of DNA between two sister chromatids of a duplicating chromosome. An increase in SCE levels is used as an index of mutagenic potential. The test can be performed with *in vitro* exposure using cultured mammalian cells (CHO or peripheral blood lymphocytes) or following *in vivo* exposure using bone marrow or lymphocytes from mammalian species such as mice, rats, or hamsters. Human lymphocytes may also be used. The assay involves the growth of cells in the presence of 5′-bromo-deoxyuridine followed by collecting metaphase spreads on glass slides. Sister chromatid exchanges (SCEs) are visualized cytologically.

(Continued)

TABLE 99 (*Continued*)

Genetic Toxicology Assays

Assay	Evaluates	Comments
In vitro chromosome aberration	Chromosome aberration/ damage	Chromosome aberration assays are used to measure genotoxicity at the chromosome level. Test systems used include permanent cell lines, such as: Chinese hamster cells ovary (CHO), Chinese hamster lung (V79 or CHL), Chinese hamster lymphoid cell line (TK6) or primary cells such as human peripheral blood lymphocytes (HPBL) and rat blood lymphocytes. Primary lymphocytes need to be stimulated to divide using a mitogen such as phytohemagglutin. Following exposure of cells to a test substance in the absence and presence of an exogenous source of metabolic activation, dividing cells are arrested in metaphase using a spindle inhibitor (e.g., Colcemid®) and evaluated. Endpoints measured include changes in chromosome structure (clastogenicity) and chromosome number (aneugenicity).
In vivo chromosome aberration	Chromosome aberration/ damage	This study is conducted using either mice or rats. To maximize delivery of the test article to the bone marrow, the route of animal dosing needs be justified. In the bone marrow analysis assay, dividing bone marrow cells are arrested in metaphase by an intraperitoneal injection of colchicine. Sample times are selected to assure analysis of first-division metaphase cells, both non-delayed and delayed. Cells are evaluated microscopically for chromosome aberrations as is done in the *in vitro* assay.

(Continued)

TABLE 99 (Continued)

Genetic Toxicology Assays

Assay	Evaluates	Comments
In vitro micronucleus	Chromosome aberration/ damage	This assay is used for the detection of micronuclei in the cytoplasm of interphase cells. Micronuclei originate from acentric chromosome fragments (*i.e.*, lacking a centromere) or whole chromosomes that are unable to migrate with the rest of the chromosomes during the anaphase stage of cell division. The assay detects the activity of clastogenic and aneugenic test chemicals or their metabolites in cells that have undergone cell division after exposure to the test substance. Various cell lines are utilized such as Chinese hamster ovary (CHO), Chinese hamster lung (V79 or CHL) or human lymphoid cell (TK6).
In vivo micronucleus	Chromosome aberration/ damage	The *in vivo* micronucleus test is routinely performed in rodents (mice or rats) and may be applied to other species including dogs and monkeys with appropriate validation. The *in vivo* micronucleus test can be applied to animals with acute treatment or integrated in subacute general toxicology studies such as 28 day repeat dose rodent toxicity studies. To maximize delivery of the test article to the bone marrow, the route of animal dosing needs to be justified. An evaluation of peripheral blood erythrocytes detects accumulated micronuclei in blood. One confounding factor is that in some species, the spleen gradually filters micronucleated reticulocytes (mnRETs) from the peripheral blood. This can be overcome using special techniques. Mice do not exhibit splenic filtration of mnRETs. In the event of negative findings, it may be necessary to demonstrate that the target cells (i.e., bone marrow) were exposed to the test article. This may be achieved by a measure of bone marrow toxicity (mitotic inhibition or depressed polychromatic erythrocyte (PCE)/total erythrocyte ratio), or in its absence, tissue distribution data. In some cases, blood serum levels may be sufficient to document exposure.

(Continued)

TABLE 99 (*Continued*)

Genetic Toxicology Assays

Assay	Evaluates	Comments
Unscheduled DNA synthesis (UDS)	DNA damage repair	The most frequently previously used methods for the detection of DNA repair synthesis are the *in vitro* and *in vivo/in vitro* unscheduled DNA synthesis (UDS) assays. Hepatocytes exposed *in vitro* or *in vivo* to the test substance are evaluated for replication of DNA during excision repair of certain types of DNA lesions by assessing the incorporation of tritiated thymidine into the DNA repair sites. As compared to exposure of cells directly *in vitro* to a substance, in the *in vivo/in vitro* UDS assay, direct exposure of animals is designed to account for complex patterns of metabolic activation, detoxification, uptake distribution, and excretion of chemicals. Hepatocytes are isolated following the administration of the test substance (by gavage, intravenous injection, or intraperitoneal injection). Hepatocytes exposed *in vitro* or *in vivo* are plated in medium containing ^{3}H-TdR. After incubation, the cells are processed for autoradiography to assess DNA synthesis that occurred. The assay is considered to be not very sensitive and has fallen out of favor for this reason.
Alkaline elution (AE)	DNA damage and repair	This assay detects increases in DNA single- and double-strand breaks and repair induced breaks, a basic site, DNA cross-linking agents and alkali labile sites either *in vitro* or *in vivo*. Several cell lines have been utilized in *in vitro* alkaline elution studies including V-79 and CHO cells. *In vivo*, the test substance is given in single or 2–3 daily doses with organ harvests at appropriate times to capture breaks before they are repaired. An advantage is that any target tissue can be used for analysis. Cells are passed through micropore filter membranes with the elution rate proportional to the number of DNA strand breaks. The alkaline elution method is considered to be sensitive in the detection of DNA damage, but it has high variability and inconsistency.

(*Continued*)

TABLE 99 (Continued)

Genetic Toxicology Assays

Assay	Evaluates	Comments
Cell transformation	Tumorigenic alterations	The *in vitro* cell transformation assays assess induction of phenotypic alterations in cultured cells that are characteristic of tumorigenic cells. Transforming activity of the test substance as measured by morphological transformation is evaluated in cryopreserved Syrian hamster embryo (SHE) cells or in cell lines such as Balb/c 3T3 or Balb/c 3T3 cells that have been transfected with v-Ha-*ras*, called Bhas 42. The SHE cell transformation assay evaluates the phenotypic changes in the colonies, which have originated from a single cell, whereas in the Balb/c 3T3 and the newer version of Balb/c 3T3 using the Bhas 42 cells, evaluate foci formation on top of the cell monolayer.
Pig-a mutation	Gene mutation	The Pig-a assay is a newer *in vivo* forward gene mutation assay undergoing intensive research and validation. It relies on immunologic detection, by flow cytometry, of cells deficient in certain surface markers that are presumed to arise due to loss of glycosylphosphatidylinositol (GPI) anchor proteins. Like the more familiar *hprt* gene, the *Pig-a* gene is X-linked and a single mutational event can lead to the mutant phenotype. The Pig-a assay offers three main benefits: use of small (~50 μL) blood samples; accumulation of mutant reticulocytes (RET) and erythrocytes (RBC) with repeat dosing, allowing detection and characterization of weak mutagens and low dose effects; and the ability to compare mutant induction across all species of toxicological concern, including man. Currently, these analyses have largely been limited to rodent blood (peripheral, splenic, or bone marrow). Analysis is typically conducted as part of a 28-day repeated dose study.

(Continued)

TABLE 99 (*Continued*)

Genetic Toxicology Assays

Assay	Evaluates	Comments
Comet	DNA damage, mutation	This *in vivo* method evaluates the genotoxic potential of a test substance to induce DNA damage in selected organ cells of rodents. The Comet Assay (alkaline single cell gel electrophoresis assay) employs a microgel electrophoretic technique that detects DNA damage in individual cells at pH ≥13. Organs to be evaluated are harvested from rats or mice dosed with the test substance. Liver and stomach/intestines are the most commonly used organs but the assay can easily be performed on any organ that can be used to make a single cell suspension. These include kidney, heart, lungs, uterus, peripheral blood, skin, brain, spleen, and bone marrow. Microscopic slides are prepared from single cell suspensions prepared from each organ and electrophoresed. During electrophoresis, DNA fragments migrate in the direction of the electric current while high molecular weight undamaged DNA does not migrate. "Comet" resembling electrophoretic tail migration, % tail DNA (also known as % tail intensity) and tail moment are determined and serve as parameters of DNA damage. The Comet assay can be easily combined with other *in vivo* assays such as a micronucleus assay, a 14-day, or a 28-day mammalian toxicology study.

(Continued)

TABLE 99 (Continued)

Genetic Toxicology Assays

Assay	Evaluates	Comments
Transgenic rodent (TGR) mutation	Gene mutation	The *in vivo* TGR mutation assay utilizes transgenic rats and mice that contain multiple copies of chromosomally integrated plasmid and phage shuttle vectors that harbor reporter genes for the detection of mutation. Mutagenic events arising in a rodent are scored by recovering the shuttle vector and analyzing the phenotype of the reporter gene in a bacterial host. Mutagenesis is normally assessed as a mutant frequency. The TGR assay evaluates tissue-specific mutagenic potential under *in vivo* exposure conditions. Bioavailability (TK) measurements can be added to demonstrate exposure. Multiple copies of the transgene are carried in every cell of each animal, permitting *in vivo* mutation analysis of almost any organ in the body. TGR assays have value in detecting induced mutagenesis in specific target organs due to high, local concentration of DNA-damaging agents, in an organ of first contact, or unique metabolism in a specific organ that otherwise is missed by other *in vivo* genetic toxicology assays. The most commonly used transgenic rodents for *in vivo* mutation analysis are the *lacZ* transgenic mouse (Muta™Mouse) and the Lambda LIZ *lacI* (BigBlue™) mouse and rat. Other TGR models have been developed including the *lacZ* plasmid mouse, gpt delta mouse and rat, rpsL, PhiX174, and sipF models. The use of a specific model has depended upon availability of animals and extent of validation.

TABLE 100

Characteristics of Initiation, Promotion, and Progression

Initiation	Promotion	Progression
Irreversible	Reversible	Irreversible
Additive	Nonadditive	Karyotypic abnormalities appear accompanied by increase growth rate and invasiveness
Dose response can be demonstrated; does not exhibit a readily measurable threshold	Dose response having a measurable threshold can be demonstrated	Benign and/or malignant tumor observed
No measurable maximum response	Measurable maximum effect	Environmental factors influence early stage of progression
Initiators are usually genotoxic	Promoters are usually not mutagenic	Progressors may not be initiators
One exposure may be sufficient	Prolonged and repeated exposure to promoters required	Progressors act to advance promoted cells to a potentially malignant stage
Must occur prior to promotion	Promoter effective only after initiation has occurred	Spontaneous progression can occur
Requires fixation through cell division	Promoted cell population dependent on continued presence of promoter	
Initiated cells are not identifiable except as foci lesions following a period of promotion	Causes expansion of the progeny of initiated cells producing foci lesions	
"Pure" initiation does not result in neoplasia without promotion	"Pure" promoters not capable of initiation	
Spontaneous (fortuitous) initiation can occur	Sensitive to hormonal and dietary factors	

Source: Adapted from Pitot, H.C. (1991) and Maronpot, R.R. (1991).

TABLE 101

Classification of Carcinogenic Chemicals Based on Mode of Action

Classification[a]	Mode of Action	Examples
I. Genotoxic	Agents which interact with DNA	
1. Direct acting (primary carcinogen; activation-independent)	Organic chemicals; direct alteration of DNA, chromosome structure, or number; metabolic conversion not required; generation of reactive electrophiles and covalent binding to DNA	Bis-chloromethylether, β-propiolactone, ethylene imine
2. Procarcinogen (secondary carcinogen; activation-dependent)	Organic chemicals; requires biotransformation to a direct acting carcinogen (proximate carcinogen)	Nitrosamines, ethylene dibromide, vinyl chloride
3. Inorganic carcinogen	Direct effects on DNA may occur through interference with DNA replication	Nickel, cadmium
II. Epigenetic	Agents for which there is no direct evidence of interaction with DNA	
4. Cytotoxin	Cytolethal; induction of regenerative cell proliferation; mutations may occur secondarily through several mechanisms including release of nucleases, generation of reactive oxygen radicals, DNA replication before adduct repair; preferential growth of preoplastic cells may be caused by selective killing of normal cells or expression of growth control genes (oncogenes)	Nitrilo triacetic acid, chloroform
5. Mitogen	Stimulation of mitogenic cell proliferation directly or via a cellular receptor; mutations may occur secondarily as a result of increased cell proliferation; preferential growth of preneoplastic cells may be caused through alteration of rates of cell birth or death	Phenobarbital, α-hexachlorocyclohexane

(Continued)

TABLE 101 (*Continued*)

Classification of Carcinogenic Chemicals Based on Mode of Action

Classification[a]	Mode of Action	Examples
6. Peroxisome proliferator	Generation of reactive oxygen radicals through pertubation of lipid metabolism; growth control genes may be activated directly or via a cellular receptor	Fenofibrate, diethylhexyl phthalate, clofibrate
7. Immunosuppressor	Enhancement of the development of virally induced, transplanted, and metastatic neoplasms possibly through impairment or loss of natural and acquired tumor resistance	Azathioprine, cyclosporin A, 6-mercaptopurine
8. Hormones and hormonal altering agents	Chronic stimulation of cell growth through activation of regulatory genes; other potential modes of action include promotional effects resulting from alteration of hormonal homeostasis, inhibition of cell death (apoptosis), generation of reactive radicals	Estrogens, diethylstilbestrol, synthetic androgens
9. Solid-state carcinogen	Generally only mesenchymal cells/tissues affected; physical size and shape of agent is critical; mechanism of action uncertain	Polymers (plastic), metal foils (gold), asbestos

(*Continued*)

TABLE 101 (*Continued*)

Classification of Carcinogenic Chemicals Based on Mode of Action

Classification[a]	Mode of Action	Examples
10. Cocarcinogen	*Simultaneous* administration enhances the carcinogenic process caused by a genotoxic carcinogen; possible mechanisms include enhanced biotransformation of a procarcinogen, inhibition of detoxification of a primary carcinogen, enhanced absorption or decreased elimination of a genotoxic carcinogen	Phorbol esters, catechol, ethanol
11. Promoter	Administration *subsequent* to a genotoxic agent promotes tumor formation through enhancement of the clonal expansion of preneoplastic cells; multiple and diverse mechanisms proposed	Phorbol esters, saccharin, croton oil
12. Progressor	Development of initiated/promoted cells influenced; associated with alterations in biochemical and morphological characteristics, increased growth rate, invasiveness, and metastases; direct or indirect induction of structural (karyotypic) changes to chromosomes	Arsenic salts, benzene, hydroxyurea

Source: Adapted from Weisburger, J.H. and Williams, G.M. (1980).

Additional source: Pitot, H.C. and Dragon, Y.P. (1993); Pitot, H.C. (1993); Maronpot, R.R. (1991); and Butterworth, B.E. and Goldsworthy, T.L. (1991).

[a] Classifications shown are not rigid. For example, a chemical may be both genotoxic and mitogenic or cytotoxic; phorbol ester can be both a promoter and a cocarcinogen.

TABLE 102

Reported Percentage Incidence (Range) of Spontaneous Tumor
Formation in Various Mouse Strains

	CD-1		B6C3F1	
Organ/Tissue	**Male**	**Female**	**Male**	**Female**
Adrenal	0–27.9 (%)	0–3.8	<1.0–1.4	<1.0
Body cavities	—	—	<1.0	<1.0
Brain	—	0–2.0	<0.1–0.1	0–0.1
Circulatory system	—	—	<1.0–2.9	<1.0–2.4
Heart	—	—	0.1–<1.0	0–0.1
Intestines	—	—	<1.0	<1.0
Kidney	0–2.8	0–1.4	<1.0	<0.1–<1.0
Leukemia/lymphoma	0–8.6	1.4–25.0	1.6–19.0	1.7–33.2
Liver	0–17.3	0–7.1	15.6–40.1	2.5–10.5
Lung/trachea	0–26.0	0–38.6	9.2–22.5	3.5–7.1
Mammary gland	—	0–7.3	—	<1.0–1.3
Ovary	NA	0–4.8	NA	<1.0
Pancreas	—	—	0.1–2.1	<0.1–<1.0
Pancreatic islets	0–2.1	0–1.4	<1.0	<1.0
Pituitary	0–0.8	0–10.0	<1.0	3.2–13.1
Skin/subcutaneous	0–2.8	0–2.0	<0.1–1.9	0.1–1.6
Stomach	0–4.9	0–3.8	0.3–1.1	<1.0
Testes[a]	0–2.0	NA	<1.0	NA
Thyroid	0–2.0	—	1.0–1.1	<1.0–1.7
Urinary bladder	0–2.0	0–1.4	0–0.1	<0.1–1.0
Uterus/vagina	NA	0–13.3	NA	1.2–1.9

Source: Adapted from Gad, S.C. and Weil, C.S. (1986).

Additional source: Chu, K. (1977), Fears, T.R. et al. (1977), Page, N.P. (1977), Gart, J.J. et al. (1979), Tarone, R.E. et al. (1981), Rao, G.N. et al. (1990), and Lang, P.L. (1987).

[a] Includes prostate and seminal vesicles.

TABLE 103

Reported Percentage Incidence (Range) of Spontaneous Tumor Formation in Various Rat Strains

Organ/Tissue	F344 Male	F344 Female	Sprague-Dawley Male	Sprague-Dawley Female	Wistar Male	Wistar Female
Adrenal	2.4–38.1 (%)	4.0–12.0	1.4–7.6	2.7–4.3	0–48.6	0–57.1
Body cavities	<1.0–9.0	0.3–1.9	1.1–1.4	1.8	—	—
Brain	0.8–8.1	<1.0	1.4–2.7	0.9–1.6	0–8.0	0–6.0
Circulatory system	0.4–3.8	<1.0	0.5	—	0–3.3	0–2.5
Heart	<1.0	<1.0	—	—	0	0
Intestines	<1.0	<1.0	—	0.5	0–3.1	0–3.8
Kidney	<1.0	<1.0	1.6	0.9	0–2.5	0–2.0
Leukemia/lymphoma	6.5–48.0	2.1–24.6	1.9–2.2	1.4–1.6	0–12.0	0–16.0
Liver	0.5–3.4	0.5–3.9	1.1	0.5–2.2	0–5.0	0–12.0
Lung/trachea	<1.0–3.0	<1.0–2.0	1.6	2.2	0–5.7	0–2.1
Mammary gland	0–1.5	8.5–41.0	0.5–2.3	36.4–45.1	0–6.7	1.3–45.0
Ovary	NA	<1.0	NA	1.1	NA	0–4.3
Pancreas	0.2–6.0	0	—	—	0–51.7	0–1.7
Pancreatic islets	0.8–4.9	0.8–1.3	0.9–2.7	0.5	0–25.0	0–4.0
Pituitary	4.7–34.7	0.3–58.6	11.2–33.2	37.3–57.6	2.3–58.3	6.7–68.0
Preputial gland	1.4–2.4	1.2–1.8	—	—	—	—

(Continued)

TABLE 103 (Continued)

Reported Percentage Incidence (Range) of Spontaneous Tumor Formation in Various Rat Strains

Organ/Tissue	F344 Male	F344 Female	Sprague-Dawley Male	Sprague-Dawley Female	Wistar Male	Wistar Female
Skin/subcutaneous	5.7–7.8	2.5–3.2	2.8–6.5	3.2–3.8	0–21.9	0–5.0
Stomach	<1.0	<1.0	—	—	0	0–2.2
Testes[a]	2.3–90.0	NA	4.2–4.3	NA	0–22.0	NA
Thyroid	3.6–12.0	4.7–10.0	1.9–3.8	1.8	0–21.7	2.5–22.4
Urinary bladder	<1.0	<1.0	0.5	—	0–2.0	0–2.0
Uterus/vagina	NA	5.5–24.6	NA	3.3–4.5	NA	1.1–25.3

Source: Adapted from Gad, S.C. and Weil, C.S. (1986).
Additional source: Chu, K. (1977), Fears, T.R. et al. (1977), Page, N.P. (1977), Gart, J.J. et al. (1979), Tarone, R.E. et al. (1981), Rao, G.N. et al. (1990), Goodman, D.G. et al. (1979), Bombard, E. et al. (1986), Walsh, K.M. and Poteracki, J. (1994), Haseman, J.K. (1983), Poteraki, J. and Walsh, K.M. (1998).

[a] Includes prostate and seminal vesicles.

TABLE 104

Frequency of Carcinogenic Response to Chemicals by Organ/System—
Rats and Mice[a]

	Number Positive at Site (%)[b]	
	Chemicals Evaluated as Carcinogenic in Rats ($n = 354$)[c]	Chemicals Evaluated as Carcinogenic in Mice ($n = 299$)[c]
Liver	143 (40%)	171 (57%)
Lung	31 (9%)	83 (28%)
Mammary gland	73 (21%)	14 (5%)
Stomach	60 (17%)	42 (14%)
Vascular system	26 (7%)	47 (16%)
Kidney/ureter	45 (13%)	12 (4%)
Hematopoietic system	35 (10%)	39 (13%)
Urinary bladder/urethra	37 (10%)	12 (4%)
Nasal cavity/turbinates	33 (9%)	4 (1%)
Ear/Zymbal's gland	30 (9%)	2
Esophagus	29 (8%)	7 (2%)
Small intestine	21 (6%)	3 (1%)
Thyroid gland	20 (6%)	10 (3%)
Skin	20 (6%)	1
Peritoneal cavity	17 (5%)	7 (2%)
Oral cavity	16 (5%)	1
Large intestine	15 (4%)	
Central nervous system	15 (4%)	2
Uterus	11 (3%)	5 (2%)
Subcutaneous tissue	10 (3%)	1
Pancreas	9 (3%)	
Adrenal gland	7 (2%)	4 (1%)
Pituitary gland	7 (2%)	4 (1%)
Clitoral gland	7 (2%)	2

(Continued)

TABLE 104 (*Continued*)

Frequency of Carcinogenic Response to Chemicals by Organ/System—
Rats and Mice[a]

	Number Positive at Site (%)[b]	
	Chemicals Evaluated as Carcinogenic in Rats ($n = 354$)[c]	Chemicals Evaluated as Carcinogenic in Mice ($n = 299$)[c]
Preputial gland	2	7 (2%)
Testes	6 (2%)	1
Harderian gland		6 (2%)
Spleen	6 (2%)	
Ovary		4 (1%)
Gall bladder		3 (1%)
Bone	3	
Mesovarium	2	
Myocardium		2
Prostate	2	
Vagina	1	

Source: From Gold, L.S. et al. (1991).

[a] Based on 354 and 299 chemicals considered carcinogenic to rats and mice, respectively, in long-term chemical carcinogenesis studies from the carcinogenic potency database (CPDB).

[b] Percentages not given when fewer than 1% of the carcinogens were active at a given site.

[c] Chemicals have been excluded for which the only positive results in the CPDB are for "all tumor bearing animals," that is, there is no reported target site.

TABLE 105

Capacity of Tissues to Undergo Hyperplasia

High capacity
  Surface epithelium
  Hepatocytes
  Renal tubules
  Fibroblasts
  Endothelium
  Mesothelium
  Hematopoietic stem cells
  Lymphoid cells
Moderate capacity
  Glandular epithelium
  Bone
  Cartilage
  Smooth muscle of vessels
  Smooth muscle of uterus
Low capacity
  Neurons
  Skeletal muscle
  Smooth muscle of GI tract

Source: From Maronpot, R.R. (1991). With permission.

TABLE 106

Selected Taxonomy of Neoplasia

Tissue	Benign Neoplasia[a]	Malignant Neoplasia[b]
Epithelium		
Squamous	Squamous cell papilloma	Squamous cell carcinoma
Transitional	Transitional cell papilloma	Transitional cell carcinoma
Glandular		
Liver cell	Hepatocellular adenoma	Hepatocellular carcinoma
Islet cell	Islet cell adenoma	Islet cell adenocarcinoma
Connective tissue		
Adult fibrous	Fibroma	Fibrosarcoma
Embryonic fibrous	Myxoma	Myxosarcoma
Cartilage	Chondroma	Chondrosarcoma
Bone	Osteoma	Osteosarcoma
Fat	Lipoma	Liposarcoma
Muscle		
Smooth muscle	Leiomyoma	Leiomyosarcoma
Skeletal muscle	Rhabdomyoma	Rhabdomyosarcoma
Cardiac muscle	Rhabdomyoma	Rhabdomyosarcoma
Endothelium		
Lymph vessels	Lymphangioma	Lymphangiosarcoma
Blood vessels	Hemangioma	Hemangiosarcoma
Lymphoreticular		
Thymus	(Not recognized)	Thymoma
Lymph nodes	(Not recognized)	Lymphosarcoma (malignant lymphoma)
Hematopoietic		

Bone marrow	(Not recognized)	Leukemia
		  Granulocytic
		  Monocytic
		  Erythroleukemia

(Continued)

TABLE 106 (*Continued*)

Selected Taxonomy of Neoplasia

Tissue	Benign Neoplasia[a]	Malignant Neoplasia[b]
Neural tissue		
Nerve sheath	Neurilemmoma	Neurogenic sarcoma
Glioma	Glioma	Malignant glioma
Astrocytes	Astrocytoma	Malignant astrocytoma
Embryonic cells	(Not recognized)	Neuroblastoma

Source: From Maronpot, R.R. (1991). With permission.

[a] "-oma," benign neoplasm.

[b] "Sarcoma," malignant neoplasm of mesenchymal origin; "carcinoma," malignant neoplasm of epithelial origin.

TABLE 107

Selected Examples of Presumptive Preneoplastic Lesions

Tissue	Presumptive Preneoplastic Lesion[a]
Mammary gland	Hyperplastic alveolar nodules (HANs), atypical epithelial proliferation, lobular hyperplasia, intraductal hyperplasia, hyperplasic terminal duct
Liver	Foci of cellular alteration, hepatocellular hyperplasia, oval cell proliferation, cholangiofibrosis
Kidney	Karyocytomegaly, atypical tubular dilation, atypical tubular hyperplasia
Skin	Increase in dark basal keratinocytes, focal hyperplasia/hyperkeratosis
Pancreas (exocrine)	Foci of acinar cell alteration, hyperplastic nodules, atypical acinar cell nodules

Source: From Maronpot, R.R. (1991). With permission.

[a] Many of these presumptive preneoplastic lesions are seen in carcinogenicity studies utilizing specific animal model systems. Generalizations about these presumptive preneoplastic lesions are inappropriate outside the context of the specific animal model system being used.

TABLE 108

Animal Neoplastic Lesions of Questionable Significance to Humans

- Male rat renal tumors with β_2-globulin nephropathy
- Rodent urinary bladder neoplasia associated with calcium phosphate precipitates
- Rodent hepatocellular neoplasia with peroxisome proliferation
- Adenohypophysis neoplasia in rats with dopamine inhibitors
- Rodent thyroid follicular cell tumors resulting from adaptive hormonal mechanisms
- Splenic sarcomas in rats
- Pancreatic islet cell neoplasia in rats with neuroleptics
- Rodent stomach carcinoid tumors associated with prolonged acid secretion suppression
- Forestomach neoplasia in rats and mice
- Osteomas in mice
- Mononuclear cell leukemia in F344 rats
- Canine mammary neoplasia related to progestagen administration
- Rodent mammary neoplasia related to adaptive hormonal responses
- Rat uterine endometrial carcinomas related to dopamine agonists
- Uterine leiomyoma in mice with β_1-antagonists
- β_2-Receptor stimulant-induced rat mesovarian leiomyomas
- Mouse ovarian tubulostromal adenomas
- Leydig cell tumors in rat and mice testes

Source: Alison, R.H. et al. (1994); Williams, G.M. and Iatropooulos, M.J. (2001).

TABLE 109

Comparative Features of Benign and Malignant Neoplasms

	Benign	Malignant
General effect on the host	Little; usually do not cause death	Will almost always kill the host if untreated
Rate of growth	Slow; may stop or regress	More rapid (but slower than "repair" tissue); autonomous; never stop or regress
Histological features	Encapsulated; remain localized at primary site	Infiltrate or invade; metastasize
Mode of growth	Usually grow by expansion, displacing surrounding normal tissue	Invade, destroy, and replace surrounding normal tissue
Metastasis	Do not metastasize	Most can metastasize
Architecture	Encapsulated; have complex stroma and adequate blood supply	Not encapsulated; usually have poorly developed stroma; may become necrotic at center
Danger to host	Most without lethal significance	Always ultimately lethal unless removed or destroyed *in situ*

(Continued)

TABLE 109 (Continued)

Comparative Features of Benign and Malignant Neoplasms

	Benign	Malignant
Injury to host	Usually negligible but may become very large and compress or obstruct vital tissue	Can kill host directly by destruction of vital tissue
Radiation sensitivity	Radiation sensitivity near that of normal parent cell; rarely treated with radiation	Radiation sensitivity increased in rough proportion to malignancy; often treated with radiation
Behavior in tissue	Cells cohesive and inhibited by mutual contact	Cells do not cohere; frequently not inhibited by mutual contact
Resemblance to tissue of origin	Cells and architecture resemble tissue of origin	Cells atypical and pleomorphic; disorganized bizarre architecture
Mitotic figures	Mitotic figures rare and normal	Mitotic figures may be numerous and abnormal in polarity and configuration
Shape of nucleus	Normal and regular; show usual stain affinity	Irregular; nucleus frequently hyperchromatic
Size of nucleus	Normal; ratio of nucleus to cytoplasm near normal	Frequently large; nucleus to cytoplasm ratio increased
Nucleolus	Not conspicuous	Hyperchromatic and larger than normal

Source: From Maronpot, R.R. (1991). With permission.

TABLE 110

Criteria for Determining the Human Relevance of Animal Bioassay Results

Supportive	Not Supportive
Same exposure route as humans	Different exposure route than humans
Tumors with several types of exposure	Tumors with only one type of exposure (not relevant to humans)
Tumors in several species	Tumors in only one species
Tumor site correspondence	No site correspondence across species
Tumors at multiple sites	Tumors at only one site
Tumors at sites of low spontaneous occurrence	Tumors at sites with high background incidence
Tumors in tissues analogous to human tissues	Tumors in animal tissues not relevant to humans
No evidence of cellular toxicity at the target site	Tumors only in organs displaying cellular toxicity
Tumors appear early in life	Tumors detectable only late in life
Tumors progress rapidly (benign to malignant)	Benign tumors only
Tumors usually fatal	Tumors not fatal
Similar metabolism (biotransformation) in animals and humans	Metabolic pathways differ in humans and animals
Genotoxic	Nongenotoxic
DNA-reactive	No reaction with DNA
Mechanism of tumorogenesis relevant to humans	Mechanism of tumorogenesis does not occur in humans
Structural similarity to known human carcinogens	Little structural similarity to known human carcinogens
No evidence for disruption of homeostasis	Homeostasis disrupted

Source: Modified from Ashby et al. (1990). With permission.

TABLE 111

Known Human Carcinogens

Aflatoxins

Alcoholic beverage consumption

4-Aminobiphenyl

Aristolochic acids

Analgesic mixtures containing phenacetin

Arsenic and inorganic arsenic compounds

Asbestos

Azathioprine

Benzene

Benzidine

Beryllium and beryllium compounds

1,3-Butadiene

1,4-Butanediol dimethanesulfonate (Busulfan)

Cadmium and cadmium compounds

Chlorambucil

1-(2-Chloroethyl)-3-(4-methylcyclohexyl)-1-nitrosourea (MeCCNU)

Bis(chloromethyl) ether and technical-grade chloromethyl methyl ether

Chromium hexavalent compounds

Coal tar and coal tar pitches

Coal tar

Coke oven emissions

Cyclophosphamide

Cyclosporin A

Diethylstilbestrol

Dyes that metabolize to benzidine

Epstein-Barr virus (EBV)

Environmental tobacco smoke

Erionite

Estrogens, steroidal

Ethylene oxide

Formaldehyde

(Continued)

TABLE 111 *(Continued)*

Known Human Carcinogens

Hepatitis B virus

Hepatitis C virus

Human immunodeficiency virus type 1 (HIV-1)

Human papilloma viruses: some genital–mucosal types

Human T-cell lymphotropic virus type 1 (HTLV-1)

Melphalan

Merkel cell polyomavirus (MCV)

Methoxsalen with ultraviolet A therapy (PUVA)

Mineral oils: (untreated and mildly treated)

Mustard gas

2-Naphthylamine

Neutrons

Nickel compounds

Oral tobacco products

Radon

Silica, crystalline (respirable size)

Solar radiation

Soots

Strong inorganic acid mists containing sulfuric acid

Sunlamps or sunbeds, exposure to

Tamoxifen

2,3,7,8-Tetrachlorodibenzo-p-dioxin (TCDD)

Thiotepa

Thorium dioxide

Tobacco smoke, environmental

Tobacco-smokeless

Tobacco smoking

o-Toluidine

Trichloroethylene (TCE)

Vinyl chloride

Ultraviolet radiation, broad spectrum UV radiation

Wood dust

X-radiation and γ radiation

Source: National Toxicology Program (2016).

References

Alison, R.H., Capen, C.-C., and Prentice, D.E. (1994), Neoplastic lesions of questionable significance to humans. *Toxicol. Pathol.* 22,179.

Ashby, J., Doerrer, N.G., Flamm, F.G., Harris, J.E., Hughes, D.H., Johannsen, F.R., Lewis, S.C., Krivanek, N.D., McCarthy, J.F., Moolenaar, R.J., Raabe, G.K., Reynolds, R.C., Smith, J.M., Stevens, J.T., Teta, M.J., and Wilson, J.D. (1990), A scheme for classifying carcinogens. *Regul. Toxicol. Pharmacol.* 12, 270.

Bomhard, E., Karbe, E., and Loeser, E. (1986), Spontaneous tumors of 2000 Wistar TNO/W.70 rats in two-year carcinogenicity studies, *J. Environ. Pathol. Toxicol. Oncol.* 7, 35.

Butterworth, B.E. and Goldsworthy, T.L. (1991), The role of cell proliferation in multistage carcinogeneis, *Proc. Soc. Exp. Biol. Med.* 198, 683.

Chu, K. (1977), *Percent Spontaneous Primary Tumors in Untreated Species Used at NCI for Carcinogen Bioassays*, NCI Clearing House, cited in Gad and Weil (1986).

Fears, T.R., Tarone, R.E., and Chu, K.C. (1977), False-positive and false-negative rates for carcinogenicity screens, *Cancer Res.* 27, 1941, cited in Gad and Weil (1986).

Gad, S.C. and Weil, C.S. (1986), *Statistics and Experimental Design for Toxicologists*, Telford Press, New Jersey, 1986.

Gart, J.J., Chu, K.C., and Tarone, R.E. (1979), Statistical issues in interpretation of chronic bioassay tests for carcinogenciity, *J. Natl. Cancer Inst.* 62, 957, cited in Gad and Weil (1986).

Gold, L.S., Slone, T.H., Manley, N.B., and Bernstein, L. (1991), Target organs in chronic bioassays of 533 chemical carcinogens, *Environ. Health Perspect.* 93, 233.

Goodman, D.G., Ward, J.M., Squire, R.A., Chu, K.C., and Linhart, M.S. (1979), Neoplastic and nonneoplastic lesions in aging F344 rats, *Toxicol. Appl. Pharmacol.* 48, 237, cited in Gad and Weil (1986).

Haseman, J.K. (1983), Patterns of tumor incidence in two-year cancer bioassay feeding studies in Fischer 344 rats, *Fundam. Appl. Toxicol.*, 3, 1.

Lang, P.L. (1987), Spontaneous neoplastic lesions in the Crl:CD-1®(ICR) BR mouse, Charles River Laboratories, Wilmington, MA.

Maronpot, R.R. (1991), Chemical carcinogenesis, in *Handbook of Toxocologic Pathology*, Haschek, W.M. and Rousseaux, G.G., Eds., Academic Press, San Diego, CA, chap.7.

National Toxicology Program (2016), The 14th Report on Carcinogens, U.S. Department of Health and Human Services, Public Health Service.

Page, N.P. (1977), Concepts of a bioassay program in environmental carcinogenesis, in *Environmental Cancer*, Kraybill, H.F. and Mehlman, M.A., Eds., Hemisphere Publishing, New York, 1977, pp. 87–171, cited in Gad and Weil (1986).

Pitot, H.C. and Dragon, Y.P. (1993), Stage of tumor progression, progressor agents, and human risk, *Proc. Soc. Exp. Biol. Med.*, 202, 37.

Pitot, H.C. (1991), Endogenous carcinogenesis: The role of tumor promotion, *Proc. Soc. Exp. Biol. Med.*, 198, 661.

Pitot, H.C. (1993), The dynamics of carcinogenesis: implications for human risk, *C.I.I.T. Activities*, Chemical Industry Institute of Toxicology, Vol. 13, No. 6.

Poteraki, J. and Walsh, K.M. (1998) Spontaneous neoplasms in control Wistar rats: A comparison of reviews, *Toxicol. Sci.* 45,1.

Rao, G.N., Haseman, J.K., Grumbein, S., Crawford, D.D., and Eustis, S.L. (1990), Growth, body weight, survival and tumor trends in (C57BL/6 X C3H/HeN)F1 (B6C3F1) mice during a nine-year period, *Toxicol. Pathol.*, 18, 71.

Rao, G.N., Haseman, J.K., Grumbein, S., Crawford, D.D., and Eustis, S.L. (1990), Growth, body weight, survival and tumor trends in F344/N rats during an eleven-year period, *Toxicol. Pathol.* 18, 61.

Tarone, R.E., Chu, K.C., and Ward, J.M. (1981), Variability in the rates of some common naturally occurring tumors in Fischer 344 rats and (C57BL/6NXC3H/HEN) F' (B6C3F1) mice, *J. Natl. Cancer Inst.*, 66, 1175, cited in Gad and Weil (1986).

Walsh, K.M. and Poteracki, J. (1994), Spontaneous neoplasms in control Wistar rats, *Fundam. Appl. Toxicol.*, 22, 65.

Weisburger, J.H. and Williams, G.M. (1980). Chemical carcinogens, in *Cassarett and Doull's Toxicology: The Basic Science of Poisons*, 2nd ed., Doull, J., Klaassen, C.D., and Amdur, M.O., Eds., Macmillan Publishing Co., New York, chap. 6.

Williams, G.M. and Iatropooulos, M.J. (2001). Principles of testing for carcinogenic activity, in *Principles and Methods of Toxicology*, 4th edn., Hayes, A. W., Ed., Taylor & Francis, Philadelphia, PA, chap. 20.

7

Neurotoxicology

TABLE 112

Examples of Potential Endpoints of Neurotoxicity

Behavioral Endpoints

Absence or altered occurrence, magnitude, or latency of sensorimotor reflex

Altered magnitude of neurological measurements, such as grip strength or hindlimb splay

Increases or decreases in motor activity

Changes in rate or temporal patterning of schedule-controlled behavior

Changes in motor coordination, weakness, paralysis, abnormal movement or posture, tremor, ongoing performance

Changes in touch, sight, sound, taste, or smell sensations

Changes in learning or memory

Occurrence of seizures

Altered temporal development of behaviors or reflex responses

Autonomic signs

Neurophysiological Endpoints

Change in velocity, amplitude, or refractory period of nerve conduction

Change in latency or amplitude of sensory-evoked potential

Change in EEG pattern or power spectrum

Neurochemical Endpoints

Alteration in synthesis, release, uptake, degradation of neurotransmitters

Alteration in second messenger-associated signal transduction

Alteration in membrane-bound enzymes regulating neuronal activity

Decreases in brain acetylcholinesterase

(Continued)

TABLE 112 (*Continued*)

Examples of Potential Endpoints of Neurotoxicity

Inhibition of neurotoxic esterase
Altered developmental patterns of neurochemical systems
Altered proteins (c-*fos*, substance P)

Structural Endpoints
Accumulation, proliferation, or rearrangement of structural elements
Breakdown of cells
GFAP increases (adults)
Gross changes in morphology, including brain weight
Discoloration of nerve tissue
Hemorrhage in nerve tissue

Source: From U.S. Environmental Protection Agency (1993).

TABLE 113

Examples of Parameters Recorded in Neurotoxicity Safety Studies

Clinical signs of neurotoxicity (onset and duration)
Body weight changes
Changes in behavior
Observations of skin, eyes, mucous membranes, etc.
Signs of autonomic nervous system effect (e.g., tearing, salivation, diarrhea)
Changes in respiratory rate and depth
Cardiovascular changes such as flushing
Central nervous system changes such as tremors, convulsion, or coma
Time of death
Necropsy results
Histopathological findings of the brain, spinal cord, and peripheral nerves

Source: From Abou-Donia, M.B. (1992).

TABLE 114

Summary of Measures in the Functional Observational Battery and the Type of Data Produced by Each

Home Cage and Open Field	Manipulative	Physiological
Posture (D)	Ease of removal (R)	Body temperature (I)
Convulsions, tremors (D)	Handling reactivity (R)	Body weight (I)
Palpebral closure (R)	Palpebral closure (R)	
Lacrimation (R)	Approach response (R)	
Piloerection (Q)	Click response (R)	
Salivation (R)	Touch response (R)	
Vocalizations (Q)	Tail pinch response (R)	
Rearing (C)	Righting reflex (R)	
Urination (C)	Landing foot splay (I)	
Defecation (C)	Forelimb grip-strength (I)	
Gait (D,R)	Hindlimb grip-strength (I)	
Arousal (R)	Pupil response (Q)	
Mobility (R)		
Stereotype (D)		
Bizarre behavior (D)		

Source: From U.S. Environmental Protection Agency (1993).

Note: D, descriptive data; R, rank order data; Q, quantal data; I, interval data; C, count data.

TABLE 115

Measures of the FOB, Divided by Functional Domain

Functional Domain	FOB Measures
Activity	Home cage posture, palpebral closure, locomotion and rearing in open field, automated measure of motor activity
Autonomic	Lacrimation, salivation, pupil response, palpebral closure, defecation, urination
Convulsive	Tonic and clonic movements, tremors, myoclonus
Excitability	Arousal, removal and handling reactivity, vocalizations
Neuromuscular	Gait, locomotion, forelimb and hind limb grip strength, landing foot splay, air righting
Sensorimotor	Tail-pinch response, auditory or click response, touch response, approach response
Physiological measures	Body weight, body temperature, piloerection

Source: Adapted from Moser, V.C. (1991).

TABLE 116

Chemicals Commonly Used as Positive Control Materials for the FOB

Chemical	Effect
Acrylamide	Increased landing foot splay, decreased grip strength (especially hind limb), ataxia, decreased motor activity, tremors
Amphetamine	Increased arousal, increased locomotion, increased rearing, stereotypical behavior
Carbaryl Physostigmine Parathion	Autonomic signs (salivation, lacrimation, miosis), tremor, muscle fasciculations, altered gait, decreased activity, hypothermia, chewing motions
Chlorpromazine	Decreased activity, low arousal, flattened posture, altered gait, decreased grip strength
Dichlorodiphenyltrichloroethane (DDT)	Tremors, myoclonus, convulsions, increased response to auditory stimuli, gait abnormalities, hyperthermia
Triethyltin	Altered gait, decreased grip strength, decreased righting ability, decreased activity and arousal

Source: Adapted from Moser, V.C. and Ross, J.F. (1996).

TABLE 117

Toxic Signs of Acetylcholinesterase Inhibition

Muscarinic Effects[a]	Nicotinic Effects[b]	CNS Effects[c]
Bronchoconstriction	Muscular twitching	Giddiness
Increased bronchosecretion	Fasciculation	Anxiety
Nausea and vomiting	Cramping	Insomnia
(absent in rats)	Muscular weakness	Nightmares
Diarrhea		Headache
Bradycardia		Apathy
Hypotension		Depression
Miosis		Drowsiness
Urinary incontinence		Confusion
		Ataxia
		Coma
		Depressed reflex
		Seizure
		Respiratory depression

Source: From Chan, P.K. and Hayes, A.W. (1989). With permission.

[a] Blocked by atropine.
[b] Not blocked by atropine.
[c] Atropine might block early signs.

TABLE 118

Representative Areas of the Nervous System for Histopathological
Evaluations

Brain

Section 1 (coronal incision rostral to olfactory tubercules): cerebral cortex, rhinal
fissure, olfactory tracts

Section 2 (coronal incision through optic chiasm): cerebral cortex, corpus callosum,
basal ganglia (globus pallidus, putamen, caudate nucleus), thalamus, hypothalamus,
internal capsule, external capsule, lateral ventricles, third ventricle

Section 3 (coronal incision through infundibulum): cerebral cortex, corpus callosum,
hippocampus, amygdala, thalamus, hypothalamus, lateral ventricles, third ventricle,
internal capsule, external capsule

Section 4 (coronal incision at caudal margin of the mammillary body): cerebral
cortex, hippocampus, medial geniculate nuclei, substantia nigra, cerebral
aqueduct

Section 5 (coronal incision at caudal border of the trapezoid body): cerebellum,
cerebellar peduncles, pyramidal tract, medulla, fourth ventricle

Section 6 (coronal incision through medulla immediately beneath the caudal edge of
the cerebellum): medulla, pyramidal tract, olivary nuclei, central canal

Trigeminal ganglia

Eye, with Optic Nerve and Retina

Spinal cord

    Cervical (longitudinal and cross sections)—at cervical enlargement

    Thoracic (longitudinal and cross sections)

    Lumbar (longitudinal and cross sections)—at lumbar enlargement

Dorsal root ganglia and associated dorsal and ventral root fibers

    Cervical region

    Lumbar region

Peripheral nerves

    Sciatic nerve (proximal region)

    Tibial nerve (proximal, at the knee)

    Tibial nerve and calf muscle (distal, at calf muscle)

    Sural nerve

References

Abou-Donia, M.B. (1992), Principles and methods of evaluating neuro-toxicity, in *Neurotoxicology*, Abou-Donia, M.B., Ed., CRC Press, Boca Raton, FL, p. 515.

Chan, P.K. and Hayes, A.W. (1989), Principles and methods for acute tox-icity and eye irritancy, in *Principles and Methods of Toxicology*, 2nd ed., A.W. Hayes, Ed., Raven Press, New York.

Moser, V.C. (1991), Applications of a neurobehavioral screening battery, *J. Am. Coll. Toxicol.* 10: 661.

Moser, V.C. and Ross, J.F. (1996), Training video and reference manual for a functional observational battery, U.S. Environmental Protection Agency and American Industrial Health Council.

U.S. Environmental Protection Agency (1993), Draft report: Principles of neurotoxicity risk assessment, *Chemical Regulation Reporter, Bureau of National Affairs*, Inc., Washington, D.C., pp. 900–943.

8

Immunotoxicology

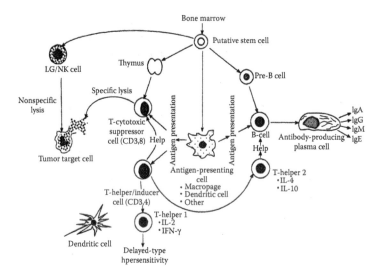

FIGURE 2
Cellular elements of the immune system.

TABLE 119

Examples of the Four Types of Hypersensitivity Responses

Agents: Clinical Manifestations	Hypersensitive Reaction	Cells Involved	Antibody	Mechanism of Cell Injury
Food additives: GI allergy Penicillin: urticaria and dermatitis Anhydrides: occupational asthma	Type I (anaphylactic)	Mast cell	IgE (and others)	Degranulation and release of inflammatory mediators such as histamine, proteolytic enzymes, chemotactic factors, prostaglandins, and leukotrienes
Cephalosporins: hemolytic anemia Aminopyrine: leukopenia Quinidine, gold: thrombocytopenia	Type II (cytotoxic)	Null (K) cells[a]	IgG, IgM	Antibody-dependent cellular cytotoxicity, or complement-mediated lysis
Hydralazine: systemic lupus erythromatosis Methicillin: chronic glomerulonephritis	Type III (immune complex)	PMNs[b]	IgG, IgM	Immune complex deposition in various tissues activates complement, which attracts PMNs causing local damage by release of inflammatory cytokines
Nickel, penicillin, dinitrochlorobenzene, phenothiasines: contact dermatitis	Type IV (delayed hypersensitivity)	T cells (sensitized); macrophages	None	Release of lymphokines activates and attracts macrophages, which release mediators that induce inflammatory reactions

Source: From Norbury, K. and Thomas, P. (1990). With permission.

Note: Defined by Coombs, R.R.A. and Gell, P.G.H. (1968).

[a] Also, T cells, monocyte/macrophages, platelets, neutrophils, and eosinophils.

[b] Polymorphonuclear leukocytes.

TABLE 120

Examples of Antemortem and Postmortem Findings That May Include Potential Immunotoxicity If Treatment Related

Parameter	Possible Observation (Cause)	Possible State of Immune Competence
Antemortem		
Mortality	Increased (infection)	Depressed
Body weight	Decreased (infection)	Depressed
Clinical signs	Rales, nasal discharge (respiratory infection)	Depressed
	Swollen cervical area (sialodacryoadenitis virus)	Depressed
Physical examinations	Enlarged tonsils (infection)	Depressed
Hematology	Leukopenia/lymphopenia	Depressed
	Leukocytosis (infection/cancer)	Enhanced/depressed
	Thrombocytopenia	Hypersensitivity
	Neutropenia	Hypersensitivity
Protein electrophoresis	Hypogammaglobulinemia	Depressed
	Hypergammaglobulinemia (ongoing immune response or infection)	Enhanced/activated
Postmortem		
Organ weights		
Thymus	Decreased	Depressed
Histopathology		
Adrenal glands	Cortical hypertrophy (stress)	Depressed (secondary)
Bone marrow	Hypoplasia	Depressed
Kidney	Amyloidosis	Autoimmunity
	Glomerulonephritis (immune complex)	Hypersensitivity
Lung	Pneumonitis (infection)	Depressed
Lymph node (see also spleen)	Atrophy	Depressed

(Continued)

TABLE 120 (*Continued*)

Examples of Antemortem and Postmortem Findings That May Include Potential Immunotoxicity If Treatment Related

Parameter	Possible Observation (Cause)	Possible State of Immune Competence
Spleen	Hypertrophy/hyperplasia	Enhanced/activated
	Depletion of follicles	Depressed B-cells
	Hypocellularity of periarteriolar sheath	Depressed T-cells
	Active germinal centers	Enhanced/activated
Thymus	Atrophy	Depressed
Thyroid	Inflammation	Autoimmunity

Source: From Norbury, K. and Thomas, P. (1990). With permission.

TABLE 121

National Toxicology Program Panel for Detecting Immune Alterations
in Rodents

Parameter	Procedures
Screen (Tier I)	
Immunopathology	• Hematology: complete blood count and differential
	• Weights: body, spleen, thymus, kidney, liver
	• Cellularity: spleen
	• Histology: spleen, thymus, lymph node
Humoral immunity	• Enumerate IgM antibody plaque-forming cells to T-dependent antigen (sRBC, KLH)
	• Lippopolysaccharide (LPS) mitogen response
Cell-mediated immunity	• Lymphocyte blastogenesis to mitogens (Con A)
	• Mixed leukocyte response against allogeneic leukocytes (MLR)
Nonspecific immunity	• Natural killer (NK) cell activity
Comprehensive (Tier II)	
Immunopathology	• Quantitation of splenic B- and T-cell numbers
Humoral-mediated immunity	• Enumeration of IgG antibody response to sRBCs
Cell-mediated immunity	• Cytotoxic T-lymphocyte (CTL) cytolysis
	• Delayed-type hypersensitivity (DTH) response
Nonspecific immunity	• Macrophage function-quantitation of resident peritoneal cells, and phagocytic ability (basal and activated by MAF)
Host resistance challenge models (endpoints)[a]	• Syngeneic tumor cells
	• PYB6 sarcoma (tumor incidence)
	• B16F10 melanoma (lung burden)
	• Bacterial models: *Listeria monocytogenes; Streptococcus species*
	• Viral models: Influenza
	• Parasite models: *Plasmodium yoelii* (Parasitaemia)

Source: Adapted from Luster, M.I. et al. (1992).
Note: The testing panel was developed using B6C3F1 female mice.
[a] For any particular chemical tested, only two or three host resistance models are
selected for examination.

TABLE 122

Immunotoxicology Functional Assays

Assay	Evaluates	Comments
Cytokine/chemokine assays	Innate immunity	Cytokines and chemokines are nonspecific immunological mediators that are important in cell–cell communication among cells of the immune system.
Natural killer (NK) cell activity	Innate immunity	NK cells have an important role in the interaction of different immunological cell types and cell functions that are important in immunological defense against viral, bacterial, parasitic, and neoplastic disease. Measurement of an antigen- or infectious microorganism-driven cytokine-enhanced immunological response is very important and measures not only the static activity, but also the ability to be stimulated by an infectious disease. NK cells connect innate and adaptive, acquired immunity. NK activity is measured *in vitro* by measuring the lysis of tumor cells sensitive to NK-mediated cytotoxicity.
Macrophage activity	Innate immunity	Macrophages are important contributors to early nonspecific innate immunity and also participate in specific immunological responses. Macrophages can initiate and modulate both specific and nonspecific immunological responses. A variety of assays can be utilized to assess macrophage activity but can pose technical issues.
Delayed-type hypersensitivity (DTH) assay	Cell-mediated immunity (CMI)	A measure of almost exclusively T-lymphocyte function. Response of animals sensitized dermally to a strong contact sensitizer is assessed. Not considered as sensitive as the MLR or CTL assays.

(Continued)

TABLE 122 (*Continued*)

Immunotoxicology Functional Assays

Assay	Evaluates	Comments
Lymphocyte blastogenesis	Cell-mediated immunity	Measures lymphocyte activation/cell proliferation in response to agents that can activate lymphocytes such as phytohemagglutin. The ability of lymphocytes to respond to activation signals in a physiological manner is used to assess overall immunocompetence.
Mixed lymphocyte reaction (MLR)	Cell-mediated immunity	An *in vitro* assay that measures the ability of lymphocytes to respond to the presence of allogeneic cells. This proliferation represents the initial stage of the acquisition of CTL function by CD8⁺ T cells, and thus serves as a measure of CMI. The MLR is a form of lymphoproliferation. Also referred to as mixed lymphocyte culture (MLC).
Cytotoxic T lymphocyte (CTL) activity	Cell-mediated immunity	The cytotoxic T lymphocyte (CTL) response is a component of the specific or acquired immune response and has been used to evaluate CMI following exposure to chemicals. The CTL response requires the interaction of the following categories of immune cells: 1. Professional antigen-presenting cells such as dendritic cells and/or macrophages 2. CD4⁺ T lymphocytes that produce help for response to T-dependent antigens 3. CD8⁺ T lymphocytes that develop into antigen-specific cytotoxic effector cells. Antigen presentation is by both class I and class II molecules of the major histocompatibility complex (MHC) to generate effector cytotoxic CTLs. For this reason, the CTL response is distinguished from the delayed type hypersensitivity (DTH) response or the T-dependent antibody response (TDAR) both of which require only class II presentation of antigens.

(Continued)

TABLE 122 (Continued)

Immunotoxicology Functional Assays

Assay	Evaluates	Comments
Antibody-forming cell assay (AFC)/ Plaque-forming cell assay(PFC)	Humoral-mediated immunity (HMI)	This assay measures the ability of animals to produce either IgM or IgG antibodies against a T-dependent or T-independent antigen following *in vivo* (or less frequently *in vitro*) immunization. Because of the involvement of multiple cellular and humoral elements in mounting an antibody response, the assay evaluates several immune parameters simultaneously.
T-dependent antibody response (TDAR)	Humoral-mediated immunity	The TDAR response requires and measures the functionality of three major immune cells: T cells, B cells, and the antigen processing and presentation ability of dendritic cells and macrophages. The measurement of TDAR is important in assessing the ability of the host to produce antibody. TDAR may be measured by evaluating the number of antibody forming cells (AFC) in the spleen following immunization with sheep red blood cells (SRBC). TDAR may also be measured by immunizing with keyhole limpet hemocyanin (KLH) or SRBC and measuring anti-KLH or anti-SRBC in the serum by ELISA.
T-independent antibody response (TIAR)	Humoral-mediated immunity	The TIAR response is an important antibody response to polysaccharide antigens such as those on the encapsulated bacteria that cause bacterial pneumonia. This antibody response occurs in the absence of T cell help and requires the presence of marginal zone B cells.

(Continued)

TABLE 122 (*Continued*)

Immunotoxicology Functional Assays

Assay	Evaluates	Comments
Host resistance assay-influenza	Overall function of the immune system	Overall health of the immune system in rats or mice is evaluated in response to influenza virus exposure. The following are mechanistic immunological function endpoints evaluated in this model:
		• Cytokines-innate immunity
		• Interferon activity—innate immunity
		• Macrophage activity—innate immunity
		• NK cell activity—innate immunity
		• CTL activity—cell-mediated immunity
		• Influenza-specific IgM, IgG (IgG1 and IgG2a)—TDAR—humoral-mediated immunity (TDAR)
		• Immunophenotyping
		• Histopathology
		Clearance of the infectious agent is the cumulative effect of the orchestrated immune response and is the best method for evaluating the overall health of the immune system.
Targeted host resistance assay-marginal zone B (MZB) cell assay	Effect of chemicals on the spleen marginal zone	Addresses concern arising from histopathology results indicating an effect of the test article on the spleen marginal zone.

(*Continued*)

TABLE 122 (*Continued*)

Immunotoxicology Functional Assays

Assay	Evaluates	Comments
Targeted host resistance assay—*Streptococcus pneumoniae* pulmonary assay	Innate immunity	Addresses concern arising from a defect in innate immunity parameters, especially macrophages and neutrophils. This assay measures bacterial clearance at 24 hours, a time when acquired immune functions have not yet developed.
Targeted host resistance assay—*Listeria monocytogenes* systemic assay	Cell-mediated immunity	Addresses concern arising from a defect in cell-mediated immunity. This assay measures bacterial clearance at multiple time points over 7 days. Intracellular Gram-positive bacterial assay to evaluate liver and splenic macrophages and neutrophils.
Targeted host resistance assay—Murine cytomegalovirus (MCMV) latent viral reactivation assay	Viral reactivation	Addresses concern arising from a decrease in cell-mediated immunity that could result in reactivation of latent viral infection.
Targeted host resistance assay—*Candida albicans* assay	Antifungal immunity	Addresses concern arising from defects in antifungal immunity by measuring clearance of infectious Candida albicans.

References

Coombs, R.R.A. and Gell, P.G.H. (1968), Classification of allergic reactions responsible for clinical hypersensitivity and disease, in *Clinical Aspects of Immunology*, Gell, P. and Coombs, R., Eds., Blackwell Scientific Publications, Oxford, pp. 121–137.

Luster, M.I., Portier, C., Pait, D., Whilte, K., Genning, C., Munson, A., and Rosenthal, G. (1992), Risk assessment in immunotoxicology I. Sensitivity and predictability of immune tests. *Fundam. Appl. Toxicol.* 18, 200–210.

Norbury, K. and Thomas, P. (1990), Assessment of immunotoxicity, in *In vivo Toxicity Testing: Principles, Procedures and Practices*, Arnold, D.L., Grice, H., and Krewski, D., Eds., Academic Press, New York, pp. 410–448.

Sjoblad, R. (1988), Potential future requirements for immunotoxicology testing of pesticides, *Toxicol. Ind. Health*, 4, 391–395.

9

Reproductive/Developmental Toxicology

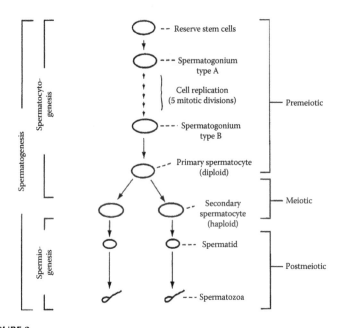

FIGURE 3

A general scheme of mammalian spermatogenesis. Each cycle is completed in 35 to 64 days, depending on the species, with a new cycle being initiated at the Type A spermatogonium level every 12 to 13 days. (From Ecobichon, D.J. (1992).)

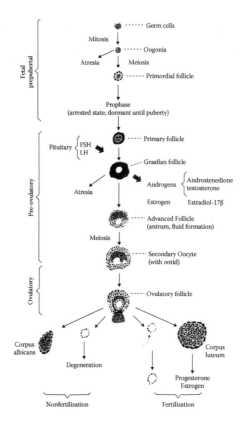

FIGURE 4
A general scheme of mammalian oogenesis. (From Ecobichon, D.J. (1992).)

TABLE 123

Reproductive Parameters for Various Species

Species	Age at Puberty	Sexual Cycle Type[a]	Sexual Cycle Duration (Days)	Ovulation Time[b]	Ovulation Type[c]	Copulation Time[b]	Copulation Length	Implantation (Days)	Gestation Period (Days)
Mouse	5–6 weeks	PE	4	2–3 h	S	Onset of estrus		4–5	19 (19–21)
Rat	6–11 weeks	PE	4–6	8–11 h	S	1–4 h		5–6	21–22
Rabbit	6–7 months	PE	Indefinite	10 h	I	Anytime[f]	Sec	7–8	31 (30–35)
Hamster	5–8 weeks	PE	4	Early estrus	S	Estrus[g]	Sec	5+	16 (15–18)
Guinea Pig	8–10 weeks	PE	16–19	10 h	S	Estrus		6	67–68
Ferret	8–12 months	ME	Seasonal	30–36 h	I	Estrus	1–3 h	12–13	42
Cat	6–15 months	PE	Seasonal[d]	24–56 h	I	3rd day[h]	1–2 h	13–14	63 (52–69)
Dog	6–8 months	ME	9	1–3 days	S	Estrus	1–2 h	13–14	61 (53–71)
Monkey	3 years	PE	28	9–20 days	S	Anytime[i]		9	168 (146–180)
Human	12–16 years	PE	27–28	14 day (13–15)	S	Anytime	15–30 min	7.5	267 (ovulation)

Source: Modified from Spector, W.S. (1956).

[a] PE, polyestrous; ME, monoestrus.
[b] Time from start of estrous cycle.
[c] I, induced ovulation; S, spontaneous ovulation.
[d] March to August.
[e] After mating.
[f] Most receptive when in estrus.
[g] 8–10 pm.
[h] Of estrus, most receptive.
[i] Most receptive 2 days before ovulation.

TABLE 124

Breeding Characteristics of Female Laboratory Mammals Compared with the Human

Parameters	Monkey[a]	Dog	Cat	Rabbit	Mouse	Guinea Pig	Hamster	Rat	Human
Age at puberty	36 months	6–8 months	6–15 months	5.5–8.5 days	35 days	55–70 days	35–56 days	37–67 days	12–15 years
Breeding season	All year	Spring–Fall	Feb–July	All year	All year	All year	All year	All year	All year
Breeding life (years)	10–15	5–10	4	1–3	1	3	1	1	35
Breeding age (months)	54	9	10	6–7	2	3	2	2–3	180
Litter size (number)	1	3–6	1–6	1–13	1–12	1–5	1–12	6–9	1
Birth weight (g)	500–700	1100–2200	125	100	1.5	75–100	2.0	5.6	1
Opening of eyes (days)	At birth	8–12	8–12	10	11	At birth	15	11	At birth
Weaning age (weeks)	16–24	6	6–9	18	3	2	3–4	3–4	
Weight at weaning (g)	4400	5800	3000	1000	11–12	250	35	10–12	

Source: Data obtained from various sources, including the following: Ecobichon, D.J. (1992); Spector, W.S. (1956); Altman, P.L. and Dittmer, D.S. (1972).

[a] Monkey = *Macaca mulatta or fascicularis.*

TABLE 125

Species Variability in Parameters Involving Spermatogenesis

Parameter	Mouse	Rat	Hamster	Rabbit	Dog	Monkey	Human
Spermatogenesis duration (days)	26–35	48–53	35	28–40	13.6	9.5	74
Duration of cycle of seminiferous epithelium (days)	8.6	12.9		10.7	13.6	9.5	16
Life span of							
B-type spermatogonia (days)	1.5	2.0		1.3	4.0	2.9	6.3
L + Z spermatocytes (days)	4.7	7.8		7.3	5.2	6.0	9.2
P + D spermatocytes (days)	8.3	12.2		10.7	13.5	9.5	15.6
Golgi spermatids (days)	1.7	2.9		2.1	6.9	1.8	7.9
Cap spermatids (days)	3.5	5.0		5.2	3.0	3.7	1.6
Testis weight (g)[a]	0.2	3.7	1.8	6.4	12.0	4.9	34.0
Daily sperm production							
Per gram testis (×10⁶)	54	14–22	22	25	20	23	4.4
Per individual (×10⁶)	5–6	80–90	70	160	300	1100	125
Sperm reserve in cauda at sexual rest (×10⁶)	49	440	575	1600		5700	420
Sperm storage in epididymal tissue (×10⁶)							
Caput	20		200				
Corpus	7	300	175				
Cauda	40–50	400	200				420

(Continued)

TABLE 125 (*Continued*)

Species Variability in Parameters Involving Spermatogenesis

Parameter	Mouse	Rat	Hamster	Rabbit	Dog	Monkey	Human
Transit time through epididymis at sexual rest (days)							
Caput and corpus	3.1	3.0		3.0		4.9	1.8
Cauda	5.6	5.1		9.7		5.6	3.7
Ejaculate volume (mL)	0.04	0.2	0.1	1.0			3.0
Ejaculated sperm (10^6/mL)	5.0			150			80.0
Sperm transit time from vagina to tube	15–60 min	30–60 min		3–4 h	20 min		15–30 min

Source: Data obtained from various sources, including: Altman, P.L. and Dittmer, D.S. (1972); Eddy, E.M. and O'Brien, D.A. (1989); Blazak, W.F. (1989); Zenick, H. and Clegg, E.D. (1989) and Spector, W.S., Ed. (1956).

[a] Combined weight of both testes.

TABLE 126

Species Variability in Parameters Involving Oogenesis

Parameter	Mouse	Rat	Guinea Pig	Hamster	Rabbit	Cat	Dog	Monkey	Human
Sexual maturity (days)	28	46–53	84	42–54	120–240	210–245	270–425	1642	
Duration of estrus (days)	9–20 h	9–20 h	6–11 h	1	30	4	9	4–6	
Ovulation time (days)	2–3 h	9–20 h	10 h		10 h	24–56 h	1–3	9–20	15
Ovulation type[a]	S	S	S	S	I	I	S	S	S
No. ova released	8	10		7	10	4–6	8–10	1	1
Follicle size (mm)	0.5	0.9	0.8		1.8		10		
Ovum diameter (mm)	0.07–0.087	0.07–0.076	0.075–0.107		0.110–0.146	0.12–0.13	0.135–0.145	0.109–0.173	0.089–0.091
Zona pellucida (mm membrane thickness)			0.012		0.011–0.023	0.012–0.115	0.135	0.012–0.034	0.019–0.035
Transport time (to reach site of implantation) (days)	4.5	3.0	3.5	3.0	2.5–4	4–8	6–8	3.0	3.0
Implantation (days)	4.5–5.0	5.5–6.0	6.0	4.5–5.0	7–8	13–14	13–14	9–11	8–13
Rate of transport of sperm to oviduct (min)	15	15–30	15		5–10				5–60
Rate of transport of embryo to uterus (h)	72	95–100	80–85		60				80
Fertile life of spermatozoa in female tract (h)	6	14	21–22		30–32				24–48
Rate of transport of ova in female tract (h)	8–12	12–14	20	5–12	6–8				24

(Continued)

TABLE 126 (*Continued*)

Species Variability in Parameters Involving Oogenesis

Parameter	Mouse	Rat	Guinea Pig	Hamster	Rabbit	Cat	Dog	Monkey	Human
Segmentation (to form blastocele) (days)	2.5–4.0	4.5	5–6	3.25	3–4				5–8
Primitive streak (days)	7.0	8.5	10.0	6.0	6.5	13.0	13.0	18.0	
Duration of organogenesis (days)	7.5–16	9–17	11–25	7–14	7–20	14–26	14–30	20–45	
Gestational length (days)	20–21	21–22	65–68	16–17	31–32	58–71	57–66	164–168	

Source: Data obtained from various sources, including the following: Ecobichon, D.J. (1992); Spector, W.S., Ed. (1956); Altman, P.L. and Dittmer, D.S. (1972); Eddy, E.M. and O'Brien, D.A. (1989); Manson, J.M. and Kang, Y.S. (1989).

[a] Ovulation type: I, induced; S, spontaneous.

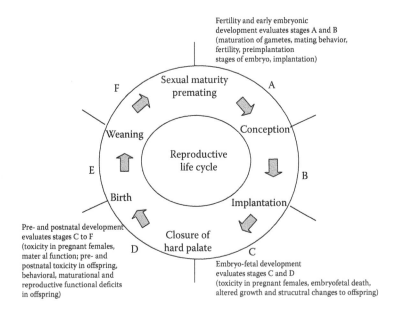

FIGURE 5

Graphic representation of an animal's reproductive life-cycle and corresponding relationship to the ICH reproductive life stages indicated by the letters A through F. Also shown are the specific stages evaluated by the standard segmented reproductive study designs-Fertility and Early Embryonic Development/Seg. I, Embryo-Fetal Development/Seg. II, and Pre- and Postnatal Development/Seg. III.

TABLE 127

Fertility and Reproductive Indices Used in Single and Multigeneration Studies

Index	Derivation
Mating	$= \dfrac{\text{No. confirmed copulations}}{\text{No. of estrous cycles required}} \times 100$
Male fertility	$= \dfrac{\text{No. males impregnating females}}{\text{No. males exposed to fertile, nonpregnant females}} \times 100$
Female fertility	$= \dfrac{\text{No. of females confirmed pregnant}}{\text{No. of females housed with fertile male}} \times 100$
Female fecundity	$= \dfrac{\text{No. of females confirmed pregnant}}{\text{No. of confirmed copulations}} \times 100$
Implantation	$= \dfrac{\text{No. of implantations}}{\text{No. of pregnant females}} \times 100$
Preimplantation loss	$= \dfrac{\text{Corpora lutea} - \text{No. of implants}}{\text{No. of Corpora lutea}} \times 100$
Parturition incidence	$= \dfrac{\text{No. of females giving birth}}{\text{No. of females confirmed pregnant}} \times 100$
Live litter size	$= \dfrac{\text{No. of litters with live pups}}{\text{No. of females confirmed pregnant}} \times 100$
Live Birth	$= \dfrac{\text{No. viable pups born/litter}}{\text{No. pups born/litter}} \times 100$
Viability	$= \dfrac{\text{No. of viable pups born}}{\text{No. of dead pups born}} \times 100$
Survival	$= \dfrac{\text{No. of pups viable on day 1}}{\text{No. of viable pups born}} \times 100$

(Continued)

TABLE 127 (*Continued*)

Fertility and Reproductive Indices Used in Single and Multigeneration Studies

Index	Derivation
Pup death (day 1–4)	$= \dfrac{\text{No. of pups dying, postnatal days 1–4}}{\text{No. of viable pups born}} \times 100$
Pup death (days 5–21)	$= \dfrac{\text{No. of pups dying, postnatal days 5–21}}{\text{No. of viable pups born}} \times 100$
Sex ratio (at birth)	$= \dfrac{\text{No. of male offspring}}{\text{No. of female offspring}} \times 100$
Sex ratio (day 4) (day 21)	$= \dfrac{\text{No. of male offspring}}{\text{No. of female offspring}} \times 100$

Source: From Ecobichon, D.J. (1992).

TABLE 128

Basic Developmental Toxicity Testing Protocol

Phase	Time	Developmental Toxicity Testing[a]
Acclimation period	Variable number of weeks	No exposure of the animals to the test agent
Cohabitation period	Day of mating determined (Day 0)	No exposure of the animals to the test agent
Preembryonic period	Day of mating through day 5,[b] 6,[c] 7[d] of pregnancy	
Period of major embryonic organogenesis	Day 5, 6, or 7 through day 15,[b,c] or 18[d] of pregnancy	Groups of pregnant animals exposed to the test agent
Fetal period	Day 15 or 18 through day 18,[b] 21,[c] or 30[d] of pregnancy	No exposure of the pregnant animals to the test agent
Term	Day 18,[b] 22,[c] or 31[d] of pregnancy	Females sacrificed (to preclude cannibalization of malformed fetuses), cesarean section performed, and young examined externally and internally

Source: Adapted from Johnson, E.M. (1990).

[a] Usually a sham-treated control group and three agent-treated groups are used with 20 to 25 mice or rats and 15 to 18 rabbits per group. The dose levels are chosen with the goal of no maternal or developmental effects in the low-dose group and at least maternal toxicity in the high-dose group (failure to gain or loss of weight during dosing, reduced feed and/or water consumption, increased clinical signs, or no more than 10% maternal death).

[b] Mice.

[c] Rats.

[d] Rabbits.

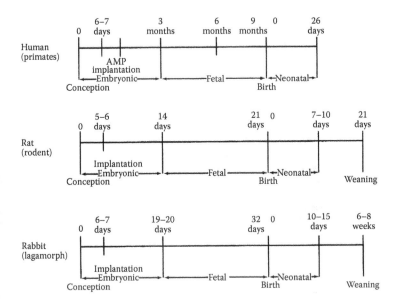

FIGURE 6

Developmental stages and timelines in the human, rat, and rabbit. AMP: Anticipated menstrual period. Average human menstrual cycle is 28 days, with ovulation occurring about 14 days. Rabbit ovulates following coitus. (Adapted from Miller, R.K. et al. (1987).)

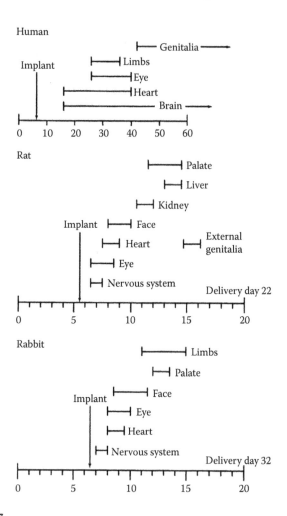

FIGURE 7
Critical periods of embryogenesis in the human, rat, and rabbit. (Adapted from Ecobichon, D.J. (1992).)

TABLE 129

Signs of Overt Maternal Toxicity

1. Daily (or isolated) body weight changes and/or effects on food and/or water consumption during the dosing period[a]
2. Changes in respiration, alertness, posture, spontaneous motor activity, color of mucous membranes, behavior (aggressive, depressed, lethargic, sedated), hair and coat appearance, color of urine, frequency of urination, and number and consistency of fecal pellets
3. Other signs such as nasal discharge, chromodacryoarrhea, salivation, vaginal bleeding, tumor, convulsions, and coma
4. Death and necropsy findings

Source: Modified from Khera, K.S. et al. (1989).

[a] Weight loss or failure to gain weight at any time during the dosing period may be followed by a rebound weight gain of sufficient magnitude to obfuscate an effect on maternal weight; therefore, maternal body weights should be determined daily.

TABLE 130

Signs of Overt Embryofetal Toxicity

Mortality: Resorptions (early and late) and dead fetuses

Dysmorphogenesis (structural alterations): malformations or variations of the offspring

Alterations to growth: growth retardation, excessive growth or precocious maturation

Functional alterations: persistent change in normal physiologic or biochemical function, or reproductive function and developmental neurobehavioral effects.

Source: Modified from Wilson, J.G. (1973).

TABLE 131

End Points of Developmental Toxicity in Female Rodents and Rabbits

Postconception evaluation

    Maternal weight and body weight gain (daily during treatment; not in rabbits)

    Clinical observations

    Food consumption

Cesarean evaluations

    Implantation number

    Corpora lutea number (not in mice)

    Gravid uterine weight

    Litter size

    Live fetuses and fetal weight (male, female, and total)

    Deaths (embryonic, fetal)

    Resorptions

    Pup weight, crown-rump length

    Incidence of malformations (external, visceral, skeletal, ossification sites)

    Increased incidence of variations (external, visceral, skeletal)

Source: Modified from Parker, R.M. (2011).

References

Altman, P.L. and Dittmer, D.S. (1972), *Biology Data Book*, 2nd ed., Vol. 1, Federation of American Societies for Experimental Biology, Bethesda, MD.

Blazak, W.F. (1989), Significance of cellular endpoints in assessment of male reproductive toxicity, in *Toxicology of the Male and Female Reproductive Systems*, Working, P.K., Ed., Hemisphere Publishing Corp., Washington, pp. 157–172, chap. 6.

Ecobichon, D.J. (1992), Reproductive toxicology, in *The Basis of Toxicity Testing*, CRC Press, Boca Raton, FL, pp. 83–112, chap. 5.

Eddy, E.M. and O'Brien, D.A. (1989), Biology of the gamete: maturation, transport, and fertilization, in *Toxicology of the Male and Female Reproductive Systems*, Working, P.K., Ed., Hemisphere Publishing Corp., Washington, pp. 31–100, chap. 3.

Johnson, E.M. (1990), The effects of riboviron on development and reproduction: A critical review of published and unpublished studies in experimental animals, *J. Am. Coll. Toxicol.* 9, 551.

Khera, K.S., Grice, H.C., and Clegg, D.J. (1989), *Current Issues in Toxicology: Interpretation and Extrapolation of Reproductive Data to Establish Human Safety Standards*, Springer-Verlag, New York.

Manson, J.M. and Kang, Y.S. (1989), Test methods for assessing female reproductive and developmental toxicology, in *Principles and Methods of Toxicology*, 2nd edition, Hayes, A.W., Ed., Raven Press, New York, pp. 311–359, chap. 11.

Miller, R.K., Kellogg, C.K., and Saltzman, R.A. (1987), Reproductive and perinatal toxicology, in *Handbook of Toxicology*, Haley, T.J. and Berndt, W.O., Eds., Hemisphere Publishing Corp., Washington.

Parker, R.M. (2011), Testing for reproductive toxicity, in *Handbook of Reproductive Toxicology*, 3rd edn., Hood, R., Ed., CRC Press, Boca Raton, FL.

Spector, W.S., Ed. (1956), *Handbook of Biological Data*, W.B. Saunders Company, Philadelphia.

Wilson, J.G. (1973), *Environment and Birth Defects*, Academic Press, New York.

Zenick, H. and Clegg, E.D. (1989), Assessment of male reproductive toxicity: a risk assessment approach, in *Principles and Methods of Toxicology, Hayes*, A.W., Ed., Raven Press, New York, pp. 275–309, chap. 10.

10

Clinical Pathology

TABLE 132

Approximate Blood Volumes in Animals Typically Used in Nonclinical Toxicology Research

| Species | Typical Body Weight (kg) | Blood Volume (mL) | | | |
		Total Volume (mL)	Weekly Sampling	Monthly Sampling	At Necropsy
Mouse	0.03	2	0.075	0.2	1
Rat	0.3	20	1	2	10
Dog	12.0	1000	50	100	400
Monkey[a]	3.0	200	10	20	100
Rabbit	3.0	200	10	20	100

Source: Adapted from Loeb, W.F. and Quimby, F.W. (1989).

[a] Rhesus or cynomolgus.

TABLE 133

Recommended Maximum Allowable Blood Collection Volumes for Animals[a]

One-Time Collection	Total for Multiple Collections over 1 Day	Total for Multiple Collections over 1 wk	Total for Multiple Collections over 1 mo	Total for Collections Done on a Weekly Basis
1 mL/100 g body weight	1 mL/100 g body weight	1 mL/100 g body weight	2 mL/100 g body weight	0.5 mL/100 g body weight per week

[a] Higher collection volumes acceptable where animals are bled under anesthesia at termination or when an equivalent volume of blood is immediately replaced.

TABLE 134

Common Sites for Blood Collection and Blood Volumes in Various Species

Species	Site	Approximate Blood Volume (mL/kg)	Comments
Mouse	Tail clip, tail vein, orbital sinus, and cardiac puncture	72	Anesthesia should be given for tail clip, orbital sinus, and cardiac puncture collection.
Rat	Tail clip, tail vein, jugular vein, orbital sinus, sublingual vein, and cardiac puncture	64	Anesthesia should be given for tail clip, sublingual, orbital sinus, and cardiac puncture collection. Tail clip not suitable for older animals.
Rabbit	Marginal ear vein, central ear artery, jugular vein, and cardiac puncture	56	Anesthesia should be given for cardiac puncture collection. Compression of central ear artery for a few minutes may be necessary to stop bleeding.
Dog	Cephalic vein, saphenous vein, and jugular vein	85	
Minipig	Anterior vena cava/jugular vein, marginal ear vein	65	
Primate	Cephalic vein, saphenous vein, femoral vein, and jugular vein	60	
Guinea pig	Marginal ear vein, jugular vein, and cardiac puncture	80	Anesthesia should be given for cardiac puncture collection.
Hamster	Orbital sinus, jugular vein, femoral vein, and cardiac puncture	78	Anesthesia should be given for orbital sinus and cardiac puncture collection.
Ferret	Cephalic vein, saphenous vein, jugular vein, and cardiac puncture	75	Anesthesia should be given for cardiac puncture collection.

Source: BVA/Frame/RSPCA/UFAW Joint Working Group (1993), Diehl et al. (2001).

TABLE 135

Erythrocyte Life Span in Various Animals[a]

Species	Mean Life Span[b] (Days)
Man	117–127 (120)[c]
Dog	90–135
Cat	66–79
Pig	62–86
Rabbit	50–80
Guinea pig	70–90
Hamster	60–70
Rat	50–68 (60)[c]
Mouse	41–55

[a] Determined by use of isotopes.
[b] Range of means from various studies.
[c] Most often cited.

TABLE 136

Mean Control Ranges of Typical Serum Clinical Chemistry Measurements in B6C3F₁ Mice

Parameter	12–14 Weeks Old	18–20 Weeks Old	32–34 Weeks Old	58–60 Weeks Old	84–86 Weeks Old	110–112 Weeks Old
Alanine aminotransferase (IU/L)	20–50	25–100	22–90	20–50	23–60	23–60
Albumin (g/dL)	2.3–4.4	2.5–4.2	2.7–3.8	3.0–4.0	3.0–3.9	3.0–4.1
Albumin/globulin ratio	1.0–2.0	0.8–2.0	1.2–1.9	1.3–1.9	1.3–2.0	1.3–2.0
Alkaline phosphatase (IU/L)	30–80 (M) 40–140 (F)	20–55 (M) 45–85 (F)	—	—	—	—
Aspartate aminotransferase (IU/L)	40–110	64–180	—	—	—	—
Bilirubin, total (mg/dL)	—	0.1–0.5	0.1–0.5	0.1–0.5	0.1–0.5	0.1–0.5
Calcium (mg/dL)	—	8.2–11.8	—	—	—	—
Chloride (mEq/L)	—	110–128	—	—	—	—
Cholesterol, total (mg/dL)	90–160	80–130	85–150	80–150	90–160	90–175
Creatine kinase (IU/L)	50–300	—	—	—	—	—
Creatinine (mg/dL)	0.3–0.8	0.2–0.8	—	—	—	—
Globulin (g/dL)	1.5–2.5	1.0–2.7	1.6–2.4	1.8–3.1	1.6–3.0	1.8–3.0
Glucose (mg/dL)	125–200	81–165	115–170	115–170	115–170	115–170
Phosphorus, inorganic (mg/dL)	—	—	—	—	—	—

(Continued)

TABLE 136 (Continued)

Mean Control Ranges of Typical Serum Clinical Chemistry Measurements in B6C3F$_1$ Mice

Parameter	12–14 Weeks Old	18–20 Weeks Old	32–34 Weeks Old	58–60 Weeks Old	84–86 Weeks Old	110–112 Weeks Old
Potassium (mEq/L)	—	3.6–7.3	—	—	—	—
Protein, total (g/dL)	4.5–5.5	4.0–6.0	4.2–6.2	4.8–6.5	4.8–6.6	5.4–6.5
Sodium (mEq/L)	—	147–163	—	—	—	—
Sorbitol dehydrogenase (IU/L)	15–50	18–57	—	—	—	—
Triglycerides (mg/dL)	75–175	75–130	100–173	90–190	110–160	90–175
Urea nitrogen (BUN) (mg/dL)	15–35	12–34	12–27	12–24	10–24	15–28

Source: Adapted from Levine, B.S. (1979–1993) and NIEHS (1985).
Note: —, data unavailable.

TABLE 137

Mean Control Ranges of Typical Serum Clinical Chemistry
Measurements in CD-1 and BALB/c Mice of Various Ages

Parameter	<1-Year-Old CD-1	>1-Year-Old CD-1	1–3-Month-Old BALB/c	6–12-Month-Old BALB/c
Alanine aminotransferase (IU/L)	30–250 (M)	20–200 (M)	—	—
	30–100 (F)	20–80 (F)	—	—
Albumin (g/dL)	—	—	1.6–2.6	1.3–2.6
Albumin/globulin ratio	—	—	—	—
Alkaline phosphatase (IU/L)	30–70	20–75	75–275	47–102
Aspartate aminotransferase (IU/L)	75–300	75–300	40–140	70–110
Bilirubin, total (mg/dL)	0.2–0.8	0.2–0.8	0.5–1.2	0.4–1.0
Calcium (mg/dL)	8.5–11.5	6.7–11.5	7.8–10.8	6.5–9.6
Chloride (mEq/L)	110–125	110–135	—	—
Cholesterol, total (mg/dL)	90–170 (M)	60–170 (M)	165–295	100–300
	60–125 (F)	50–100 (F)		
Creatine kinase (IU/L)	—	—	—	—
Creatinine (mg/dL)	0.3–1.0	0.2–2.0	—	—
Globulin (g/dL)	—	—	—	—
Glucose (mg/dL)	75–175	60–150	75–150	40–160
Phosphorus, inorganic (mg/dL)	7.5–11.0	6.0–10.0	4.5–8.9	4.7–7.2
Potassium (mEq/L)	6.5–9.0	6.6–9.0	—	—
Protein, total (g/dL)	4.5–6.0	3.5–5.6	4.4–6.0	4.4–6.4
Sodium (mEq/L)	145–160	155–170	—	—
Triglycerides (mg/dL)	60–140 (M)	40–150 (M)	—	—
	50–100 (F)	25–75 (F)	—	—
Urea nitrogen (BUN) (mg/dL)	20–40	20–70	10–30	10–30

Source: Adapted from Frithe, C.H. et al. (1980) and Wolford, S.T. et al. (1986).
Note: —, data unavailable.

TABLE 138

Mean Control Ranges of Typical Serum Clinical Chemistry Measurements in CD® Rats

Parameter	10–12 Weeks Old	18–20 Weeks Old	32–34 Weeks Old	58–60 Weeks Old	84–86 Weeks Old	108–110 Weeks Old
Alanine aminotransferase (IU/L)	10–40	10–50	10–50	20–60	20–60	20–60
Albumin (g/dL)	3.4–4.1 (M)	3.3–4.2 (M)	3.5–4.0 (M)	3.0–3.8 (M)	3.0–4.0 (M)	2.7–3.5 (M)
	3.5–4.5 (F)	3.5–4.7 (F)	4.0–5.0 (F)	3.5–4.5 (F)	3.7–4.5 (F)	3.3–3.7 (F)
Albumin/globulin ratio	1.0–1.5	1.0–1.5	1.0–1.5	0.75–1.75	0.75–1.75	0.75–1.5
Alkaline phosphatase (IU/L)	140–300 (M)	50–150 (M)	50–150 (M)	50–150 (M)	50–150 (M)	50–100 (M)
	80–100 (F)	25–150 (F)	25–100 (F)	25–100 (F)	25–100 (F)	25–100 (F)
Aspartate aminotransferase (IU/L)	45–90	45–100	45–120	60–120	75–150	75–150
Bile acids, total (μmol/L)	20–60	20–60	—	—	—	—
Bilirubin, total (mg/dL)	0.2–0.4	0.1–0.5	0.1–0.5	0.1–0.5	0.1–0.5	0.1–0.4
Calcium (mg/dL)	9.8–12.0	9.8–12.0	9.8–12.0	9.8–12.0	9.8–12.0	9.8–12.0
Chloride (mEq/L)	97–105	97–105	95–105	97–105	97–105	95–105
Cholesterol, total (mg/dL)	50–85	50–100	70–140	60–150	130–180 (M)	130–180 (M)
					100–150 (F)	90–150 (F)
Creatine kinase (IU/L)	50–400	50–300	50–500	—	—	—
Creatinine (mg/dL)	0.3–0.8	0.3–0.9	0.3–1.0	0.4–0.8	0.4–0.8	0.4–1.3
γGT (IU/L)	0–2	0–2	0–3	0–5	0–7	0–5
Globulin (g/dL)	2.5–4.0	2.5–4.0	2.0–4.5	2.0–4.5	2.0–4.5	2.0–4.5
Glucose (mg/dL)	90–175	100–175	100–200	100–200	100–175	100–175

(*Continued*)

TABLE 138 (Continued)

Mean Control Ranges of Typical Serum Clinical Chemistry Measurements in CD® Rats

Parameter	10–12 Weeks Old	18–20 Weeks Old	32–34 Weeks Old	58–60 Weeks Old	84–86 Weeks Old	108–110 Weeks Old
LDH (IU/L)	50–400	50–400	50–500	—	—	—
Phosphorus, inorganic (mg/dL)	7.0–10.0	4.0–8.5	4.0–8.0	3.5–7.0	3.5–8.0	4.0–7.0
Potassium (mEq/L)	5.5–8.0	4.0–7.0	4.0–7.0	4.0–7.0	3.5–6.0	3.5–6.0
Protein, total (g/dL)	6.2–7.6 (M) 6.3–8.2 (F)	6.2–7.8 (M) 6.5–8.5 (F)	6.2–8.0 (M) 7.0–9.0 (F)	6.0–8.0 (M) 6.5–8.5 (F)	6.3–7.6 (M) 6.7–8.0 (F)	5.7–6.5 (M) 6.3–7.1 (F)
Sodium (mEq/L)	140–153	140–153	140–153	140–153	140–153	140–145
Sorbitol dehydrogenase (IU/L)	10–30	10–30	10–30	—	—	—
Triglycerides (mg/dL)	50–125	50–200	50–200	50–300	75–400	50–300
Urea nitrogen (BUN) (mg/dL)	12–18	12–20	12–20	12–18	12–18	12–30

Source: Adapted from Levine, B.S. (1979–1993) and Charles River Laboratories (1993b).
Note: —, data unavailable.

TABLE 139

Mean Control Ranges of Typical Serum Clinical Chemistry Measurements in F-344 Rats

Parameter	12–14 Weeks Old	18–20 Weeks Old	32–34 Weeks Old	58–60 Weeks Old	84–86 Weeks Old	110–112 Weeks Old
Alanine aminotransferase (IU/L)	25–45	30–62	20–40	56–100 (M) 33–65 (F)	41–80 (M) 32–50 (F)	25–60
Albumin (g/dL)	3.8–4.7	3.0–5.0	4.0–5.0	3.8–5.0	3.8–5.0	3.5–5.0
Albumin/globulin ratio	1.5–2.3	1.1–2.5	1.5–2.0	1.4–1.9	1.4–2.0	1.2–1.8
Alkaline phosphatase (IU/L)	200–300 (M) 150–250 (F)	58–154 (M) 45–120 (F)	45–80	31–68	—	—
Aspartate aminotransferase (IU/L)	50–90	50–100	—	—	—	—
Bile acids, total (μmol/L)	10–50	—	—	—	—	—
Bilirubin, total (mg/dL)	—	0.1–0.5	0.1–0.4	0.1–0.5	0.1–0.5	0.1–0.4
Calcium (mg/dL)	—	9.5–12.0	9.5–11.2	9.5–11.5	9.5–11.5	9.8–11.7
Chloride (mEq/L)	—	97–115	98–110	100–112	97–100	104–113
Cholesterol, total (mg/dL)	70–110 (M) 90–135 (F)	50–80 (M) 80–120 (F)	50–80 (M) 85–130 (F)	68–125 (M) 110–150 (F)	100–120	125–175
Creatine kinase (IU/L)	60–300	100–400	300–700	300–500	100–500	100–400
Creatinine (mg/dL)	0.5–1.0	0.4–0.8	—	—	—	—
Globulin (g/dL)	1.5–2.5	1.2–2.8	2.0–3.0	2.3–3.5	2.0–3.0	2.2–3.2
Glucose (mg/dL)	100–180	90–170	80–130	90–140	90–140	90–140

(Continued)

TABLE 139 (Continued)

Mean Control Ranges of Typical Serum Clinical Chemistry Measurements in F-344 Rats

Parameter	12–14 Weeks Old	18–20 Weeks Old	32–34 Weeks Old	58–60 Weeks Old	84–86 Weeks Old	110–112 Weeks Old
LDH (IU/L)	—	500–800	400–800	150–400	—	—
Phosphorus, inorganic (mg/dL)	—	3.9–7.3	—	—	—	—
Potassium (mEq/L)	—	3.6–5.9	4.0–5.7	4.1–5.5	4.0–5.2	4.0–5.1
Protein, total (g/dL)	6.0–7.2	5.7–7.6	6.2–7.5	6.5–7.6	6.0–7.8	6.1–8.0
Sodium (mEq/L)	—	140–155	142–158	142–156	138–149	138–146
Sorbitol dehydrogenase (IU/L)	15–60	5–25	5–35	—	—	—
Triglycerides (mg/dL)	100–400 (M) 25–130 (F)	75–150 (M) 35–70 (F)	125–190 (M) 30–70 (F)	90–175 (M) 40–85 (F)	110–240 (M) 60–145 (F)	80–220
Urea nitrogen (BUN) (mg/dL)	15–25	10–26	12–24	10–20	10–20	12–25

Source: Adapted from Levine, B.S. (1979–1993), NIEHS (1985), and Burns, K.F. et al. (1971).
Note: —, data unavailable.

TABLE 140

Mean Control Ranges of Typical Serum Clinical Chemistry Measurements in Beagle Dogs

Parameter	6–8 Months Old	9–11 Months Old	12–14 Months Old	15–18 Months Old	19–30 Months Old
Alanine aminotransferase (IU/L)	20–40	20–40	20–40	20–40	20–40
Albumin (g/dL)	2.5–3.5	2.5–3.5	2.5–3.5	2.5–4.0	2.7–4.5
Albumin/globulin ratio	0.8–1.5	0.8–1.5	0.8–1.5	0.8–2.0	0.8–2.0
Alkaline phosphatase (IU/L)	120–160 (M) 100–130 (F)	70–120 (M) 60–100 (F)	50–100	35–100	35–100
Aspartate aminotransferase (IU/L)	30–45	30–50	25–50	25–50	25–50
Bilirubin, total (mg/dL)	0.1–0.7	0.1–0.7	0.1–0.7	0.1–0.3	0.1–0.3
Calcium (mg/dL)	9.0–11.5	9.0–11.5	9.0–11.5	10.0–11.3	10.0–11.5
Chloride (mEq/L)	100–115	100–115	100–115	105–119	105–115
Cholesterol, total (mg/dL)	150–250	125–250	125–250	125–250	125–225
Creatine kinase (IU/L)	100–400	100–400	100–400	—	—
Creatinine (mg/dL)	0.5–0.8	0.7–0.9	0.7–0.9	—	—
γGT (IU/L)	0–5	0–5	0–5	—	—
Globulin (g/dL)	2.5–3.5	2.5–3.5	2.5–3.5	2.5–3.5	2.5–3.5
Glucose (mg/dL)	100–130	100–130	100–130	70–110	70–110
Haptoglobin (mg/dL)	50–200	50–150	25–100	—	—
LDH (IU/L)	30–100	30–100	30–100	—	—

(Continued)

TABLE 140 (Continued)

Mean Control Ranges of Typical Serum Clinical Chemistry Measurements in Beagle Dogs

Parameter	6–8 Months Old	9–11 Months Old	12–14 Months Old	15–18 Months Old	19–30 Months Old
Phosphorus, inorganic (mg/dL)	6.0–9.0	4.0–6.0	3.0–5.0	3.0–5.0	3.0–4.7
Potassium (mEq/L)	4.2–5.0	4.2–5.0	4.2–5.0	4.1–5.1	4.2–5.2
Protein, total (g/dL)	5.5–6.5	5.5–6.5	5.5–6.5	5.5–6.5	5.7–6.6
Sodium (mEq/L)	143–147	143–147	143–147	143–153	143–153
Triglycerides (mg/dL)	30–60	30–75	30–75	—	—
Urea nitrogen (BUN) (mg/dL)	10–20	10–20	10–20	10–20	10–20

Source: Adapted from Levine, B.S. (1979–1993), Clarke, D. et al. (1992), Pickrell, J.A. et al. (1974), and Kaspar, L.V. and Norris, W.P. (1977).

Note: —, data unavailable.

TABLE 141

Mean Control Ranges of Typical Serum Clinical Chemistry Measurements in Nonhuman Primates of Various Ages

Parameter	3–7-Year-Old Cynomolgus	1–2-Year-Old Rhesus	3–7-Year-Old Rhesus	<1.5-Year-Old Marmoset	>1.5-Year-Old Marmoset	1–5-Year-Old Baboon	6–15-Year-Old Marmoset
Alanine aminotransferase (IU/L)	20–60	20–50	15–40	45–75	40–70	15–50	20–50
Albumin (g/dL)	3.5–4.8	3.0–4.5	3.2–4.5	3.5–5.8	3.5–5.8	3.1–4.5	2.0–4.5
Albumin/globulin ratio	1.0–1.5	1.0–1.5	1.0–1.5	1.0–1.5	1.0–1.5	1.0–1.5	1.0–1.5
Alkaline phosphatase (IU/L)	300–800 (M) 200–500 (F)	200–600	70–300	100–250	35–80	200–1000	100–200
Amylase (IU/L)	200–500	—	—	1000–2000	500–1500	200–400	200–500
Aspartate aminotransferase (IU/L)	25–60	25–60	15–70	100–200	100–200	18–35	20–35
Bilirubin, total (mg/dL)	0.3–0.8	0.1–0.8	0.1–0.6	0.1–0.9	0.1–0.9	0.3–0.7	0.3–0.5
Calcium (mg/dL)	9.0–11.0	8.2–10.5	8.5–10.3	8.1–12.4	8.5–11.7	8.0–9.5	7.5–10.0
Chloride (mEq/L)	100–115	103–115	97–110	80–110	93–119	105–115	100–110
Cholesterol, total (mg/dL)	90–160	90–160	90–170	90–210	105–230	75–200	70–125
Creatine kinase (IU/L)	140–200	200–1000	200–600	—	—	—	—
Creatinine (mg/dL)	0.7–1.2	0.5–0.9	0.7–1.2	0.2–1.0	0.2–1.0	0.8–1.2	1.0–1.8
γGT (IU/L)	40–90	—	10–60	—	—	—	—

(Continued)

TABLE 141 (Continued)

Mean Control Ranges of Typical Serum Clinical Chemistry Measurements in Nonhuman Primates of Various Ages

Parameter	3–7-Year-Old Cynomolgus	1–2-Year-Old Rhesus	3–7-Year-Old Rhesus	<1.5-Year-Old Marmoset	>1.5-Year-Old Marmoset	1–5-Year-Old Baboon	6–15-Year-Old Marmoset
Globulin (g/dL)	3.0–4.5	3.0–4.0	3.0–4.0	2.5–4.0	3.5–4.0	2.5–4.0	2.5–4.5
Glucose (mg/dL)	50–100	50–100	41–80	180–275	130–240	50–125	50–140
LDH (IU/L)	300–600	130–600	125–600	125–500	100–350	100–400	100–350
Phosphorus, inorganic (mg/dL)	4.0–7.0	3.2–5.0	3.0–5.3	5.5–9.8	4.0–7.5	4.7–7.5	1.3–4.5
Potassium (mEq/L)	3.0–4.5	3.0–4.2	3.1–4.1	3.5–5.0	3.0–4.8	3.2–4.3	3.7–4.8
Protein, total (g/dL)	7.0–9.0	6.7–8.0	7.0–8.3	5.5–7.5	6.0–8.0	6.0–8.0	6.0–7.5
Sodium (mEq/L)	140–153	144–150	142–148	150–170	150–170	142–158	142–158
Triglycerides (mg/dL)	30–70	50–200	50–200	75–200	75–200	25–60	30–125
Urea nitrogen (BUN) (mg/dL)	15–25	14–26	14–25	17–35	15–32	10–25	10–25

Source: Adapted from Levine, B.S. (1979–1993), Clarke, D. et al. (1992), Kapeghian, L.C. and Verlangieri, A.J. (1984), Davy, C.W. et al. (1984), Yarbrough, L.W. et al. (1984), and Hack, C.A. and Gleiser, C.A. (1982).

Note: —, data unavailable.

TABLE 142

Mean Control Ranges of Typical Serum Clinical Chemistry
Measurements in New Zealand White Rabbits

Parameter	15–20 Weeks Old	25–40 Weeks Old	1–2 Years Old
Alanine aminotransferase (IU/L)	25–70	25–70	25–70
Albumin (g/dL)	3.8–5.0	3.5–4.7	3.0–4.5
Albumin/globulin ratio	2.0–3.0	2.0–3.0	2.0–3.0
Alkaline phosphatase (IU/L)	50–120	40–120	15–90
Aspartate aminotransferase (IU/L)	20–50	10–35	10–30
Bilirubin, total (mg/dL)	0.1–0.5	0.1–0.5	0.2–0.6
Calcium (mg/dL)	12.0–14.0	11.0–14.0	12.0–15.0
Chloride (mEq/L)	97–110	96–108	100–110
Cholesterol, total (mg/dL)	20–60	20–60	20–60
Creatine kinase (IU/L)	200–800	200–1000	200–1000
Creatinine (mg/dL)	1.0–1.6	0.8–1.6	0.8–1.7
γGT (IU/L)	—	0–10	0–6
Globulin (g/dL)	1.4–1.9	1.5–2.2	1.5–2.5
Glucose (mg/dL)	100–160	100–175	80–140
LDH (IU/L)	50–200	50–200	35–125
Phosphorus, inorganic (mg/dL)	4.6–7.2	4.0–7.0	3.0–5.0
Potassium (mEq/L)	4.0–5.2	4.0–5.0	3.3–4.5
Protein, total (g/dL)	5.4–6.6	5.5–7.0	5.5–7.5
Sodium (mEq/L)	132–144	132–145	132–150
Urea nitrogen (BUN) (mg/dL)	10–20	12–22	12–25

Source: Adapted from Levine, B.S. (1979–1993), Hewett, C.D. et al. (1989), and Yu, L. et al. (1979).

Note: —, data unavailable.

TABLE 143

Control Ranges of Typical Clinical Chemistry Measurements in Minipigs (Gottingen)[a]

Parameters			
Alanine aminotransferase (IU/L)	19–331	γ-Glutamyltransferase (γGT) (IU/L)	38–108
Albumin (g/dL)	3.4–5.1	Globulin (g/dL)	0.6–3.4
Albumin/globulin ratio	1.0–6.4	Glucose (mg/dL)	53–224
Alkaline phosphatase (IU/L)	50–642	Phosphorus, inorganic (mg/dL)	4.3–11.6
Aspartate aminotransferase (IU/L)	10–799	Potassium (mEq/L)	3.1–9.8
Bilirubin, total (mg/dL)	0.0–0.3	Protein, total (g/dL)	5.2–7.4
Calcium (mg/dL)	9.7–12.8	Sodium (mEq/L)	132–159
Chloride (mEq/L)	91–115	Sorbitol dehydrogenase (IU/L)	0–6
Cholesterol, total (mg/dL)	34–144	Triglycerides (mg/dL)	7–58
Creatinine (mg/dL)	0.4–3.9	Total bile acids (µmol/L)	0–96
		Urea nitrogen (BUN) (mg/dL)	3.0–19.0

Source: Marshall Farms, Unpublished data.

[a] 4–6 months of age.

TABLE 144

Mean Control Ranges of Typical Hematology Measurements in B6C3F$_1$ Mice

Parameters	12–14 Weeks Old	18–20 Weeks Old	32–34 Weeks Old	58–60 Weeks Old	84–86 Weeks Old	110—112 Weeks Old
Erythrocyte count (10^6/mm^3)	9.0–10.2	7.5–10.5	8.0–10.4	8.0–10.0	8.6–10.4	7.7–10.4
Hematocrit (%)	44.1–49.5	36.0–48.6	40.8–46.6	38.5–45.5	40.0–46.9	36.0–43.5
Hemoglobin (g/dL)	15.0–17.1	13.1–16.5	15.2–18.2	14.5–17.9	15.0–18.2	13.0–16.8
Leukocyte count, total (10^3/mm^3)	3.0–7.8 (M) 2.5–5.0 (F)	5.5–10.9 (M) 3.2–5.2 (F)	6.1–13.3 (M) 4.2–9.3 (F)	6.1–13.2 (M) 4.6–10.5 (F)	7.0–13.4 (M) 3.9–7.9 (F)	5.0–16.5 (M) 4.2–8.8 (F)
MCH (pg)	16.6–18.8	16.9–20.2	16.4–18.9	15.8–18.0	15.9–18.3	15.7–18.7
MCHC (g/dL)	34.6–38.4	34.6–40.4	37.1–41.2	36.5–39.0	36.2–39.4	35.7–38.8
MCV (fl)	44.0–52.0	45.4–53.6	44.0–48.0	42.0–47.0	42.0–48.0	46.0–50.0
Methemoglobin (% Hb)	—	0–3.0	0–2.5	0–1.5	0–0.9	0–1.0
Platelet count (10^3/mm^3)	700–1100	500–1000	800–1200	700–1200	400–1100	400–800
Reticulocyte count (% RBC)	0.5–2.0	1.0–3.9	0.4–2.8	0.4–1.6	0.2–2.3	0.5–2.5

Source: Adapted from Levine, B.S. (1979–1993) and NIEHS (1985).

Note: —, data not available.

TABLE 145

Mean Control Ranges of Typical Hematology Measurements in CD-1 and BALB/c Mice of Various Ages

Parameters	1-3-Month-Old BALB/c	6-12-Month-Old BALB/c	<1-Year-Old CD-1	>1-Year-Old CD-1
Erythrocyte count (10^6/mm³)	8.5–10.5	8.8–10.6	8.0–10.0	6.0–9.0
Hematocrit (%)	42.5–47.9	38.3–46.9	36.9–46.9	28.2–41.1
Hemoglobin (g/dL)	14.5–16.8	14.2–17.0	13.6–16.8	10.4–14.9
Leukocyte count, total (10^3/mm³)	2.0–5.7	2.0–5.0	4.0–12.0 (M)	3.4–17.0 (M)
			3.5–9.7 (F)	2.4–13.4 (F)
MCH (pg)	15.8–18.4	15.1–17.5	16.1–18.6	15.1–18.4
MCHC (g/dL)	34.2–38.1	35.1–40.6	34.8–38.2	34.6–37.6
MCV (fl)	46.3–50.3	40.9–45.9	44.5–49.7	41.3–51.1
Platelet count (10^3/mm³)	—	—	700–1400	700–1500
Reticulocyte count (% RBC)	—	—	1.6–3.7	1.7–5.0

Source: Adapted from Frithe, C.H. et al. (1980) and Wolford, S.T. et al. (1986).
Note: —, data not available.

TABLE 146

Mean Control Ranges of Typical Hematology Measurements in CD® Rats

Parameters	10–12 Weeks Old	18–20 Weeks Old	32–34 Weeks Old	58–60 Weeks Old	84–86 Weeks Old	108–110 Weeks Old
Activated partial thromboplastin time (seconds)	14.0–20.0 (M) 12.0–18.0 (F)	14.0–20.0 (M) 13.0–18.0 (F)	14.0–17.0 (M) 13.0–16.0 (F)	16.0–19.0 (M) 15.0–18.0 (F)	—	—
Erythrocyte count (10^6/mm^3)	6.8–8.5 (M) 7.0–8.2 (F)	7.0–9.8 (M) 6.5–9.2 (F)	7.0–9.6 (M) 6.5–8.8 (F)	7.0–9.2(M) 6.5–8.5 (F)	7.0–9.2 (M) 6.0–8.5 (F)	6.2–8.2 (M) 5.8–8.0 (F)
Fibrinogen (mg/dL)	—	200–300 (M) 130–190 (F)	—	—	—	—
Hematocrit (%)	40.0–48.0	36.0–52.0	36.0–50.0	38.0–48.0	38.0–50.0	35.0–45.0
Hemoglobin (g/dL)	14.0–17.0	14.0–17.0	14.0–17.0	14.0–17.0	14.0–17.0	12.0–15.0
Leukocyte count, total (10^3/mm^3)	6.0–18.0 (M) 4.0–14.0 (F)	6.0–19.0 (M) 5.0–14.0 (F)	6.0–18.0 (M) 4.0–11.0 (F)	5.0–15.0 (M) 3.0–9.0 (F)	10.0–15.0 (M) 6.0–10.0 (F)	5.0–18.0 (M) 3.0–12.0 (F)
MCH (pg)	19.0–22.0	16.0–20.0	17.0–21.0	16.0–21.0	16.0–20.0	16.0–20.0
MCHC (g/dL)	33.0–38.0	31.0–38.0	31.0–38.0	32.0–38.0	31.0–36.0	31.0–36.0
MCV (fl)	53.0–63.0	50.0–60.0	45.0–60.0	46.0–58.0	48.0–56.0	50.0–63.0
Methemoglobin (% Hb)	0.4–1.2	0.4–1.2	0.4–1.2	—	—	—
Platelet count (10^3/mm^3)	900–1300	800–1200	700–1200	700–1200	700–1200	700–1200
Prothrombin time (seconds)	9.0–14.0	9.0–14.0	10.0–14.0	10.0–14.0	—	—
Reticulocyte count (% RBC)	0.2–1.0	0.2–0.8	0.2–0.8	—	—	—

Source: Adapted from Levine, B.S. (1979–1993) and Charles River Laboratories (1993a).
Note: —, data not available.

TABLE 147

Mean Control Ranges of Typical Hematology Measurements in F-344 Rats

Parameters	10–12 Weeks Old	18–20 Weeks Old	32–34 Weeks Old	58–60 Weeks Old	84–86 Weeks Old	108–110 Weeks Old
Erythrocyte count ($10^6/mm^3$)	7.2–8.6	7.0–10.0	8.5–9.6	7.2–9.5	7.5–9.8	6.5–9.6
Hematocrit (%)	39.5–45.5	42.0–50.0	41.4–46.7	40.0–46.6	40.3–45.5	40.0–48.5
Hemoglobin (g/dL)	15.0–17.0	15.0–17.3	15.0–17.8	15.7–17.5	15.5–17.6	13.0–18.5
Leukocyte count, total ($10^3/mm^3$)	7.1–13.5 (M) 5.4–11.7 (F)	6.5–10.7 (M) 4.5–7.0 (F)	6.5–8.7 (M) 4.4–6.5 (F)	5.8–9.0 (M) 4.5–6.2 (F)	5.7–8.5 (M) 3.2–6.0 (F)	5.0–15.0 (M) 3.5–8.0 (F)
MCH (pg)	18.5–21.0	17.5–20.8	18.5–21.0	18.1–20.7	18.0–20.5	18.5–22.0
MCHC (g/dL)	36.6–39.6	35.3–39.2	37.8–40.0	36.9–40.5	37.0–40.6	36.3–40.9
MCV (fl)	48.0–58.0	48.3–56.1	48.0–56.0	47.0–56.0	47.0–56.0	50.0–58.0
Methemoglobin (% Hb)	—	0–3.0	0–4.0	0–2.5	0–2.7	0–2.0
Platelet count ($10^3/mm^3$)	400–750	350–700	400–870	450–700	450–700	200–450
Reticulocyte count (% RBC)	1.0–2.0	0.7–2.0	0.8–2.0	0.8–2.0	0.3–1.5	0.5–2.5

Source: Adapted from Levine, B.S. (1979–1993) and NIEHS (1985).

Note: —, data not available.

TABLE 148

Mean Control Ranges of Typical Hematology Measurements in Beagle Dogs

Parameters	6–8 Months Old	9–11 Months Old	12–14 Months Old	15–18 Months Old	19–30 Months Old
Activated partial thromboplastin time (seconds)	9.0–13.0	9.0–13.0	9.0–13.0	9.0–13.0	9.0–13.0
Erythrocyte count ($10^6/mm^3$)	6.0–7.3	6.2–8.0	6.2–8.2	5.8–7.3	5.8–7.3
Fibrinogen (mg/dL)	150–300	100–200	—	—	—
Hematocrit (%)	41.5–49.0	44.3–54.9	46.0–54.6	42.5–55.0	42.0–52.0
Hemoglobin (g/dL)	14.5–17.3	15.8–18.0	16.0–18.8	13.0–19.0	13.0–19.0
Leukocyte count, total ($10^3/mm^3$)	5.5–14.0	6.8–13.6	5.7–15.5	5.0–15.0	6.0–18.0
MCH (pg)	21.5–25.1	21.6–24.9	22.0–25.2	22.5–26.0	23.0–26.0
MCHC (g/dL)	33.0–37.0	33.0–36.4	34.0–36.0	30.0–34.0	30.0–34.0
MCV (fl)	65.0–71.0	64.0–73.0	64.0–72.0	65.0–78.0	65.0–78.0
Methemoglobin (% Hb)	0–2.0	0–1.5	0–1.5	—	—
Platelet count ($10^3/mm^3$)	150–400	150–400	150–400	150–400	150–400
Prothrombin time (seconds)	6.2–8.4	6.8–8.4	6.2–8.8	6.5–9.0	6.5–9.0
Reticulocyte count (% RBC)	0–0.7	0–0.7	0–0.7	0–0.7	0–0.7

Source: Adapted from Levine, B.S. (1979–1993), Bulgin, M.S. et al. (1970), and Jordan, J.E. (1977).
Note: —, data not available.

TABLE 149

Mean Control Ranges of Typical Hematology Measurements in Nonhuman Primates of Various Ages

Parameters	3–7-Year-Old Cynomolgus	1–2-Year-Old Rhesus	3–7-Year-Old Rhesus	<1.5-Year-Old Marmoset	>1.5-Year-Old Marmoset	1–5-Year-Old Baboon	6–15-Year-Old Baboon
Activated partial thromboplastin time (seconds)	15.5–22.7	15.0–22.0	15.0–22.0	—	—	—	30–60
Erythrocyte count ($10^6/mm^3$)	4.5–7.2	4.4–5.8	4.2–6.2	4.2–6.2	4.6–6.8	4.2–5.7	4.0–5.3
Hematocrit (%)	31.5–37.9	31.5–39.2	29.3–39.0	30.0–42.1	37.7–47.5	31.0–43.0	34.0–42.0
Hemoglobin (g/dL)	10.4–12.4	10.8–13.5	9.8–13.1	12.6–15.0	13.5–16.8	10.8–13.5	10.3–13.1
Leukocyte count, total ($10^3/mm^3$)	5.3–13.4	4.5–13.3	4.3–12.2	5.5–13.0	4.6–11.3	4.9–13.0	4.8–13.9
MCH (pg)	18.9–22.3	19.8–24.8	19.6–23.2	24.0–30.5	23.0–29.0	22.0–27.0	22.0–28.0
MCHC (g/dL)	32.0–35.6	31.3–35.5	31.7–37.5	32.1–42.6	32.2–42.5	28.0–34.0	30.0–35.0
MCV (fl)	57.1–63.9	66.0–74.0	56.0–70.0	66.0–76.0	68.0–77.0	63.0–80.0	75.0–91.0
Platelet count ($10^3/mm^3$)	150–400	200–600	200–500	200–500	200–500	200–500	200–500
Prothrombin time (seconds)	11.5–14.0	9.9–12.2	11.2–14.4	—	—	—	9.0–13.0
Reticulocyte count (% RBC)	0–0.5	0–1.4	0–1.5	0–5.0	0–4.7	0–2.3	0–1.9

Source: Adapted from Levine, B.S. (1979–1993), Kapeghian, L.C. and Verlangieri, A.J. (1984); Yarbrough, L.W. et al. (1984), and Hack, C.A. and Gleiser, C.A. (1982).

Note: —, data not available.

TABLE 150

Mean Control Ranges of Typical Hematology Measurements
in New Zealand White Rabbits

Parameters	15–20 Weeks Old	25–40 Weeks Old	1–2 Years Old
Activated partial thromboplastin time (seconds)	11.7–14.5	11.3–14.9	10.5–15.8
Erythrocyte count (10^6/mm³)	5.5–7.0	4.8–6.7	4.9–7.0
Fibrinogen (mg/dL)	125–300	125–300	125–400
Hematocrit (%)	37.0–44.5	37.0–44.5	37.5–44.7
Hemoglobin (g/dL)	12.0–14.7	10.9–14.4	10.5–14.8
Leukocyte count, total (10^3/mm³)	5.4–11.9	3.6–7.9	4.8–13.5
MCH (pg)	20.2–23.0	21.8–24.5	20.4–23.4
MCHC (g/dL)	32.3–34.9	32.2–34.8	30.0–34.1
MCV (fl)	61.4–68.6	64.8–69.5	64.8–72.0
Platelet count (10^3/mm³)	175–500	175–500	200–500
Reticulocyte count (% RBC)	0–2.0	0–2.0	0–3.0
Prothrombin time (seconds)	8.2–9.8	8.0–10.0	8.0–10.3

Source: Adapted from Levine, B.S. (1979–1993), Hewett, C.D. et al. (1989) and Jain (1986).

TABLE 151

Control Ranges of Typical Hematology Measurements in Minipigs (Gottingen)[a]

Parameters			
Erythrocyte count (10^6/mm^3)	6.5–10.4	Mean corpuscular hemoglobin (pg)	12.9–21.7
Hematocrit (%)	31.0–50.3	Mean corpuscular hemoglobin conc. (g/dL)	25.8–37.0
Hemoglobin (g/dL)	9.4–16.5	Mean corpuscular volume (fL)	39.8–63.9
Leukocyte count, total (10^3/mm^3)	4.5–19.7	Platelet count (10^3/mm^3)	144–1088
Neutrophils (%)	9.0–45.5	Reticulocyte count (%RBC)	0.2–5.1
Lymphocytes (%)	55.9–81.7	Prothrombin time (seconds)	10.7–13.8
Monocytes (%)	0.7–8.5	Activated partial thromboplastin time (seconds)	12.1–17.1
Eosinophils (%)	0.2–5.9		
Basophils (%)	0.1–3.7		

Source: Marshall Farms, Unpublished data.

[a] Age ~4–6 mo.

TABLE 152

24-hr Mean Urinalysis Data with Standard Deviation (SD) and Standard Error of the Mean (SEM) in Adult Male Rats: Fischer-344, Sprague-Dawley, and Wistar

	Strain								
	F-344			Sprague-Dawley			Wistar		
Parameters	Mean	SD	SEM	Mean	SD	SEM	Mean	SD	SEM
Volume (mL)	5.92	2.15	0.88	14.83	7.63	3.12	12.68	4.06	1.66
Volume (mL/100 g body weight)	1.78	0.556	0.227	2.824	1.339	0.547	2.453	0.761	0.311
Sodium (μEq/mL)	62.7	20.3	8.3	54.3	32.7	13.4	41.67	16.27	6.64
Potassium (μEq/mL)	197.67	32.87	13.42	168.0	75.2	30.7	146.0	37.5	15.3
Chloride (μEq/mL)	105.0	54.5	22.3	64.7	47.5	19.4	60.0	24.9	10.2
Protein (g/dL)	0.4833	0.0983	0.0401	0.5167	0.1941	0.0792	0.3667	0.0816	0.0333
Glucose (mg/dL)	7.33	17.96	7.33	0.00	0.00	0.00	0.00	0.00	0.00
Alkaline phosphatase (ALP) (IU)	154.2	54.4	22.2	87.1	53.7	21.9	141.4	43.4	17.8
Lactate dehydrogenase (LDH) (IU)	3.83	9.39	3.83	34.17	83.69	34.17	0.00	0.00	0.00
Osmolality (mOsm/kg)	1312.3	210.5	86.0	1206	497	203	1197	325	133
pH	6.18	0.41	0.17	6.83	0.75	0.31	6.167	0.406	0.17
Creatinine (Cr) (mg/dL)	144.2	22.8	9.3	142.0	61.9	25.3	165.7	60.7	24.8
Sodium/Cr (μEq/mg Cr)	43.2	124	5.07	35.78	7.88	3.22	25.37	7.55	3.08
Potassium/Cr (μEq/mg Cr)	137.2	11.6	4.72	117.29	15.55	6.35	91.10	16.1	6.59
Chloride/Cr (μEq/mg Cr)	70.3	35.3	14.40	39.5	25.0	10.20	36.9	14.7	5.99
Protein/Cr (g/mg Cr)	0.0039	0.00058	0.00024	0.0038	0.0012	0.00047	0.0023	0.0004	0.00016
Glucose/Cr (mg/mg Cr)	0.05	0.123	0.05	0	0	0	0	0	0

TABLE 153

24-hr Mean Urinalysis Data with Standard Deviation (SD) and Standard Error of the Mean (SEM) in Adult Female Rats: Fischer-344, Sprague-Dawley, and Wistar

	Strain								
	F-344			Sprague-Dawley			Wistar		
Parameters	Mean	SD	SEM	Mean	SD	SEM	Mean	SD	SEM
Volume (mL)	8.82	4.32	1.76	8.43	3.43	1.40	18.22	8.07	3.29
Volume (mL/100 g body weight)	4.93	2.41	0.98	2.839	1.119	0.457	6.48	2.46	1.00
Sodium (µEq/mL)	152.0	41.4	16.9	155.2	16.2	6.6	81.7	63.5	25.9
Potassium (µEq/mL)	304.0	91.2	36.8	324.2	50.8	20.7	179.7	97.9	40.0
Chloride (µEq/mL)	205.5	61.6	25.2	249.7	78.7	32.1	104.7	79.5	32.5
Protein (g/dL)	0.3333	0.0516	0.0211	0.4667	0.1211	0.0494	0.1833	0.0983	0.0401
Glucose (mg/dL)	9.3	14.5	5.9	8.33	13.05	5.33	0.00	0.00	0.00
Alkaline phosphatase (ALP) (IU)	25.22	11.14	4.55	16.1	11.1	4.5	32.0	24.9	10.2
Lactate dehydrogenase (LDH) (IU)	0.00	0.00	0.00	13.83	19.02	7.76	2.50	6.12	2.50
Osmolality (mOsm/kg)	1764	520	212	2286	650	266	1083	428	175
pH	7.00	0.632	0.26	7.83	1.17	0.48	7.50	1.23	0.50
Creatinine (Cr) (mg/dL)	91.8	27.8	11.4	161.5	54.2	22.1	71.50	23.98	9.79
Sodium/Cr (µEq/mg Cr)	169.9	27.7	11.32	105.5	34.6	14.13	120.9	85.0	34.72
Potassium/Cr (µEq/mg Cr)	334.6	27.7	11.32	213.5	47.5	19.39	264.1	129.3	52.77
Chloride/Cr (µEq/mg Cr)	225.6	24.6	10.06	162.8	43.8	17.90	149.9	106.8	43.62
Protein/Cr (g/mg Cr)	0.0040	0.0015	0.00062	0.0031	0.00086	0.00035	0.0027	0.0015	0.0006
Glucose/Cr (mg/mg Cr)	0.0817	0.1266	0.0517	0.0417	0.0646	0.0264	0	0	0

TABLE 154

Comparison of Biochemical Components in Urine of Normal Experimental Animals

Component (mg/kg body wt/day) or Property	Rat	Rabbit	Cat	Dog	Goat	Sheep
Volume (mL/kg body wt/day)	15.0–35.0	20.0–350	10.0–30.0	20.0–167	7.0–40.0	10.0–40.0
Specific gravity	1.040–1.076	1.003–1.036	1.020–1.045	1.015–1.050	1.015–1.062	1.015–1.045
pH	7.30–8.50	7.60–8.80	6.00–7.00	6.00–7.00	7.5–8.80	7.50–8.80
Calcium	3.00–9.00	12.1–19.0	0.20–0.45	1.00–3.00	1.00–3.40	1.00–3.00
Chloride	50.0–75.0	190–300	89.0–130	5.00–15.0	186–376	—
Creatinine	24.0–40.0	20.0–80.0	12.0–30.0	15.0–80.0	10.0–22.0	5.80–14.5
Magnesium	0.20–1.90	0.65–4.20	1.50–3.20	1.70–3.00	0.15–1.80	0.10–1.50
Phosphorous, inorganic	20.0–40.0	10.0–60.0	39.0–62.0	20.0–50.0	0.5–1.6	0.10–0.50
Potassium	50.0–60.0	40.0–55.0	55.0–120	40.0–100	250–360	300–420
Protein, total	1.20–6.20	0.74–1.86	3.10–6.82	1.55–4.96	0.74–2.48	0.74–2.17
Sodium	90.4–110.0	50.0–70.0	—	2.00–189	140–347	0.80–2.00
Urea nitrogen (g/kg/day)	1.00–1.60	1.20–1.50	0.80–4.00	0.30–0.50	0.14–0.47	0.11–0.17
Uric acid	8.00–12.0	4.00–6.00	0.20–13.0	3.1–6.0	2.00–5.00	2.00–4.00

(Continued)

TABLE 154 (*Continued*)

Comparison of Biochemical Components in Urine of Normal Experimental Animals

	Swine	Cattle	Horse	Monkey
Volume (mL/kg body wt/day)	5.00–30.0	17.0–45.0	3.0–18.0	70.0–80.0
Specific gravity	1.010–1.050	1.025–1.045	1.020–1.050	1.015–1.065
pH	6.25–7.55	7.60–8.40	7.80–8.30	5.50–7.40
Calcium	—	0.10–3.60	—	10.0–20.0
Chloride	—	10.0–140	81.0–120	80.0–120
Creatinine	20.0–90.0	15.0–30.0	—	20.0–60.0
Magnesium	—	2.00–7.00	—	3.20–7.10
Phosphorous, inorganic	—	0.01–6.20	0.05–2.00	9.00–20.6
Potassium	—	240–320	—	160–245
Protein, total	0.33–1.49	0.25–2.99	0.62–0.99	0.87–2.48
Sodium	—	2.00–40.0	—	—
Urea nitrogen (g/kg/day)	0.28–0.58	0.05–0.06	0.20–0.80	0.20–0.70
Uric acid	1.00–2.00	1.00–4.00	1.00–2.00	1.00–2.00

Source: From Mitruka, B.M. and Rawnsley, H.M. (1977). With permission.

TABLE 155

Normal Human Laboratory Values[a]

	Blood, Plasma, or Serum	
	Reference Value	
Determination	**Conventional Units**	**SI Units**
Ammonia (NH_3)-diffusion	20–120 mcg/dL	12–70 mcmol/L
Ammonia nitrogen	15–45 µg/dL	11–32 µmol/L
Amylase	35–118 IU/L	0.58–1.97 mckat/L
Anion gap (Na^+-[Cl^- + HCO_3^-]) (P)	7–16 mEq/L	7–16 mmol/L
Antinuclear antibodies	Negative at 1:10 dilution of serum	Negative at 1:10 dilution of serum
Antithrombin III (AT III)	80–120 U/dL	800–1200 U/L
Bicarbonate: Arterial	21–28 mEq/L	21–28 mmol/L
Venous	22–29 mEq/L	22–29 mmol/L
Bilirubin: Conjugated (direct)	≤0.2 mg/dL	≤4 mcmol/L
Total	0.1–1 mg/dL	2–18 mcmol/L
Calcitonin	<100 pg/mL	<100 ng/L
Calcium: Total	8.6–10.3 mg/dL	2.2–2.74 mmol/L
Ionized	4.4–5.1 mg/dL	1–1.3 mmol/L
Carbon dioxide content (plasma)	21–32 mmol/L	21–32 mmol/L
Carcinoembryonic antigen	<3 ng/mL	<3 mcg/L
Chloride	95–110 mEq/L	95–110 mmol/L
Coagulation screen		
Bleeding time	3–9.5 min	180–570 seconds
Prothrombin time	10–13 seconds	10–13 seconds
Partial thromboplastin time (activated)	22–37 seconds	22–37 seconds
Protein C	0.7–1.4 µ/mL	700–1400 U/mL
Protein S	0.7–1.4 µ/mL	700–1400 U/mL
Copper, total	70–160 mcg/dL	11–25 mcmol/L
Corticotropin (adrenocorticotropic hormone [ACTH])—0800 h	<60 pg/mL	<13.2 pmol/L

(Continued)

TABLE 155 (*Continued*)

Normal Human Laboratory Values[a]

	Blood, Plasma, or Serum			
		Reference Value		
Determination	**Conventional Units**		**SI Units**	
Cortisol: 0800 h	5–30 mcg/dL		138–810 nmol/L	
1800 h	2–15 mcg/dL		50–410 nmol/L	
2000 h	≤50% of 0800 h		≤50% of 0800 h	
Creatine kinase (CK): Female	20–170 IU/L		0.33–2.83 mckat/L	
Male	30–220 IU/L		0.5–3.67 mckat/L	
CK isoenzymes, MB fraction	0–12 IU/L		0–0.2 mckat/L	
Creatine	0.5–1.7 mg/dL		44–150 mcmol/L	
Fibrinogen (coagulation factor 1)	150–360 mg/dL		1.5–3.6 g/L	
Follicle-stimulating hormone (FSH)				
Female	2–13 mIU/mL		2–13 IU/L	
Midcycle	5–22 mIU/mL		5–22 IU/L	
Male	1–8 mIU/mL		1–8 IU/L	
Glucose, fasting	65–115 mg/dL		3.6–6.3 mmol/L	
Glucose Tolerance Test (Oral)	mg/dL		mmol/L	
	Normal	**Diabetic**	**Normal**	**Diabetic**
Fasting	70–105	>140	3.9–5.8	>7.8
60 min	120–170	≥200	6.7–9.4	≥11.1
90 min	100–140	≥200	5.6–7.8	≥11.1
120 min	70–120	≥140	3.9–6.7	≥7.8
Gamma glutamyl transferase: Male	9–50 units/L		9–50 units/L	
Female	8–40 units/L		8–40 units/L	
Haptoglobin	44–303 mg/dL		0.44–3.03 g/L	
Hematologic tests				
Fibrinogen	200–400 mg/dL		2–4 g/L	
Hematocrit: Female	36%–44.6%		0.36–0.446 fraction of 1	
Male	40.7%–50.3%		0.4–0.503 fraction of 1	

(*Continued*)

TABLE 155 (*Continued*)

Normal Human Laboratory Values[a]

	Blood, Plasma, or Serum	
	Reference Value	
Determination	**Conventional Units**	**SI Units**
Hemoglobin A_{1C}	5.3%–7.5% of total Hb	0.053–0.075
Hemoglobin: Female	12.1–15.3 g/dL	121–153 g/L
Male	13.8–17.5 g/dL	138–175 g/L
Leukocyte count (WBC)	3800–9800/mcL	$3.8–9.8 \times 10^9$/L
Erythrocyte count (RBC): Female	$3.5–5 \times 10^6$/mcL	$3.5–5 \times 10^{12}$/L
Male	$4.3–5.9 \times 10^6$/mcL	$4.3–5.9 \times 10^{12}$/L
MCV	80–97.6 mcm^3	80–97.6 fl
MCH	27–33 pg/cell	1.66–2.09 fmol/cell
MCHC	33–36 g/dL	20.3–22 mmol/L
Erythrocyte sedimentation rate (ESR) (sed rate)	≤30 mm/h	≤30 mm/h
Erythrocyte enzymes Glucose-6-phosphate dehydrogenase (G-6-PD)	250–5000 units/10^6 cells	250–5000 mcunits/cell
Ferritin	10–383 ng/mL	23–862 pmol/L
Folic acid: normal	>3.1–12.4 ng/mL	7–28.1 nmol/L
Platelet count	$150–450 \times 10^3$/mcL	$150–450 \times 10^9$/L
Reticulocytes	0.5%–1.5% of erythrocytes	0.005–0.015
Vitamin B_{12}	223–1132 pg/mL	165–835 pmol/L
Iron: Female	30–160 mcg/dL	5.4–31.3 mcmol/L
Male	45–160 mcg/dL	8.1–31.3 mcmol/L
Iron-binding capacity	220–420 mcg/dL	39.4–75.2 mcmol/L
Isocitrate dehydrogenase	1.2–7 units/L	1.2–7 units/L
Isoenzymes		
Fraction 1	14%–26% of total	0.14–0.26 fraction of total

(*Continued*)

TABLE 155 (*Continued*)

Normal Human Laboratory Values[a]

	Blood, Plasma, or Serum	
	Reference Value	
Determination	**Conventional Units**	**SI Units**
Fraction 2	29%–39% of total	0.29–0.39 fraction of total
Fraction 3	20%–26% of total	0.20–0.26 fraction of total
Fraction 4	8%–16% of total	0.08–0.16 fraction of total
Fraction 5	6%–16% of total	0.06–0.16 fraction of total
Lactate dehydrogenase	100–250 IU/L	1.67–4.17 mckat/L
Lactic acid (lactate)	6–19 mg/dL	0.7–2.1 mmol/L
Lead	≤50 mcg/dL	≤2.41 mcmol/L
Lipase	10–150 units/L	10–150 units/L
Lipids		
Total cholesterol		
Desirable	<200 mg/dL	<5.2 mmol/L
Borderline-high	200–239 mg/dL	<5.2–6.2 mmol/L
High	>239 mg/dL	>6.2 mmol/L
LDL		
Desirable	<130 mg/dL	<3.36 mmol/L
Borderline-high	130–159 mg/dL	3.36–4.11 mmol/L
High	>159 mg/dL	>4.11 mmol/L
HDL (low)	<35 mg/dL	<0.91 mmol/L
Triglycerides		
Desirable	<200 mg/dL	<2.26 mmol/L
Borderline-high	200–400 mg/dL	2.26–4.52 mmol/L
High	400–1000 mg/dL	4.52–11.3 mmol/L
Very high	>1000 mg/dL	>11.3 mmol/L

(*Continued*)

TABLE 155 (*Continued*)

Normal Human Laboratory Values[a]

	Blood, Plasma, or Serum	
	Reference Value	
Determination	**Conventional Units**	**SI Units**
Magnesium	1.3–2.2 mEqL	0.65–1.1 mmol/L
Osmolality	280–300 mOsm/kg	280–300 mmol/kg
Oxygen saturation (arterial)	94%–100%	0.94–1 fraction of 1
PCO_2, arterial	35–45 mmHg	4.7–6 kPa
PH, arterial	7.35–7.45	7.35–7.45
PO_2, arterial: breathing room air[b]	80–105 mmHg	10.6–14 kPa
On 100% O_2	>500 mmHg	
Phosphatase (acid), total at 37°C	0.13–0.63 IU/L	2.2–10.5 IU/L or 2.2–10.5 mckat/L
Phosphatase alkaline[c]	20–130 IU/L	20–130 IU/L or 0.33–2.17 mckat/L
Phosphorus, inorganic[d] (phosphate)	2.5–5 mg/dL	0.8–1.6 mmol/L
Potassium	3.5–5 mEq/L	3.5–5 mmol/L
Progesterone		
Female	0.1–1.5 ng/mL	0.32–4.8 nmol/L
Follicular phase	0.1–1.5 ng/mL	0.32–4.8 nmol/L
Luteal phase	2.5–28 ng/mL	8–89 nmol/L
Male	<0.5 ng/mL	<1.6 nmol/L
Prolactin	1.4–24.2 ng/mL	1.4–24.2 mcg/L
Prostate specific antigen	0–4 ng/mL	0–4 ng/mL
Protein		
Total	6–8 g/dL	60–80 g/L
Albumin	3.6–5 g/dL	36–50 g/L
Globulin	2.3–3.5 g/dL	23–35 g/L
Rheumatoid factor	<60 IU/mL	<60 kIU/L
Sodium	135–147 mEq/L	135–147 mmol/L

(*Continued*)

TABLE 155 (*Continued*)

Normal Human Laboratory Values[a]

	Blood, Plasma, or Serum	
	Reference Value	
Determination	**Conventional Units**	**SI Units**
Testosterone: Female	6–86 ng/dL	0.21–3 nmol/L
Male	270–1070 ng/dL	9.3–37 nmol/L
Thyroid hormone function tests		
Thyroid-stimulating hormone (TSH)	0.35–6.2 mcU/mL	0.35–6.2 mU/L
Thyroxine-binding globulin capacity	10–26 mcg/dL	100–260 mcg/L
Total triiodothyronine (T_3)	75–220 ng/dL	1.2–3.4 nmol/L
Total thyroxine by RIA (T_4)	4–11 mcg/dL	51–142 nmol/L
T_3 resin uptake	25%–38%	0.25–0.38 fraction of 1
Aspartate aminotransferase (AST)	11–47 IU/L	0.18–0.78 mckat/L
Alanine aminotransferase (ALT)	7–53 IU/L	0.12–0.88 mckat/L
Transferrin	220–400 mg/dL	2.20–4.00 g/L
Urea nitrogen (BUN)	8–25 mg/dL	2.9–8.9 mmol/L
Uric acid	3–8 mg/dL	179–476 mcmol/L
Vitamin A (retinol)	15–60 mcg/dL	0.52–2.09 mcmol/L
Zinc	50–150 mcg/dL	7.7–23 mcmol/L

(*Continued*)

TABLE 155 (*Continued*)

Normal Human Laboratory Values[a]

	Urine	
	Reference Value	
Determination	**Conventional Units**	**SI Units**
Calcium[e]	50–250 mcg/day	1.25–6.25 mmol/day
Catecholamines: Epinephrine	<20 mcg/day	<109 nmol/day
Norepinephrine	<100 mcg/day	<590 nmol/day
Catecholamines, 24 h	<110 µg	<650 nmol
Copper[e]	15–60 mcg/day	0.24–0.95 mcmol/day
Creatinine		
Child	8–22 mg/kg	71–195 µmol/kg
Adolescent	8–30 mg/kg	71–265 µmol/kg
Female	0.6–1.5 g/day	5.3–13.3 mmol/day
Male	0.8–1.8 g/day	7.1–15.9 mmol/day
pH	4.5–8	4.5–8
Phosphate[e]	0.9–1.3 g/day	29–42 mmol/day
Potassium[e]	25–100 mEq/day	25–100 mmol/day
Protein		
Total	1–14 mg/dL	10–140 mg/L
At rest	50–80 mg/day	50–80 mg/day
Protein, quantitative	<150 mg/day	<0.15 g/day
Sodium[e]	100–250 mEq/day	100–250 mmol/day
Specific gravity, random	1.002–1.030	1.002–1.030
Uric acid, 24 h	250–750 mg	1.48–4.43 mmol

Source: From © 2000 by Facts and Comparisons, reprinted with permission from *Drug Facts and Comparisons. 2000 Loose Leaf Edition*, St. Louis, MO: Facts and Comparisons, a Wolters Kluwer Company.

[a] In this table, normal reference values for commonly requested laboratory tests are listed in traditional units and in SI units. The table is a guideline only. Values are method dependent and "normal values" may vary between laboratories. The SI unit katal (kat) is the amount of enzyme generating 1 mol of product per second.

[b] Age dependent.

[c] Infants and adolescents up to 104 IU/L.

[d] Infants in the 1st year up to 6 mg/dL.

[e] Diet dependent.

References

Bulgin, M.S., Munn, S.L., and Gee, S. (1970), Hematologic changes to 4 ½ years of age in clinically normal Beagle dogs, *J. Am. Vet. Met. Assoc.,* 157, 1004.

Burns, K.F., Timmons, E.H., and Poiley, S.M. (1971), Serum chemistry and hematological values for axenic (germfree) and environmentally associated inbred rats, *Lab. Anim. Sci.,* 21, 415.

BVA/Frame/RSPCA/UFAW Joint Working Group (1993), Removal of blood from laboratory mammals and birds, First Report of the BVA/Frame/RSPCA/UFAW Joint Working Group on Refinement, *Lab. Anim.* 27, 1–22, 1993.

Charles River Laboratories (1993a), Hematology parameters for the Crl:CD®BR rat, Wilmington, MA.

Charles River Laboratories (1993b), Serum Chemistry Parameters for the Crl: CD®BR rat, Wilmington, MA.

Clarke, D., Tupari, G., Walker, R., and Smith, G. (1992), Stability of serum biochemical variables from Beagle dogs and Cynomologus monkeys, *Am. Assoc. Clin. Chem. Special Issue* 17, October.

Davy, C.W., Jackson, M.R., and Walker, S. (1984), Reference intervals for some clinical chemical parameters in the marmoset (*Callithrix jacchus*): effect of age and sex, *Lab. Anim.* 18, 135.

Diehl, K.-H, Hull, R., Morton, D., Pfister, R., Rabemampianina, Y., Smith, D., Vidal, J-M., and van de Vortenbosch, C. (2001), A good practice guide to the administration of substances and removal of blood, including routes and volumes. *J. Appl. Toxicol.* 21:15–23.

Facts and Comparisons (2000), *Drug Facts and Comparisons, 2000 Loose Leaf Edition*, Wolters Kluwer Company, St Louis, MO.

Frithe, C.H., Suber, R.L., and Umholtz, R. (1980), Hematologic and clinical chemistry findings in control BALB/c and C57BL/6 mice, *Lab. Anim. Sci.,* 30, 835.

Hack, C.A. and Gleiser, C.A. (1982), Hematologic and serum chemical reference values for adult and juvenile baboons (Papio sp), *Lab. Anim. Sci.,* 32, 502.

Hewett, C.D., Innes, D.J., Savory, J., and Wills, M.R. (1989), Normal biochemical and hematological values in New Zealand White rabbits, *Clin. Chem.*, 35, 1777.

Jain, N.C. (1986), *Schalm's Veterinary Hematology*, Lea & Febiger, Philadelphia.

Jordan, J.E. (1977), Normal laboratory values in beagle dogs at twelve to eighteen months of age, *Am. J. Vet. Res.*, 38, 409, 1c7.

Kapeghian, L.C. and Verlangieri, A.J. (1984), Effects of primaquine on serum biochemical and hematological parameters in anesthetized Macaca fasicularis, *J. Med. Primatol.*, 13, 97.

Kaspar, L.V. and Norris, W.P. (1977), Serum chemistry values of normal dogs (Beagles): associations with age, sex and family line, *Lab. Anim. Sci.*, 27, 980.

Levine, B.S., Unpublished data, 1979–1993.

Loeb, W.F. and Quimby, F.W. (1989), *The Clinical Chemistry of Laboratory Animals*, Pergamon Press, New York.

Marshall Farms, *Gottingen Minipig Reference Data*, Marshall Farms, New York.

Mitruka, B.M. and Rawnsley, H.M. (1977), *Clinical Biochemical and Hematological Reference Values in Normal Experimental Animals*, Masson Publishing, New York.

NIEHS (1985), A Summary of Control Values for F344 rats and B6C3F1 Mice in 13 week Subchronic Studies, Program Resources, Inc., Research Triangle Park, NC.

Pickrell, J.A., Schluter, S.J., Belasich, J.J., Stewart, E.V., Meyer, J., Hubbs, C.H., and Jones, R.K. (1974), Relationship of age of normal dogs to blood serum constituents and reliability of measured single values, *Am. J. Vet. Res.*, 35, 897.

Wolford, S.T., Schroer, R.A., Gohs, F.X., Gallo, P.P., Brodeck, M., Falk, H.B., and Ruhren, R.J. (1986), Reference range data base for serum chemistry and hematology values in laboratory animals, *Toxicol. Environ. Health*, 18, 161.

Yarbrough, L.W., Tollett, J.L., Montrey, R.D., and Beattie, R.J. (1984), Serum biochemical, hematological and body measurement data for common marmosets (Callithrix jacchus), *Lab. Anim. Sci.*, 34, 276.

Yu, L., Pragay, D.A., Chang, D., and Wicher, K. (1979), Biochemical parameters of normal rabbit serum, *Clin. Biochem.*, 12, 83.

11

Risk Assessment

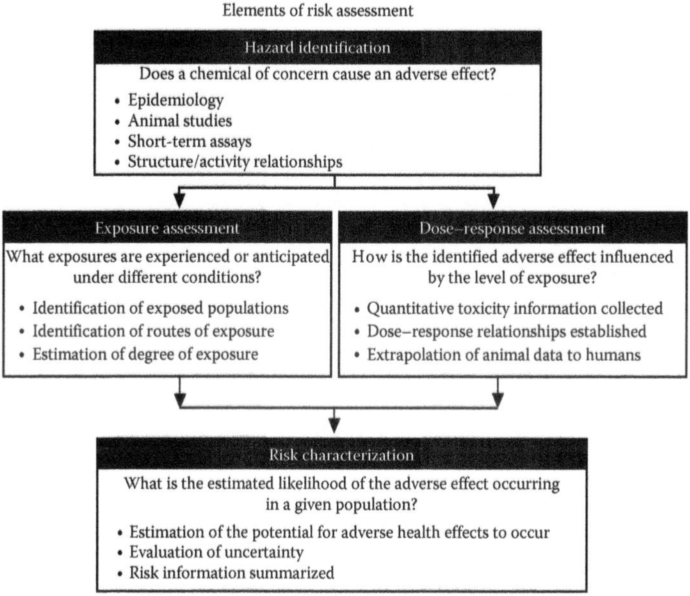

Elements of risk assessment

Hazard identification

Does a chemical of concern cause an adverse effect?

- Epidemiology
- Animal studies
- Short-term assays
- Structure/activity relationships

Exposure assessment

What exposures are experienced or anticipated under different conditions?

- Identification of exposed populations
- Identification of routes of exposure
- Estimation of degree of exposure

Dose–response assessment

How is the identified adverse effect influenced by the level of exposure?

- Quantitative toxicity information collected
- Dose–response relationships established
- Extrapolation of animal data to humans

Risk characterization

What is the estimated likelihood of the adverse effect occurring in a given population?

- Estimation of the potential for adverse health effects to occur
- Evaluation of uncertainty
- Risk information summarized

FIGURE 8

The four major elements of risk assessment. The information developed in the risk assessment process is utilized in risk management where decisions are made as to the need for, the degree of, and the steps to be taken to control exposures the chemical of concern. (From National Research Council (1983); U.S. Environmental Protection Agency (1989a); and Hopper, L.D. et al. (1992).)

TABLE 156

Typical Factors Considered in a Risk Assessment

Physical and chemical properties of the chemical

Patterns of use

Handling procedures

Availability and reliability of control measures

Source and route of exposure under ordinary and extraordinary conditions

Potential for misuse

Magnitude, duration, and frequency of exposure

Nature of exposure (oral, dermal, inhalation)

Physical nature of the exposure (solid, liquid, vapor, etc.)

Influence of environmental conditions of exposure

Population exposed

    Number

    Sex

    Health status

    Personal habits (e.g., smoking)

    Lifestyles (e.g., hobbies, activities)

Source: From Ballantyne, B. and Sullivan, J.B. (1992a).

TABLE 157

Major Factors That Influence a Risk Assessment

Factor	Effect
Low-dose extrapolation	Can involve as many as 50 or more assumptions each of which introduce uncertainty Often considered the greatest weakness in risk assessment
Population variation	The use of standard exposure factors can underestimate actual risk to hypersensitive individuals. Addressing the risk assessment to the most sensitive individuals can overestimate risk to the population as a whole
Exposure variation	The use of modeling and measurement techniques can provide exposure estimates that diverge widely from reality
Environmental variation	Can affect actual exposures to a greater or lesser degree than assumed to exist
Multiple exposures	Risk assessments generally deal with one contaminant for which additive, synergistic, and antagonistic effects are unaccounted. Can result in underestimate or overestimate of risk
Species differences	It is generally assumed that humans are equivalent to the most sensitive species. Can overestimate or underestimate risk
Dose based on body weight	Toxicity generally does not vary linearly with body weight but exponentially with body surface area
Choice of dose levels	Use of unrealistically high-dose levels can result in toxicity unlikely to occur at actual exposure levels. The number of animals being studied may be insufficient to detect toxicity at lower doses
UFs	The use of UFs in attempting to counter the potential uncertainty of a risk assessment can overestimate risk by several orders of magnitude
Confidence intervals	The upper confidence interval does not represent the true likelihood of an event and can overestimate risk by an order of magnitude or more
Statistics	Experimental data may be inadequate for statistical analysis. Statistical significance may not indicate biological significance, and a biologically significant effect may not be statistically significant. Statistical significance does not prove causality. Conversely, lack of statistical significance does not prove safety

TABLE 158

Human Data Commonly Used in Risk Assessment

Study Type	Alternative Terms	Comments on Use
Cross-sectional	Prevalence, survey	Sampling of a population at a given point in time to assess prevalence of a disease. Most useful for studying chronic diseases of high frequency. Cannot measure incidence. Although associations may be drawn with prevalent cases, the temporal and causal order of such associations cannot be determined
Case–control	Retrospective, dose or case–referent, case history	Compares previous exposure in subjects with disease with one or more groups of subjects without disease. Selection of cases and noncases can be controlled. Exposures cannot be controlled. If exposure data available, a NOEL may be identified. Exposure history may be difficult to reconstruct outside of an occupational setting. Recall and other biases are possible due to retrospective evaluation. Allows estimation of relative odds of exposure in cases and controls but not absolute risk
Cohort	Longitudinal, prospective, incidence	Population or sample of subjects at risk of disease observed through time for outcome of interest. May fail to detect rare outcome. Many factors can be controlled for reduced bias (prospective design). Dose–response curves may be constructed if dose or exposure data available. Allows estimation of absolute and relative risk
Clinical trials		Type of cohort study in which investigator controls treatment (exposure). Generally not applicable to environmental issues. Intervention trials in which an exposure is removed or changed (e.g., medication, smoking, diet) are useful for evaluating causality
Experimental studies		Controlled human exposures generally of low dose and limited exposure time. Used for hazard identification, dose–response, and risk characterization
Case reports		Suggests nature of acute end points. Cannot be used to support absence of hazard

Source: From Piantadose, S. (1992) and U.S. Environmental Protection Agency (1989b).

TABLE 159

Epidemiological Measures (Rates and Ratios)

Annual crude death rate	$= \dfrac{\text{Total number of deaths during a given year}}{\text{Total population at midyear}}$
Annual specific death rate	$= \dfrac{\text{Total number of deaths in a specific group during a given year}}{\text{Total population in the specific group at mid year}}$
Proportional mortality rate	$= \dfrac{\text{Total number of deaths in a specific group}}{\text{Total number of deaths}}$
Infant mortality rate (IMR)	$= \dfrac{\text{Infant deaths}}{\text{Total lives births}}$
Standard mortality ratio (SMR)	$= \dfrac{\text{Observed deaths}}{\text{Expected deaths}}$
Cause-of-death ratio	$= \dfrac{\text{Deaths from a specific cause over a period of time}}{\text{Total deaths due to all causes in the same time period}}$
Incidence rate	$= \dfrac{\text{Number of new cases over a period of time}}{\text{Population at risk over the same time period}}$
Prevalence rate	$= \dfrac{\text{Number of existing cases at a point in time}}{\text{total population}}$
Relative risk (risk ratio)	$= \dfrac{\text{Incidence among the exposed}}{\text{Incidence among the nonexposed}}$
Attribural risk (risk difference)	$= \text{Incidence among the exposed} - \text{Incidence among the nonexposed}$
Relative odds ratio	$= \dfrac{\text{Number of exposed individuals with disease}}{\text{Number of exposed individuals without disease}}$ $\times \dfrac{\text{Number of nonexposed individuals without disease}}{\text{Number of nonexposed individuals with disease}}$

Source: From Selevan, S.G. (1993), Hallenbeck, W.H. and Cunningham, K.M. (1986), and Gamble and Battigelli (1978).

Note: Rate is a proportion with change over time considered. In practice, the term "rate" is often used interchangeably with "ratio" without reference to time. Often a rate is reported as an incidence per 100, 1000 or 100,000. For example, an incidence of $0.001 \times 10,000 = 10$ per 10,000.

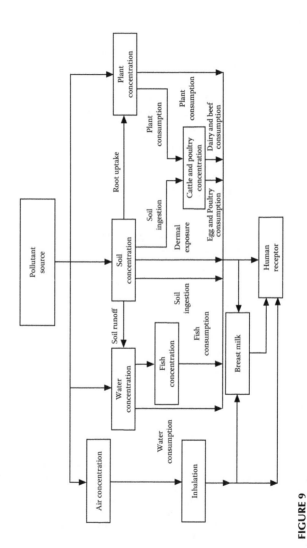

FIGURE 9
Exposure pathways to an environmental pollutant. (Modified from Lowe, J.A. et al. (1990).)

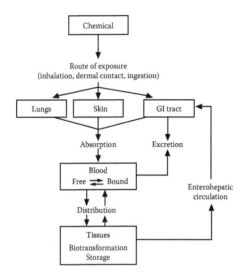

FIGURE 10

Diagrammatic representation of the possible pharmacokinetic fate of a chemical after exposure by inhalation, dermal contact, and ingestion. The lungs and skin also have enzyme systems capable of biotransformation (not shown). The fate of a chemical after exposure can vary considerably between species. Pharmacokinetic information is essential for accurate risk assessments. Such information can be obtained from animal studies/or physiologically-based pharmacokinetic models. In the absence of such data, assumptions are often made that introduce a great degree of uncertainty into the risk assessment process. (Modified from Ballantyne and Sullivan (1992a).)

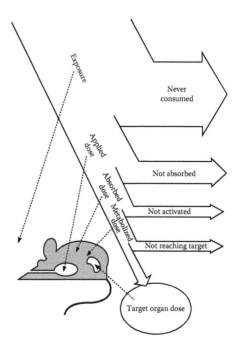

FIGURE 11

Representation of the relationships between ambient exposure and critical target dose and the progressive decrease in effective exposure due to various biological barriers. (From ILSI (1995). *Low-Dose Extrapolation of Cancer Risks: Issues and Perspectives*. Used with permission. 1995 International Life Sciences Institute, Washington, D.C.)

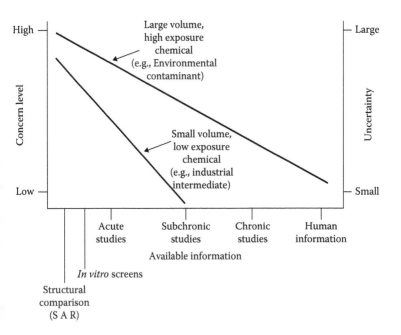

FIGURE 12

Relationship between the degree of uncertainty associated with the risk assessment of a chemical, the concern for human exposure, and the toxicological information available on the chemical. In practice, the larger the toxicological database available on a chemical of concern ("weight-of-evidence"), the greater the certainty (less uncertainty) that the estimated "safe" exposure level will be protective of individuals exposed to the chemical. Similarly, the concern that the risk assessment will underestimate the risk decreases with a larger toxicological database. Generally, less toxicological information will be required to reduce the concern level and uncertainty associated with a small-volume, low-exposure chemical (for which the exposed population is well characterized and the exposures can be controlled) as compared with a large-volume, high-exposure chemical.

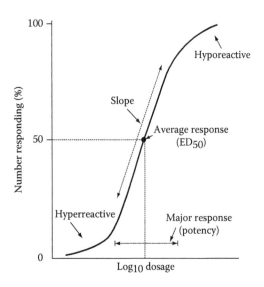

FIGURE 13

Typical sigmoid cumulative dose-response curve for a toxic effect which is symmetrical about the average (50% response) point. Note, dosage is presented on a log scale. The major response (potency) occurs around the average response. The midpoint of the curve is referred to as the median effective dose for the effect being considered (ED_{50}). If mortality is the endpoint, this point is referred to as the median lethal dose (LD_{50}). The 95% confidence limits are the narrowest at the midpoint which makes this the point most useful for comparison of toxicity between chemicals. The slope of the curve is determined by the increase in response as a function of incremental increases in dosage. A steep slope indicates a majority of a population will respond within a narrow dose range, while a flatter curve indicates that a much wider dose range is required to affect a majority of the exposed population. Hyperreactive and hyporeactive individuals are at the extreme left and right sides of the curve, respectively. (From Ballantyne, B. (1992b).)

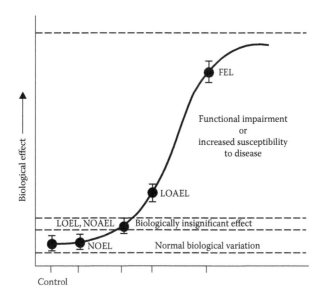

FIGURE 14

A dose-response curve from a typical toxicology study showing dose-related indices commonly used in risk assessment. A well-designed study should include dose levels that produce a Frank Effect (FEL), a Lowest Observable Adverse Effect (LOAEL), and either a Lowest Observable Effect (LOEL), a No Observable Adverse Effect (NOAEL), or a No Observable Effect (NOEL). A FEL is a dose or exposure level that produces unmistakable adverse health effects that cause functional impairment or increased susceptibility to disease; a LOAEL is the lowest dose or exposure level that produces and adverse health effect; a LOEL is the lowest dose or exposure level that produces an observable effect, but not to a degree which would be expected to have a significant impact on the health of the animal (the LOEL is sometimes confused with a LOAEL); a NOAEL is the highest dose or exposure level at which no adverse health effects are observed which are capable of functional impairment or increased susceptibility to disease (the NOAEL can be equivalent to the LOEL); a NOEL is the highest dose or exposure level at which no effects are observed outside of the range of normal biological variation for the species and strain under study. The effect, if any, observed at the NOEL should not be statistically significant when compared with the control group. (Adapted from Ecobichon, D.J. (1992a).)

TABLE 160

The Duration of Studies in Experimental Animals and Time Equivalents in Humans

Species	Duration of Study in Months				
	1	3	6	12	24
Percent of Life Span					
Rat	4.1	12	25	49	99
Rabbit	1.5	4.5	9	18	36
Dog	0.82	2.5	4.9	9.8	20
Pig	0.82	2.5	4.9	9.8	20
Monkey	0.55	1.6	3.3	6.6	13
Human Equivalents (In Months)					
Rat	34	101	202	404	808
Rabbit	12	36	72	145	289
Dog	6.5	20	40	81	162
Pig	6.5	20	40	81	162
Monkey	4.5	13	27	61	107

Source: Adapted from Paget, G.E. (1970). From Ecobichon, D.J. (1992b). With permission.

TABLE 161

Comparative Mammalian Reference Values for Relative Dose Calculations

Species	Average Life Span (Years)	Body Weight (kg)	Food Consumption (g/Day)	Food Consumption Factor[a]	Water Consumption (mL/Day)	Inhalation Rate (m³/Day)
Human	70	70	2000	0.028	1200	16
Mouse	1.5–2	0.03	4	0.13	6	0.052
Rat	2	0.35	18	0.05	50	0.29
Hamster	2.4	0.14	12	0.083	27	0.13
Guinea pig	4.5	0.84	34	0.040	200	0.40
Rabbit	7.8	3.8	186	0.049	410	2
Cat	17	3	90	0.030	220	1.2
Dog	12	12.7	318	0.025	610	4.3
Monkey (Rhesus)	18	8	320	0.040	530	5.4

Source: Adapted from U.S. EPA (1985).

[a] Fraction of body weight consumed per day as food.

TABLE 162

Reference Comparative Physiological Values[a,b]

Parameter	Mouse	Rat	Human
Tissue Perfusion (% of Cardiac Output)			
Brain	7.5 (2.0–13.0)	1.2	14.0 (13.0–15.0)
Heart	4.4 (2.8–6.0)	2.9	3.3 (2.6–4.0)
Kidney	24.8 (14.6–35.0)	17.8	22.0
Liver (total)	21.0	18.6 (17.0–26.0)	26.5 (26.0–27.0)
Liver (arterial only)	8.4	6.7	—[c]
Viscera	30.3	26.3	30.0
Adipose tissue	—[c]	4.5 (4.0–5.0)	4.7 (4.5–5.0)
Tissue Volume (% of Body Weight)			
Heart	0.4	0.5	0.6
Kidney	1.5	0.9 (0.9–1.0)	1.1 (0.4–1.5)
Liver	5.0 (4.0–5.9)	4.0 (3.7–4.2)	3.0 (2.4–4.0)
GI tract	6.8	4.3 (3.0–5.5)	3.8 (3.0–4.5)
Fat	7.6 (4.0–9.8)	8.4 (7.0–9.0)	15.5 (9.0–23.1)
Blood	7.6	7.2 (4.9–9.0)	7.2
Muscle	59.0 (45.0–73.0)	59.0 (50.0–73.0)	52.4 (43.4–73.0)
Skin	14.5	16.0	4.3
Marrow	2.7	—[c]	2.5 (2.1–2.8)
Skeletal tissue	9.0	—[c]	—[c]

(Continued)

TABLE 162 (Continued)

Reference Comparative Physiological Values[a,b]

Parameter	Mouse	Rat	Human
Cardiac Output			
Absolute (L/min)	0.0129 (0.110–0.160)	0.1066 (0.0730–0.1340)	5.59 (4.60–6.49)
Relative (L/min · kg)	0.535 (0.440–0.711)	0.327 (0.248–0.646)	0.080 —[c]
Alveolar Ventilation (L/min)			
	0.026 (0.012–0.039)	0.080 (0.075–0.085)	4.6 (4.0–5.8)
Minute Volume			
Absolute (L/min)	0.038 (0.024–0.052)	0.169 (0.057–0.336)	7.4 (6.0–9.0)
Relative (L/min · kg)	1.533 (1.239–1.925)	0.780 (0.142–2.054)	0.089 (0.014–0.127)
Respiratory Frequency (breaths/min)			
	171 (100–213)	117 (60–153)	14 (10–16)

Source: Data derived from U.S. EPA (1988).

[a] Mean of reported values. Brackets contain range of reported values from which mean was calculated. Absence of range indicates value was from a single report.

[b] Values presented are for unanesthetized animals.

[c] No data found.

TABLE 163

Typical Human Exposure Values Useful in Risk Assessments

Body weight	
Young child (1–3 year)	13 kg
Older child (5 year)	20 kg
Typical adult	80 kg
Lifespan	75 yr
Inhalation rate	
Typical adult	16 m³/day
Child (<1 year)	4.0 m³/day
Child (1–12 year)	9.0 m³/day
Industrial worker (8 h work shift)	10 m³/shift
Industrial worker (24 h total)	20 m³/day
Reasonable worst case	30 m³/day
(see Table 169 for inhalation rates relative to age and degree of activity)	
Drinking water ingestion rate	
Adult (average)	1.2 L/day (16 mL/kg-day)
Adult (90th percentile)	3.1 L/day (42 mL/kg-day)
Infant (<1 year)	0.5 L/day
Food consumption rate (adults)	
Total average meat intake	1.8 g/kg-day
Total average vegetable intake	2.5 g/kg-day
Total average fruit intake	1.1 g/kg-day
Total average dairy intake	3.5 g/kg-day
Total average dairy intake (child 1–3 years)	43.2 g/kg-day
Total grain intake	2.2 g/kg-day
Total average fish intake	0.62 g/kg-day
Adult total food intake	2000 g/day
Breast milk intake rate	
Average	657 mL/day
Upper percentile	977 mL/day

(Continued)

TABLE 163 (*Continued*)

Typical Human Exposure Values Useful in Risk Assessments

Exposed skin surface	
Typical adult	0.20 m²
Reasonable worst case	0.53 m²
Swimming or bathing (average)	
Male	1.94 m²
Female	1.69 m²
Soil ingestion rate (soil + dust)	
Children	
Average	100 mg/day
Upper percentile	200 mg/day
Pica child (soil)	1000 mg/day
Adult (average)	50 mg/day
Activities	
a. Showering (typically one event/day)	
Average	10 min/day
95th percentile	35 min/day
(a 5 min shower is estimated to use 40 gallons of water)	
b. Bathing (typically one event/day)	
Median	20 min/event
90th percentile	45 min/event
c. Time indoors	
All ages	20 h/day
Average residence volume	492 m³ (0.45 air changes/h)
d. Time outdoors	
Children (ages 3–6 months)	0.4 h/day
Children (ages 6–12 months)	2.3 h/day
Children	5 h/day
Adults	1.5 h/day

Source: U.S. EPA, 1989d, 1997, 2011.

TABLE 164

Comparative Mammalian Organ Weights (g/100 g Body Weight)

Species	Brain	Heart	Adrenals	Kidneys[a]	Lungs	Liver	Spleen	Testes[a]
Human	1.96	0.42	0.02	0.41	0.73	2.30	0.25	0.04
Mouse	1.35	0.68	0.02	2.60	0.66	5.29	0.32	0.62
Rat	0.46	0.32	0.01	0.70	0.40	3.10	0.20	0.92
Monkey (Rhesus)	2.78	0.38	0.02	0.54	1.89	2.09	0.14	0.03
Dog	0.59	0.85	0.01	0.30	0.94	2.94	0.45	0.15
Rabbit	0.40	0.35	0.02	0.70	0.53	3.19	0.04	0.13
Hamster	0.88	0.47	0.02	0.53	0.46	5.16	—	—
Guinea pig	1.33	0.53	0.07	1.17	1.18	5.14	0.21	0.65
Cat	0.77	0.45	0.02	1.07	1.04	2.59	0.29	0.07
Minipig	0.45	0.55	0.01	0.54	0.65	2.00	0.22	0.25

[a] Combined paired weight.

TABLE 165

Body Fluid Volumes for Men and Women

Parameter	Adult Male[a]		Adult Female[b]	
	Volume (L)	% of Body Weight	Volume (L)	% of Body Weight
Total body water	45.0	60	33.0	55
Extracellular water	11.25	15	9.0	15
Intracellular water	33.75	45	24.0	40
Total blood volume	5.4	7.2	4.3	7.2
Plasma volume	3.0	—	2.6	—
Erythrocyte volume	2.4	—	1.7	—

Source: Adapted from Plowchalk, D. et al. (1993).

[a] Volumes calculated for an adult male with a body weight of 75 kg and a hematocrit of 45%.

[b] Volumes calculated for an adult female with a body weight of 60 kg and a hematocrit of 40%.

TABLE 166

Relationship between Body Weight and Body Surface Area in a Number of Vertebrates

Species	Weight (g)	Surface Area (cm²)
Mouse	20	46
Rat	200	325
Guinea pig	400	565
Rabbit	1500	1270
Cat	2000	1380
Monkey	4000	2980
Minipig	8000	3333
Dog	12,000	5770
Man	70,000	18,000

Source: From Niesink, R.J.M., deVries, J., and Hollinger, M.A. (1996) and unpublished data.

TABLE 167

Constants for Estimating Surface Area of Mammals[a]

Species	Constant (K)
Rat	9.6
Mouse	9.0
Rabbit	10.0
Guinea pig	9.0
Monkey	11.8
Dog	11.0
Cat	8.7

Source: Data derived from Spector, W.S., Ed. (1956).

[a] $A = KW^{2/3}$ where A = surface area (cm²); K = constant; W = body weight (g).

TABLE 168

Median Total Body Surface Area (m²) for Humans by Age

Age (Years)	Males	Females
3–5	0.728	0.711
6–8	0.931	0.919
9–11	1.16	1.16
12–14	1.49	1.48
15–17	1.75	1.60
Adult	2.09	1.84

Source: Adapted from U.S. EPA (1989d).

TABLE 169

Summary of Human Inhalation Rates for Men, Women, and Children by Activity Level (m³/h)[a]

	Resting[b]	Light[c]	Moderate[d]	Heavy[e]
Adult male	0.7	0.8	2.5	4.8
Adult female	0.3	0.5	1.6	2.9
Average adult[f]	0.5	0.6	2.1	3.9
Child, age 6	0.4	0.8	2.0	2.4
Child, age 10	0.4	1.0	3.2	4.2

Source: From U.S. EPA (1989d).

[a] Values of inhalation rates for males, females, and children presented in this table represent the mean of values reported for each activity level in USEPA (1985).

[b] Includes watching television, reading, and sleeping.

[c] Includes most domestic work, attending to personal needs and care, hobbies, and conducting minor indoor repairs and home improvements.

[d] Includes heavy indoor cleanup, performance of major indoor repairs and alterations, and climbing stairs.

[e] Includes vigorous physical exercise and climbing stairs carrying a load.

[f] Derived by taking the mean of the adult male and adult female values for each activity level.

TABLE 170

Summary of Drug Absorption in Neonates, Infants, and Children

	Neonate	Infants	Children
Physiological Alteration			
Gastric emptying time	Irregular	Increased	Slightly increased
Gastric pH	>5	4 to 2	Normal (2–3)
Intestinal motility	Reduced	Increased	Slightly increased
Intestinal surface area	Reduced	Near adult	Adult pattern
Microbial colonization	Reduced	Near adult	Adult pattern
Biliary function	Immature	Near adult	Adult pattern
Muscular blood flow	Reduced	Increased	Adult pattern
Skin permeability	Increased	Increased	Near adult pattern
Possible Pharmacokinetic Consequences			
Oral absorption	Erratic—reduced	Increased rate	Near adult pattern
I.M. absorption	Variable	Increased	Adult pattern
Percutaneous absorption	Increased	Increased	Near adult pattern
Rectal absorption	Very efficient	Efficient	Near adult pattern
Pre-systemic clearance	<adult	>adult	>adult

Source: From Ritschel, W.A. and Kerns, G.L. (1999), with permission, Table 24-1 "Summary of Drug Absorption in Neonates, Infants and Children" of the *Handbook of Basic Pharmacokinetics*, 5th Edition, p 307. c 1999 by the American Pharmaceutical Association. Originally adapted from Morselli, P.L., 1983.

Note: Direction of alteration given relative to expected normal adult pattern. Data contained in the table reflect developmental differences that might be expected in healthy pediatric patients. Certain conditions/disease states might modify the function and/or structure of the absorptive surface area, GI motility, and/or systemic blood flow impacting on the rate or extent of absorption. Generally, neonate ≤ 1 month of age, infants = 1–24 months of age, children = 2–12 years of age. As the age limits defining these developmental stages are somewhat arbitrary, some overlap in the functional capacity between these stages should be expected. Because physiological development is a dynamic process, it should be kept in mind that functional changes occur incrementally over time and do not abruptly change from one age group to another.

TABLE 171

Drug Metabolism in the Neonate, Infant, and Child

	Neonate	Infants	Children
Physiological Alteration			
Liver/body weight ratio	Increased	Increased	Slightly increased
Cytochromes P450 activity	Reduced	Increased	Slightly increased
Blood esterase activity	Reduced	Normal (by 12 months)	Adult pattern
Hepatic blow flow	Reduced	Increased	Near adult pattern
Phase II enzyme activity	Reduced	Increased	Near adult pattern
Possible Pharmacokinetic Consequences			
Metabolic rates	Reduced	Increased	Near adult pattern[a]
Pre-systemic clearance	Reduced	Increased	Near adult pattern
Total body clearance	Reduced	Increased	Near adult pattern[a]
Inducibility of enzymes	More evident	Slightly increased	Near adult pattern[a]

Source: From Ritschel, W.A. and Kerns, G.L. (1999), with permission, Table 24-5 "Drug Metabolism in the Neonate, Infant and Child" of the *Handbook of Basic Pharmacokinetics*, 5th Edition, p. 314. c 1999 by the American Pharmaceutical Association. Originally adapted from Morselli, P.L., 1983.

Note: Direction of alteration given relative to expected normal adult patterns. Generally, neonate ≤ 1 month of age, infants = 1–24 months of age, children = 2–12 years of age. As the age limits defining these developmental stages are somewhat arbitrary, some overlap in the functional capacity between these stages should be expected. Because physiological development is a dynamic process, it should be kept in mind that functional changes occur incrementally over time and do not abruptly change from one age group to another.

[a] Denotes assumption of adult pattern of activity after the conclusion of puberty. The activity of all drug-metabolizing enzymes is generally higher before vs. after puberty.

TABLE 172

Renal Function in the Neonate, Infant, and Child

	Neonate	Infants	Children
Physiological Alteration			
Kidney/body weight ratio	Increased	Increased	Near adult values
Glomerular filtration rate	Reduced	Normal (by 12 months)	Normal adult values
Active tubular secretion	Reduced	Near normal	Normal adult values[a]
Active tubular reabsorption	Reduced	Near normal	Normal adult values
Proteins present in urine	Present (30%)	Low to absent	Normally absent
Urinary acidification capacity	Low	Normal (by 1 month)	Normal adult activity
Urine output (mL/h/kg)	3–6	2–4	1–3
Urine concentrating capacity	Reduced	Near normal	Normal adult values
Possible Pharmacokinetic Consequences			
Active drug excretion	Reduced	Near normal	Normal adult pattern
Passive drug excretion	Reduced to Increased	Increased	Normal adult pattern
Excretion of basic drugs	Increased	Increased	Near normal

Source: From Ritschel, W.A. and Kerns, G.L. (1999) with permission, Table 24-6 "Renal Function in the Neonate, Infant and Child" of the *Handbook of Basic Pharmacokinetics*, 5th Edition, p. 315. c 1999 by the American Pharmaceutical Association. Originally adapted from Morselli, P.L., 1983.

Note: Direction of alteration given relative to expected normal adult patterns. Generally, neonate $\leq$ 1 month of age, infants = 1–24 months of age, children = 2–12 years of age. As the age limits defining these developmental stages are somewhat arbitrary, some overlap in the functional capacity between these stages should be expected. Because physiological development is a dynamic process, it should be kept in mind that functional changes occur incrementally over time and do not abruptly change from one age group to another.

[a] Denotes slight increase in excretion rate for basic compounds.

TABLE 173

Plasma Protein Binding and Drug Distribution in Neonates, Infants, and Children

	Neonate	Infants	Children
Physiological Alteration			
Plasma albumin	Reduced	Near normal	Near adult pattern
Fetal albumin	Present	Absent	Absent
Total proteins	Reduced	Decreased	Near adult pattern
Total globulins	Reduced	Decreased	Near adult pattern
Serum bilirubin	Increased	Normal	Normal adult pattern
Serum free fatty acids	Increased	Normal	Normal adult pattern
Blood pH	7.1–7.3	7.4 (normal)	7.4 (normal)
Adipose tissue	Scarce (↑CNS)	Reduced	Generally reduced
Total body water	Increased	Increased	Near adult pattern
Extracellular water	Increased	Increased	Near adult pattern
Endogenous maternal substances (Ligands)	Present	Absent	Absent
Possible Pharmacokinetic Consequences			
Free fraction	Increased	Increased	Slightly increased
Apparent volume of distribution			
Hydrophilic drugs	Increased	Increased	Slightly increased
Hydrophobic drugs	Reduced	Reduced	Slightly decreased
Tissue/plasma ratio	Increased	Increased	Slightly increased

Source: From Ritschel, W.A. and Kerns, G.L. (1999), with permission, Table 24-2 "Plasma Protein Binding and Drug Distribution" of the *Handbook of Basic Pharmacokinetics*, 5th Edition, p. 309. c 1999 by the American Pharmaceutical Association. Originally adapted from Morselli, P.L., 1983.

Note: Direction of alteration given relative to expected normal adult pattern. Generally, neonate ≤ 1 month of age, infants = 1–24 months of age, children = 2–12 years of age. As the age limits defining these developmental stages are somewhat arbitrary, some overlap in the functional capacity between these stages should be expected. Because physiological development is a dynamic process, it should be kept in mind that functional changes occur incrementally over time and do not abruptly change from one age group to another.

TABLE 174

Developmental Patterns for the Ontogeny of Important Drug-Metabolizing Enzymes in Humans

Enzyme(s)	Known Developmental Pattern
Phase I Enzymes	
CYP2D6	Low to absent in fetal liver but present at 1 week of age. Poor activity (i.e., 20% of adult) by 1 month. Adult competence by 3–5 years of age
CYP2C19, CYP2C9	Apparently absent in fetal liver. Low activity in first 2–4 weeks of life with adult activity reached by approximately 6 months. Activity may exceed adult levels during childhood and declines to adult levels after conclusion of puberty
CYP1A2	Not present in appreciable levels in human fetal liver. Adult levels reached by approximately 4 months and exceeded in children at 1–2 years of age. Adult activity reached after puberty
CYP3A7	Fetal form of CYP3A that is functionally active (and inducible) during gestation. Virtually disappears by 1–4 weeks of postnatal when CYP3A4 activity predominates but remains present in approximately 5% of individuals
CYP3A4	Extremely low activity at birth reaching approximately 30%–40% of adult activity by 1 month and full adult activity by 6 months. May exceed adult activity between 1–4 years of age, decreasing to adult levels after puberty
Phase II Enzymes	
NAT2	Some fetal activity by 16 weeks gestation. Poor activity between birth and 2 months of age. Adult phenotype distribution reached by 4–6 months with adult activity reached by 1–3 years
TPMT	Fetal levels approximately 30% of adult values. In newborns, activity is approximately 50% higher than adults with phenotype distribution that approximates adults. Exception is Korean children where adult activity is seen by 7–9 years of age

(Continued)

TABLE 174 (*Continued*)

Developmental Patterns for the Ontogeny of Important Drug-Metabolizing Enzymes in Humans

Enzyme(s)	Known Developmental Pattern
UGT	Ontogeny is isoform specific. In general, adult activity is reached by 6 to 24 months of age
ST	Ontogeny is isoform specific and appears more rapid than that for UGT. Activity for some isoforms may exceed adult levels during infancy and early childhood

Source: From Ritschel, W.A. and Kerns, G.L. (1999), with permission, Table 24–4 "Developmental Patterns for the Ontogeny of Important Drug Metabolizing Enzymes in Man" of the *Handbook of Basic Pharmacokinetics*, 5th Edition, p. 3122. c 1999 by the American Pharmaceutical Association. Originally adapted from Leeder, J.S. and Kearns, G.L., 1997.

Abbreviations include: CYP, cytochrome P450; NAT2, N-acetyltransferase-2; TPMT, thiopurine methyltransferase; UGT, glucuronosyltransferase and ST, sulfotransferase.

TABLE 175

Frequency of Selected Adverse Pregnancy Outcomes in Humans

Event	Frequency per 100	Unit
Spontaneous abortion, 8–28 weeks	10–20	Pregnancies or women
Chromosomal anomalies in spontaneous abortions, 8–28 weeks	30–40	Spontaneous abortions
Chromosomal anomalies from amniocentesis	2	Amniocentesis specimens
Stillbirths	2–4	Stillbirths and live births
Low birth weight <2500 g	7	Live births
Major malformations	2–3	Live births
Chromosomal anomalies	0.2	Live births
Severe mental retardation	0.4	Children to 15 years of age

Source: Modified from National Foundation/March of Dimes, 1981, from Manson and Wise (1991), Casarett and Doull's *Toxicology: The Basic Science of Poisons*, 4th ed., 1991. With permission of McGraw-Hill.

TABLE 176

Probabilities of Spontaneous Abortion in Humans

Time from Ovulation	Probability of Fetal Death in Gestation Interval (%)
1–6 days	54.6
7–13 days	24.7
14–20 days	8.2
3–5 weeks	7.6
6–9 weeks	6.5
10–13 weeks	4.4
14–17 weeks	1.3
18–21 weeks	0.8
22–25 weeks	0.3
26–29 weeks	0.3
30–33 weeks	0.3
34–37 weeks	0.3
38+ weeks	0.7

Source: Modified from Kline and Stein, 1985, from McGuigan, *Hazardous Materials Toxicology*: Clinical Principles of Environmental Health, 1992. With permission.

TABLE 177

Risk Assessment Calculations

1. Human Equivalent Dose (HED)

$$HED = (Animal\ dose) \times \left[\frac{Animal\ Body\ weight}{Human\ Body\ weight}\right]^{1/3}$$

where

Dose = mg/kg; Body weight = kg

2. ppm–mg/m³ Conversion

$$PPM = \frac{(mg/m^3) \times (R)}{(MW)}$$

where

ppm, exposure concentration as ppm; mg/m³, exposure concentration as mg/m³; R, universal gas constant (24.5 at 25°C and 760 mmHg); MW, molecular weight.

3. Airborne Concentration to Equivalent Oral Dose*

$$EOD = \frac{(C) \times (EL) \times (MV) \times (AF) \times (10^{-6})}{(BW)}$$

where

EOD, equivalent oral dose (mg/kg); C, concentration of substance in air (mg/m³); EL, exposure length (min); MV, minute volume, species specific (mL/min); AF, absorption factor (fraction of inhaled substance absorbed), default = 1; 10^{-6}, conversion factor m³ ↔ mL; BW, body weight (kg)

(Continued)

TABLE 177 (Continued)

Risk Assessment Calculations

4. *Oral Dose to Equivalent Airborne Concentration**

$$EAC = \frac{(OD) \times (BW)}{(MV) \times (AF) \times (EL) \times (10^{-6})}$$

where

EAC, equivalent airborne concentration (mg/m³); OD, oral dose (mg/kg); BW, body weight (kg); MV, minute volume, species specific (mL/min); AF, absorption factor, fraction of inhaled substance absorbed; (default = 1) EL, exposure length (min); 10⁻⁶, Conversion factor m³ ↔ mL

* Caution should be exercised when using Equations 3 and 4. These give crude approximations in that the time period will be set and protracted for inhalation and may be either bolus for gavage studies or averaged over the entire day for feeding and drinking water studies. They assume that there will be no chemical reactivity associated with oral administration and no portal entry effects and that the target organ effects will be the same regardless of the route of administration

5. *Lifetime Exposure (h)*

$$\text{Lifetime exposure} = \left(\frac{\text{hours exposed}}{\text{per day}}\right) \times \left(\frac{\text{days exposed}}{\text{per week}}\right) \times \left(\frac{\text{weeks exposed}}{\text{per year}}\right) \times (\text{years exposed})$$

6. *Exposure from Ingestion of Contaminated Water*

$$LADD = \frac{(C) \times (CR) \times (ED) \times (AF)}{(BW) \times (TL)}$$

where

LADD, lifetime average daily dose (mg/kg/day); C, concentration of contaminant in water (mg/L); CR, water consumption rate (L/day); ED, exposure duration (days); AF, absorption factor (fraction of ingested contaminant absorbed) default = 1 (dimensionless); BW, body weight (kg); TL, typical lifetime (days)

(Continued)

TABLE 177 (*Continued*)

Risk Assessment Calculations

7. Exposure from Dermal Contact with Contaminated Water

$$LADD = \frac{(C)\times(SA)\times(EL)\times(AR)\times(ED)\times(SV)\times(10^{-9})}{(BW)\times(TL)}$$

where

LADD, lifetime average daily dose (mg/kg/day); C, concentration of contaminant in water (mg/L); SA, surface area of exposed skin (cm²); EL, exposure length (min/day); AR, absorption rate (µg/cm²/min); SV, specific volume of water (1 L/kg); ED, exposure duration (days); 10^{-9}, conversion factor (kg ↔ µg); BW, body weight (kg); TL, typical lifetime (days)

8. Exposure from Ingestion of Contamination Soil

$$LADD = \frac{(C)\times(CR)\times(ED)\times(AF)\times(FC)\times(10^{-6})}{(BW)\times(TL)}$$

where

LADD, lifetime average daily dose (mg/kg/day); C, concentration of contaminant in soil (mg/kg); CR, soil consumption rate (mg/day); ED, exposure duration (days); AF, absorption factor (fraction of ingested contaminant absorbed) default 10, (dimensionless); FC, fraction of total soil from contaminated source; 10^{-6}, conversion factor kg ↔ mg; BW, body weight (kg); TL, typical lifetime (days)

9. Exposure from Dermal Contact with Contaminated Soil

$$LADD = \frac{(C)\times(SA)\times(BF)\times(FC)\times(SDF)\times(ED)\times(10^{-6})}{(BW)\times(TL)}$$

where

LADD (mg/kg/day); C, concentration of contaminant in soil (mg/kg); SA, surface area of exposed skin (cm²); BF, bioavailability factor (percent absorbed/day); FC, fraction of total soil from contaminated source; SDF, soil deposition factor; amount deposited per unit area of skin (mg/cm²/day); ED, exposure duration (days); BW, body weight (kg); TL, typical lifetime (days)

(*Continued*)

TABLE 177 (Continued)
Risk Assessment Calculations

10. *Exposure from Inhalation of Contaminated Particles in Air*

$$LADD = \frac{(C) \times (PC) \times (IR) \times (RF) \times (EL) \times (AF) \times (ED) \times (10^{-6})}{(BW) \times (TL)}$$

where
LADD, lifetime average daily dose (mg/kg/day); C, concentration of contaminant on particulate (mg/kg); PC, particulate concentration in air (mg/m³); IR, inhalation rate (m³/h); RF, respirable fraction of particulates; EL, exposure length (h/day); AF, absorption factor (fraction of inhaled contaminant absorbed) default = 1; ED, exposure duration (days); 10⁻⁶, conversion factor kg ↔ mg; BW, body weight (kg); TL, typical lifetime (days)

11. *Exposure from Inhalation of Vapors*

$$LADD = \frac{(C) \times (IR) \times (EL) \times (AF) \times (ED)}{(BW) \times (TL)}$$

where
LADD, lifetime average daily dose (mg/kg/day); C, concentration of contaminant in air (mg/m³); IR, inhalation rate (m³/h); EL, exposure length (h/day); AF, absorption factor (fraction of inhaled contaminant absorbed) default = 1; ED, exposure duration (days); BW, body weight (kg); TL, typical lifetime (days)

12. *Calculation of an RfD (Reference Dose)*

$$RfD = \frac{(NOAEL)}{(UFs) \times (MF)}$$

where
RfD (mg/kg/day); UFs, uncertainty factors generally multiples of 10 (although 3 or 1 are occasionally used depending on the strength and quality of the data). The following UFs are usually used:
UF
10 accounts for variation in the general population. Intended to protect sensitive subpopulations

(Continued)

TABLE 177 (Continued)

Risk Assessment Calculations

10 used when extrapolating from animals to humans. Intended to account for interspecies variability between humans and animals

10 used when a NOAEL is derived from a subchronic rather than a chronic study in calculating a chronic RfD

10 applied when a LOAEL is used instead of a NOAEL. Intended to account for the uncertainty in extrapolating from LOAELs to NOAELs

10 used to account for uncertainty associated with extrapolation when the database is incomplete

MF; multiple of 1–10; intended to reflect a professional qualitative assessment of the uncertainty in the critical study from which the NOAEL is derived as well as the overall quality of the database. Accounts for the uncertainty not addressed by the UFs. The use of the MF was discontinued in 2004 as it was felt that the uncertainties addressed by this factor are covered by the UFs

13. *Estimating an LD₅₀ of a Mixture*

$$\frac{1}{\text{Predicted LD}_{50}} = \frac{Pa}{\text{LD}_{50} \text{ of Component a}} = \frac{Pb}{\text{LD}_{50} \text{ of Component b}} + \cdots \frac{Pn}{\text{LD}_{50} \text{ of Component n}}$$

where

P, fraction of components in the mixture

14. *Estimation of Maximal Attainable Air Concentration of a Chemical*

$$\text{MAAC} = \frac{(vp) \times (mw) \times (10^6)}{(760) \times (R)}$$

where

MAAC, maximal attainable air concentration (mg/m³); vp, vapor pressure of the chemical (mmHg) at 25°C; mw, molecular weight in grams; 760, atmospheric pressure at 25°C; R, 24.5 (universal gas constant at 25°C and 760 mmHg)

Note: In the absence of analytical measurements, this equation can give a worst-case estimate of the theoretically achievable air concentration of a volatile chemical at equilibrium in any size room with no ventilation and an infinite source of chemical. Actual air concentrations could be lower depending on physical properties of the chemical, the amount of chemical being used, room ventilation, and other handling practices but not exceed the MAAC

(Continued)

TABLE 177 (*Continued*)

Risk Assessment Calculations

15. *Haber's Rule*

$$C \times t = k$$

where

C, exposure concentration; t, time; and k, a constant

Note: Haber's rule has been historically used to relate exposure concentration and duration to a toxic effect. Basically, this concept states that exposure concentration and exposure duration (ED) may be reciprocally adjusted to maintain a cumulative exposure constant (k) and that this cumulative exposure constant will always reflect a specific toxic response. In general terms, it states that the shorter the time of exposure, the higher the concentration that will be needed to achieve the same toxic effect as occurs with a longer period of exposure at lower concentrations. The inverse relationship of concentration and time may be valid when the toxic response to a chemical is equally dependent upon the concentration and the ED. However, work by ten Berg and co-workers in 1986 with acutely toxic chemicals revealed chemical-specific relationships between exposure concentration and exposure time that were often exponential rather than linear. This relationship can be expressed by the Equation $C^n \times t = k$, where n represents a chemical specific, and even toxic end point specific, exponent. The relationship described by this equation is basically the form of a linear regression analysis of the log–log transformation of a plot of C vs. t. ten Berge et al. examined the airborne concentration (C) and short-term exposure time (t) relationship relative to lethal responses for approximately 20 chemicals and found that the empirically derived value of n ranged from 0.8 to 3.5 among this group of chemicals. Hence, these workers showed that the value of the exponent (n) in the equation $C^n \times t = k$ quantitatively defines the relationship between exposure concentration and ED for a given chemical and for a specific toxic or health effect end point. Haber's rule is the special case where $n = 1$. As the value of n increases, the plot of concentration vs. time yields a progressive decrease in the slope of the curve. In short, the best expression for extrapolation over several time points is $C^n \times t = k$, where the value for n is derived from existing data (Standing Operating Procedures of the National Advisory Committee on Acute Exposure Guideline Levels for Hazardous Substances, 2001)

(*Continued*)

TABLE 177 (Continued)

Risk Assessment Calculations

16. *Time-Weighted Average (TWA) for an 8 h Workday*

$$TWA = \frac{C_1 T_1 + C_2 T_2 + \ldots C_n T_n}{8}$$

where

C_n, concentration measured during a period of time (<8 h); T_n, duration of the period of exposure in hours at concentration C_n ($\Sigma T = 8$)

17. *Risk for Noncarcinogens (Hazard Index)*

$$Risk = \frac{MDD}{ADI}$$

If: Risk > 1, a potential risk exists, which may be significant. Risk < 1, risk is insignificant.

where

MDD, maximum daily dose; ADI, acceptable daily intake

18. *Lifetime Risk for Carcinogens*

$$Risk = (LADD) \times (SF)$$

If: Risk = 10^{-6}, risk is insignificant; 10^{-6}–10^{-4}, possible risk; 10^{-4}, risk may be significant.

where

LADD, lifetime average daily dose (mg/kg/day); SF, slope factor or cancer potency factor (mg/kg/day)$^{-1}$ (chemical and route specific)

19. *Total Risk from a Single Contaminant via Multiple Exposure Pathways*

Total = Σ risks from all exposure pathways

Example: Total risk (from a contaminant in water) = (risk from ingestion) + (risk from showering) + (risk from swimming)

(Continued)

TABLE 177 (Continued)

Risk Assessment Calculations

20. *Total Risk from Multiple Contaminants via a Single Exposure Pathway*

Total risk = Σ risks from all contaminants in the media

Example:

Total risk from contaminants A, B, and C in water = total risk from contaminant A + total risk from contaminant B + total risk from contaminant C

For calculations 19 and 20: total risk < 1 is insignificant; total risk > 1 may be significant. Both of these methods are extremely conservative and can greatly overestimate risk.

Source: From U.S. EPA (1989d); Paustenbach, D.J. and Leung, H.-W. (1993); Environ Corporation (1990); U.S. EPA (1989c); and Lynch, J.R. (1979).

References

Ballantyne, B. and Sullivan, J.B. (1992a), Basic principles of toxicology. In: *Hazardous Materials Toxicology: Clinical Principles of Environmental Health*, Sullivan, J.B. and Krieger, G.R., Eds., Williams and Wilkins, Baltimore, chap. 2.

Ballantyne, B. (1992b), Exposure-dose-response relationships. In: *Hazardous Materials Toxicology: Clinical Principles of Environmental Health*. Sullivan, J.B. and Krieger, G.R., Eds., Williams and Wilkins, Baltimore, MD, chap 3.

Ecobichon, D.J. (1992a), *The Basis of Toxicity Testing*, CRC press, Boca Raton, FL, chap. 7.

Ecobichon, D.J. (1992b), *The Basis of Toxicity Testing*, CRC Press, Boca Raton, FL, chap. 4.

Environ Corporation (1990), *Risk Assessment Guidance Manual*, AlliedSignal Inc., Morristown, NJ.

Gamble, J.F. and Battigelli, M.C. (1978), Epidemiology, in *Patty's Industrial Hygiene and Toxicology, 3rd Revised Edition*, Vol. 1, Clayton, G.D. and Clayton, F.E., Eds., John Wiley and Sons, New York, chap. 5.

Hallenbeck, W.H. and Cunningham K.M. (1986), *Qualitative evaluation of human and animal studies, in Quantitative Risk Assessment for Environmental and Occupational Health*, Lewis Publishers, Chelsea, MI, chap. 3.

Hooper, L.D., Oehme, F.W., and Krieger, G.R. (1992), Risk assessment for toxic hazards, in *Hazardous Materials Toxicology: Clinical Principles of Environmental Health*, Sullivan, J.B. and Krieger, G.R., Eds., Williams and Wilkins, Baltimore, MD, chap. 7.

ILSI (1995), *Low-Dose Extrapolation of Cancer Risks: Issues and Perspectives*, International Life Sciences Institute, Washington, D.C., p. 188.

Klein, J. and Stein, Z. (1985), Very early pregnancy, in *Reproduction Toxicology*, Dixon, R.L., Ed., Raven Press, New York, 259.

Leeder, J.S. and Kerns, G.L. (1997), Pharmacogenetics in pediatrics: Implications for practice, *Pediatr. Clin. North Am.* 44, 55.

Lowe, J.A. et al. (1990), *In Health Effects of Municipal Waste Incineration*, CRC Press, Boca Raton, FL.

Lynch, J.R. (1979), Measurement of worker exposure, In: *Patty's Industrial Hygiene and Toxicology, Vol. III. Theory and Rationale of Industrial Hygiene Practice*, Cralley, L.V. and Cralley, L.J., Eds., John Wiley and Sons, New York, chap. 6.

Manson, J.M. and Wise, L.D. (1991), Teratogens. In: *Cassarret and Doull's Toxicology: The Basic Science of Poisons*, 4th ed., Amdur, M.O., Doull, J. and Klaassen, C.D., Eds., Pergamon Press, New York, chap. 7.

McGuigan, M.A. (1993), Teratogenesis and reproductive toxicology, in *Hazardous Materials Toxicology: Clinical Principles of Environmental Health*, Sullivan, J.B. and Krieger, G.R., Eds., Williams and Wilkins, Baltimore, MD, chap. 16.

Morselli, P.L. (1989), Clinical pharmacology of the prenatal period and early infancy, *Clin. Pharmacokinetics* 17 (Suppl. 1) 13.

National Foundation/March of Dimes (1981), Report of Panel II. *Guidelines for reproductive studies in exposed human populations, in Guidelines for Studies of Human Populations Exposed to Mutagenic and Reproductive Hazards*, Bloom, A.D., Ed., The Foundation, New York, 37.

National Research Council (1983), *Risk Assessment in the Federal Government*, National Academy Press, Washington, D.C.

Niesink, R.J.M., deVries, J., and Hollinger, M.A. (1996), *Toxicology: Principles and Applications*, CRC Press, Boca Raton, FL.

Paget, G.E. (1979), *Methods in Toxicology*, Blackwell Scientific Publishers, Oxford, 49.

Paustenbach, D.J. and Leung, H.W. (1993), Techniques for assessing the health risk of dermal contact with chemicals in the environment, In: *Health Risk Assessment: Dermal and Inhalation Exposure and Absorption of Toxicants*, Wang, R.G.M., Knaak, J.B., and Maibach, H.I., Eds., CRC Press, Boca Raton, FL, chap. 23.

Piantadose, S. (1992), Epidemiology and principles of surveillance regarding toxic hazards in the environment, in *Hazardous Materials Toxicology: Clinical Principles of Environmental Health*, Sullivan, J.B. and Krieger, G.R., Eds., Williams and Wilkins, Baltimore, MD, chap. 6.

Plowchalk, D., Meadows, M.J., and Mattinson, D.R. (1993), Comparative approach to toxicokinetics, In: *Occupational and Environmental*

Reproductive Hazards, A Guide for Clinicians, Paul, M., Ed., Williams and Wilkins, Baltimore, MD, chap. 3.

Ritschel, W.A. and Kerns, G.L. (1999), *Handbook of Basic Pharmacokinetics*, 5th ed., American Pharmaceutical Association, Washington, D.C.

Selevan, S.G. (1993), Epidemiology. In: *Occupational and Environmental Reproductive Hazards: A Guide for Clinicians*, Paul, M. Ed., Williams and Wilkins, Baltimore, MD, chap. 9.

Spector, W.S. (1956), *Handbook of Biological Data*, W.B. Saunders, Philadelphia.

U.S. Environmental Protection Agency (1984), *Techniques for the Assessment of the Carcinogenic Risk to the U.S. Population Due to Exposure from Selected Volatile Organic Compounds from Drinking Water via the Ingestion, Inhalation and Dermal Routes*, Cothern, C.R., Coniglio, W.A., and Marcus, W.L., Office of Drinking Water, NTIS, PB-84-213941.

U.S. Environmental Protection Agency (1985), *Development of Statistical Distributions or Ranges of Standard Factors Used in Exposure Assessments*, Office of Health and Environmental Assessments, EPA No. 600/8-85-010, NTIS, PB85-242667.

U.S. Environmental Protection Agency (1988), *Reference Physiological Parameters in Pharmacokinetic Modeling*, Arms, A.D. and Travis, C.C., Office of Risk Analysis, EPA No., 600/6-88/004.

U.S. Environmental Protection Agency (1989a), *General Quantitative Risk Assessment Guidance for Non-Cancer Health Effects*, ECAP-CIN-538M (cited in Hooper et al. 1992).

U.S. Environmental Protection Agency (1989b), *Interim Methods for the Development of Inhalation Reference Doses*, Blackburn, K., Dourson, M., Erdreich, L., Jarabek, A.M., and Overton, J., Jr., Environmental Criteria and Assessment Offices, EPA/600/8-88/066F.

U.S. Environmental Protection Agency (1989c), *Risk Assessment Guidance for Superfund, Vol. 1: Health Evaluation Manual*, Office of Emergency and Remedial Response, EPA No. 540/1-89/002.

U.S. Environmental Protection Agency (1989d), *Exposure Factors Handbook*. Konz, J.J., Lisi, K., Friebele, E., and Dixon, D.A., Office of Health and Environmental Assessments, EPA No. 600/8-89/043

U.S. Environmental Protection Agency (1997), *Exposure Factors Handbook*. Wood, P., Phillips, L., Adenuga, A., Koontz, M., Rector, H., Wilkes, C., and Wilson, M., National Center for Environmental Assessments, EPA No. 600/p-95/002Fa.

U.S. Environmental Protection Agency (2011), *Exposure Factors Handbook: 2011 Edition*, National Center for Environmental Assessment, Washington, DC, EPA/600/R-09/052F.

12

Human Clinical Toxicology

TABLE 178

Some Common Clinical Presentations and Differential Diagnoses in Overdose

Presentation	Toxicological Causes	Other Medical Examples
Asymptomatic with history	Almost any drug	Not applicable
Gastrointestinal complaints	Salicylate, theophylline, iron, colchicine quinidine, almost any drug	Food poisoning, allergy, ulcer, pancreatitis, obstruction, gallstones, genitourinary
Coma	Narcotics, sedatives, antipsychotics, alcohol, tricyclics, long-lasting benzodiazepines	Infectious and metabolic encephalopathy, trauma, anoxia, cerebrovascular accident, brain death
Seizures	Theophylline, tricyclics, isoniazid, stimulants, camphor, carbon monoxide, hypoglycemic agents, alcohol withdrawal	Idiopathic, arteriovenous malformation, tumor, trauma, hypoxia, febrile, inborn errors
Psychosis and altered mental status	Anticholinergics, stimulants, withdrawal	Psychiatric, infection, metabolic/inborn errors
Acidosis	Salicylate, ethanol, methanol, ethylene glycol, cyanide drugs causing seizures	Shock, diabetes, uremia, lactic acidosis
Respiratory depression (usually with coma)	Narcotics, sedatives, benzodiazepines	Cerebrovascular accident, metabolic coma, tumor
Pulmonary edema	Salicylates, narcotics, iron, paraquat (initially)	Heart failure, disseminated intravascular coagulation

(Continued)

TABLE 178 (*Continued*)

Some Common Clinical Presentations and Differential Diagnoses in Overdose

Presentation	Toxicological Causes	Other Medical Examples
Arrhythmias	Tricyclics, quinidine, anticholinergics, β-blockers, digoxin, lithium, antipsychotics, organophosphates	Atherosclerotic heart disease
Hypotension	Narcotics, sedatives, tricyclics, antipsychotics, β-blockers, theophylline, iron	Heart failure, shock, hypovolemia, disseminated intravascular coagulation
Hypertension	Cocaine, amphetamines, cyanide, nicotine, clonidine (initially)	Essential hypertension, pheochromocytoma, carcinoid, hyperrenin states, renal failure
Ataxia	Antiepileptics, barbiturates, alcohol, lithium, organomercury	Cerebellar degeneration

Source: From Osterloh, J. (1990). Reprinted with permission.

TABLE 179

Toxicological Syndromes by Class of Drugs

Narcotics

Heroin, morphine, codeine, oxycodone, hydromorphone, hydrocodenone, propoxyphene, pentazocine, meperidine, diphenoxylate, fentanyl and derivatives, buprenorphine, methadone

CNS depression (somnolent → coma)	If BP decreases, pulse does not increase
Slowed respiratory rate	Pinpoint pupils
T° normal or low	DTR usually decrease

Alcohols—Barbiturates

Ethanol, methanol, isopropanol, ethylene glycol, amo-, pento-, seco-, buta-, phenobarbital, butalbital, glutethimide, methaqualone, ethchlorvynol, phenytoin

CNS depression (stuporous → coma)	DTR decreases
Ataxia	Metabolic acidosis with alcohols and
T° usually decreases	ethylene glycol except isopropanol
	If BP decreases, pulse may increase

Anticholinergics

Atropine, scopolamine, antihistamines, phenothiazines, tricyclics, quinidine, amantadine, jimson weed, mushrooms

Delirious	Decreased bowel sounds
Increased pulse, increased T°	Urinary retention
Skin flushed, warm, pink	Blurred vision
Dry (no sweating)	Arrhythmias, prolonged QT

Stimulants

Cocaine, amphetamines, and derivatives (e.g., ice, MDA, MDMA, DOB), phencyclidine, lysergic acid, psilocybin

Acute psychosis (nonreality)	Increased muscle tone/activity
Increased pulse, increased BP, increased T°	Dilated pupils
Increased respiratory rate	Sweating
Agitation	Seizures

(Continued)

TABLE 179 (*Continued*)

Toxicological Syndromes by Class of Drugs

Antidepressants

Anticholinergic syndrome	Sinus tachycardia (early)
Hypotension	Supraventricular tachycardia (early)
Coma	Widened QRS, QT
Seizures	Ventricular arrhythmias

Benzodiazepines

CNS depression	BP, pulse, T° not greatly affected
Respiratory depression	DTR intact

Phenothiazines

Decreased BP, decreased T°	Anticholinergic syndrome (see earlier)
Rigidity, dystonias, torticollis	Seizures
Pinpoint pupils	

Salicylates

Abdominal pain	Shock
Respiratory alkalosis (early)	Diaphoresis
Metabolic acidosis	Hypoglycemia

Theophylline

Tachycardia	Hypotension
Hypokalemia	Seizures

Iron

Abdominal pain	Acidosis
GI bleeding	Renal failure
Hypotension	Cardiovascular collapse
Hypovolemia	

Lithium

Tremor	Hyperreflexia
Chorea	Rigidity
Abdominal pain	Seizures

(Continued)

TABLE 179 (*Continued*)

Toxicological Syndromes by Class of Drugs

Isoniazid

Metabolic acidosis	Hepatitis
Seizures	

Oral Hypoglycemics

Hypoglycemia	Diaphoresis
Coma	

Acetaminophen

Liver necrosis

β-Blockers

Bradycardia	Hypotension with slowed cardiac
Hyperglycemia	conduction

Source: From Osterloh, J. (1990). Reprinted with permission.

Abbreviations: CNS, central nervous system; BP, blood pressure; DTR, deep tendon reflexes; GI, gastrointestinal; QRS, QT, electrocardiogram parameters; MDA, methylenedioxy amphetamine; MDMA, 3,4-methylenedioxy methamphetamine; DOB, 4-bromo-2,5-dimethoxyamphetamine.

TABLE 180

Common Drugs Included on Most Toxicology Screens

Alcohols—Ethanol, methanol, isopropanol, acetone

Barbiturates/sedatives—Amobarbital, secobarbital, pentobarbital, butalbital, butabarbital, phenobarbital, glutethimide, ethchlorvynol, methaqualone

Antiepileptics—Phenytoin, carbamazepine, primadone, phenobarbital

Benzodiazepines—Chlordiazepoxide, diazepam, alprazolam, temazepam

Antihistamines—Diphenhydramine, chlorpheniramine, brompheniramine, tripelennamine, trihexiphenidyl, doxylamine, pyrilamine

Antidepressants—Amitriptyline, nortriptyline, doxepin, imipramine, desipramine, trazedone, amoxapine, maprotiline

Antipsychotics—Trifluoperazine, perphenazine, prochlorperazine, chlorpromazine

Stimulants—Amphetamine, methamphetamine, phenylpropanolamine, ephedrine, MDA, MDMA (other phenylethylamines), cocaine, phencyclidine

Narcotics analgesics—Heroin, morphine, codeine, oxycodone, hydrocodone, hydromorphone, meperidine, pentazocine, propoxyphene, methadone

Other analgesics—Salicylates, acetaminophen

Cardiovascular drugs—Lidocaine, propranolol, metoprolol, quinidine, procainamide, verapamil

Others—Theophylline, caffeine, nicotine, oral hypoglycemics, strychnine

Source: From Osterloh, J. (1990). Reprinted with permission.

Reference

Osterloh, J. (1990), Utility and reliability of emergency toxicologic testing, *Emerg. Med. Clin. N. Am.* 8, 693.

13

Industrial Chemical Toxicology

TABLE 181

Combined Tabulation of Toxicity Classes

		Various Routes of Administration			
Toxicity Rating	Commonly Used Term	LD$_{50}$ Single Oral Dose Rats	Inhalation 4 h Vapor Exposure Mortality 2/6–4/6 Rats	LD$_{50}$ Skin Rabbits	Probable Lethal Dose for Man
1	Extremely toxic	≤1 mg/kg	<10 ppm	≤5 mg/kg	A taste, 1 grain
2	Highly toxic	1–50 mg	10–100 ppm	5–43 mg/kg	1 teaspoon, 4 cc
3	Moderately toxic	50–500 mg	100–1,000 ppm	44–340 mg/kg	1 ounce, 30 g
4	Slightly toxic	0.5–5 g	1,000–10,000 ppm	0.35–2.81 g/kg	1 cup, 250 g
5	Practically nontoxic	5–15 g	10,000–100,000 ppm	2.82–22.59 g/kg	1 quart, 1000 g
6	Relatively harmless	>15 g	>100,000 ppm	>22.6 g/kg	>1 quart

Source: Adapted from Hodge, H.C. and Sterner, J.H. (1949). With permission.

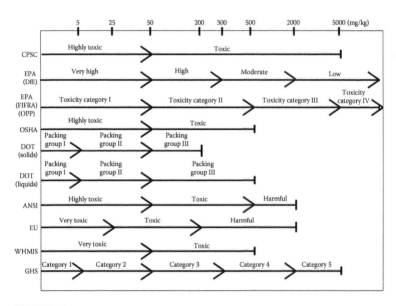

FIGURE 15

Classification based on rat acute oral LD$_{50}$. GHS Category 5 can also include chemicals for which assignment to more hazardous class is not warranted but there are indications of significant effects in humans and/or some mortality and/or significant clinical signs of toxicity within the Category 4 dose range. Refer to United Nations (2011) for specifics. CPSC, Consumer Product Safety Commission; EPA (DfE), U.S. Environmental Protection Agency (Design for the Environment Program); EPA (FIFRA), U.S. Environmental Protection Agency (Federal Insecticide, Fungicide and Rodenticide Act); EPA (OPP), U.S. Environmental Protection Agency (Office of Pesticide Programs); OSHA, U.S. Occupational Safety and Health Administration; DOT, U.S. Department of Transportation; ANSI, American National Standards Institute; EU, European Union; WHMIS, Workplace Hazardous Materials Information System (Canada); GHS, United Nations Globally Harmonized System. Use the following example of the DOT (solids) classification as an aid for interpreting the values of this figure: Packing Group I (≤5 mg/kg); Packing Group II (>5 to ≤50 mg/kg); Packing Group III (>50 to ≤200 mg/kg). (Adapted from Schurger, M.G. and McConnell, F., *Eastman Chemicals*, Kingsport, TN, 1989.)

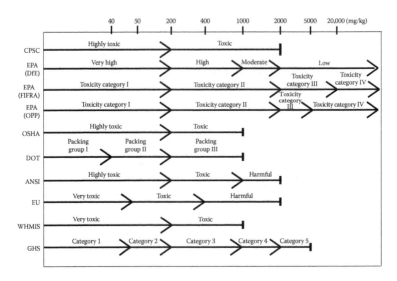

FIGURE 16

Classification based on rabbit or rat acute dermal LD_{50}. CPSC, Consumer Product Safety Commission; EPA (DfE), U.S. Environmental Protection Agency (Design for the Environment Program); EPA (FIFRA), U.S. Environmental Protection Agency (Federal Insecticide, Fungicide and Rodenticide Act); EPA (OPP), U.S. Environmental Protection Agency (Office of Pesticide Programs); OSHA, U.S. Occupational Safety and Health Administration; DOT, U.S. Department of Transportation; ANSI, American National Standards Institute; EU, European Union; WHMIS, Workplace Hazardous Materials Information System (Canada); GHS, United Nations Globally Harmonized System. Refer to the legend for Figure 15 for an aid to interpreting the values of this figure. (Adapted from Schurger, M.G. and McConnell, F., *Eastman Chemicals*, Kingsport, TN, 1989.)

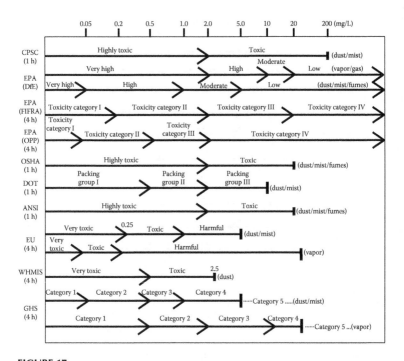

FIGURE 17

Classification based on rat acute inhalation LC_{50} (mg/L). CPSC, Consumer Product Safety Commission; EPA (DfE), U.S. Environmental Protection Agency (Design for the Environment Program); EPA (FIFRA), U.S. Environmental Protection Agency (Federal Insecticide, Fungicide and Rodenticide Act); EPA (OPP), U.S. Environmental Protection Agency (Office of Pesticide Programs); OSHA, U.S. Occupational Safety and Health Administration; DOT, U.S. Department of Transportation; ANSI, American National Standards Institute; EU, European Union; WHMIS, Workplace Hazardous Materials Information System (Canada); GHS, United Nations Globally Harmonized System. Refer to the legend for Figure 15 for an aid to interpreting the values of this figure. GHS Category 5 can include chemicals with LC_{50}s in the range equivalent to an oral/dermal LD_{50} between 2000 and 5000 mg/kg. Refer to United Nations (2011) for specifics. (Adapted from Schurger, M.G. and McConnell, F., *Eastman Chemicals*, Kingsport, TN, 1989.)

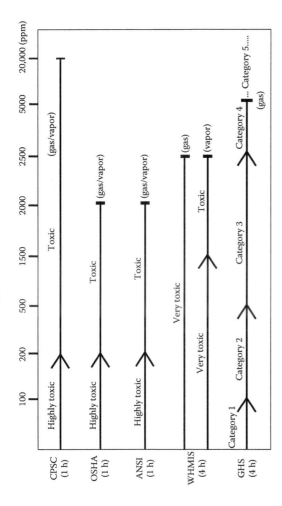

FIGURE 18

Classification based on rat acute inhalation LC$_{50}$ (ppm). CPSC, Consumer Product Safety Commission; OSHA, U.S. Occupational Safety and Health Administration; ANSI, American National Standards Institute; WHMIS, Workplace Hazardous Materials Information System (Canada); GHS, United Nations Globally Harmonized System. Refer to the legend for Figure 15 for an aid to interpreting the values of this figure. GHS Category 5 can include chemicals with LC$_{50}$s in the range equivalent to an oral/dermal LD$_{50}$ between 2000 and 5000 mg/kg. Refer to United Nations (2011) for specifics. (Adapted from Schurger, M.G. and McConnell, F., *Eastman Chemicals*, Kingsport, TN, 1989.)

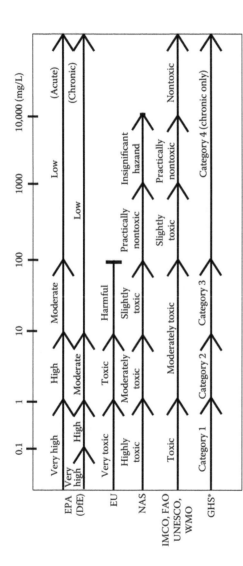

FIGURE 19

Classification based on fish acute LC_{50}. The GHS classification also applies to chronic toxicity provided certain criteria related to biodegradation and bioconcentration are met. Refer to United Nations (2011) for specifics. EPA (DfE), U.S. Environmental Protection Agency (Design for the Environment Program); EU, European Union; NAS, U.S. National Academy of Sciences; IMCO, Inter-government Maritime Consultative Organization; FAO, Food and Agriculture Organization; UNESCO, United Nations Educational, Scientific and Cultural Organization; WMO, World Meteorological Organization; GHS, United Nations Globally Harmonized System. Refer to the legend for Figure 15 for an aid to interpreting the values of this figure. (Adapted from Schurger, M.G. and McConnell, F., *Eastman Chemicals*, Kingsport, TN, 1989.)

TABLE 18.2

Classification Based on Mutagenicity/Genotoxicity

	Category	Classification	Criteria
U.S. Environmental Protection Agency Design for the Environment Program (2011)		Very high	Substances known to induce heritable mutations or to be regarded as if they induce heritable mutations in the germ cells of humans
		High	Substances which cause concern for humans owing to the possibility that they may induce heritable mutations in the germ cells of humans OR Evidence of mutagenicity supported by positive results in *in vitro* AND *in vivo* somatic cells and/or germ cells of humans or animals
		Moderate	Evidence of mutagenicity supported by positive results in *in vitro* OR *in vivo* somatic cells and/or germ cells of humans or animals
		Low	Negative for chromosomal aberrations and gene mutations, or no structural alerts
United Nations Globally Harmonized System (GHS)	1A	Known/presumed	Positive evidence from epidemiological studies
	1B	Known/presumed	Positive results in *In vivo* heritable germ cell tests in mammals Human germ cell tests *In vivo* somatic mutagenicity tests combined with some evidence of germ cell mutagenicity
	2	Suspected/possible	May include heritable mutations in human germ cells Positive evidence from tests in mammals and somatic cell tests *In vivo* somatic genotoxicity supported by *in vitro* mutagenicity

Source: U.S. Environmental Protection Agency (2011); United Nations (2011).

TABLE 183

Classification Based on Carcinogenicity

Agency	Category	Classification	Description
U.S. Environmental Protection Agency (2005)		Carcinogenic to humans	This descriptor indicates strong evidence of human carcinogenicity based on epidemiology and experimental information including exposure and mode of action associations
		Likely to be carcinogenic to humans	This descriptor is appropriate when the weight of evidence is adequate to demonstrate carcinogenic potential to humans but does not reach the weight of evidence for the above descriptor *Carcinogenic to humans*
		Suggestive evidence of carcinogenic potential	This descriptor is appropriate when the available information is suggestive of carcinogenicity and a concern for potential carcinogenic effects in humans is raised, but the information is judged not sufficient for a stronger conclusion
		Inadequate information to assess carcinogenic potential	This descriptor is appropriate when available information is judged inadequate for applying one of the other descriptors
		Not likely to be carcinogenic to humans	This descriptor is appropriate when the available information is considered robust for deciding that there is no basis for human hazard concern

(Continued)

TABLE 183 *(Continued)*
Classification Based on Carcinogenicity

Agency	Category	Classification	Description
U.S. Environmental Protection Agency (1986)	A	Carcinogenic to humans	Sufficient evidence from epidemiology studies to support a causal association
	B1	Probably carcinogenic to humans	Limited evidence in humans from epidemiology studies
	B2	Probably carcinogenic to humans	Sufficient evidence from animal studies but inadequate or no data in humans
	C	Possibly carcinogenic to humans	Limited or equivocal evidence from animal studies but inadequate or no data in humans
	D	Not classifiable as to humans carcinogenicity	Inadequate or no data from animals and inadequate or no data in humans
	E	Evidence of noncarcinogenicity for humans	No evidence of carcinogenicity in at least two animal species and no evidence in humans
U.S. Environmental Protection Agency Design for the Environment Program (2011)	Very high	Known or presumed human carcinogen	
	High	Suspected human carcinogen	
	Moderate	Limited or marginal evidence of carcinogenicity in animals (and inadequate evidence in humans)	
	Low	Negative studies or robust mechanism-based SAR	Robust mechanism-based SAR analysis which may include (i) negative studies on relevant/suitable analog(s) and/or (ii) combination of lack of structural alerts and features suggestive of potential carcinogenic activity and negative supportive, short-term predictive tests

(Continued)

TABLE 183 (*Continued*)
Classification Based on Carcinogenicity

Agency	Category	Classification	Description
International Agency for Research on Cancer (IARC)	1	Carcinogenic to humans	Sufficient epidemiological evidence for carcinogenicity in humans or sufficient evidence of carcinogenicity from animals studies with strong evidence for a carcinogenic mechanism relevant to humans
	2A	Probably carcinogenic to humans	Sufficient evidence from animal studies and limited evidence in humans
	2B	Possibly carcinogenic to humans	Sufficient evidence from animal studies but inadequate evidence in humans or limited evidence in humans and less than sufficient evidence in animals
	3	Not classifiable as to human carcinogenicity	Inadequate data to classify
	4	Not carcinogenic	Sufficient evidence of noncarcinogenicity in humans and/or animals

(*Continued*)

TABLE 183 (*Continued*)
Classification Based on Carcinogenicity

Agency	Category	Classification	Description
European Economic Community (EEC)	1	Known to be carcinogenic to humans	Sufficient evidence to establish a causal association between human exposure and cancer
	2	Regarded as if carcinogenic to humans	Sufficient evidence to provide a strong presumption that human exposure may result in cancer. Based on long-term animal studies and/or other relevant information
	3	Causes concern due to possible carcinogenic effects	Inadequate information to make a satisfactory assessment. Some evidence from animal studies but insufficient to place in Category 2
United Nations Globally Harmonized System (GHS)	1A	Known human carcinogen	Based on human evidence
	1B	Presumed human carcinogen	Based on demonstrated animal carcinogenicity
	2	Suspected carcinogen	Limited evidence of human or animal carcinogenicity

Source: Ecobichon, D.J. (1992); European Economic Community (EEC) (1993); U.S. Environmental Protection Agency (2007); U.S. Environmental Protection Agency (2011); and United Nations (2011).

TABLE 184

Classification Based on Repeated Dose Target Organ Toxicity

Category	Classification		Criteria		
			NOAEL or LOAEL (mg/kg/day or mg/L 6 h/day)		
			90 days	40–50 days	28 days
U.S. Environmental Protection Agency Design for the Environment Program (2011)	High	Oral	<10	<20	<30
		Dermal	<20	<40	<60
		Inhalation*			
		Vapor/gas	<0.2	<0.4	<0.6
		Dust/mist/fume	<0.02	<0.04	<0.06
	Moderate	Oral	10–100	20–200	30–300
		Dermal	20–200	40–400	60–600
		Inhalation*			
		Vapor/gas	0.2–1.0	0.4–2.0	0.6–3.0
		Dust/mist/fume	0.02–0.2	0.04–0.4	0.06–0.6
	Low	Oral	>100	>200	>300
		Dermal	>200	>400	>600
		Inhalation*			
		Vapor/gas	>1.0	>2.0	>3.0
		Dust/mist/fume	>0.2	>0.4	>0.6

(Continued)

TABLE 184 (*Continued*)

Classification Based on Repeated Dose Target Organ Toxicity

	Category	Classification	Criteria
United Nations Globally Harmonized System (GHS)	1	Significant toxicity in humans	Reliable, good quality human case studies or epidemiological studies
		Presumed significant toxicity in humans	Animal studies with significant and/or severe toxic effects relevant to humans at generally low exposure (as per guidance values)
	2	Presumed to be harmful to human health	Animal studies with significant toxic effects relevant to humans at generally moderate exposure (as per guidance values)
			Human evidence in exceptional cases

Source: U.S. Environmental Protection Agency (2011) and United Nations (2011).

TABLE 185

Classification Based on Reproductive and Developmental Toxicity

	Category	Classification	Criteria	
U.S. Environmental Protection Agency Design for the Environment Program (2011)			NOAEL or LOAEL (mg/kg/day or *mg/L/day)	
		High	Oral	<50
			Dermal	<100
			Inhalation*	
			Vapor/gas	<1
			Dust/mist/fume	<0.1
		Moderate	Oral	50–100
			Dermal	100–500
			Inhalation*	
			Vapor/gas	1–2.5
			Dust/mist/fume	0.1–0.5
		Low	Oral	>250–1000
			Dermal	>500–2000
			Inhalation*	
			Vapor/gas	>2.5–20
			Dust/mist/fume	>0.5–5
		Very low	Oral	>1000
			Dermal	>2000
			Inhalation*	
			Vapor/gas	>20
			Dust/mist/fume	>5
United Nations Globally Harmonized System (GHS)	1A	Known	Known to cause effects on human reproduction or on development based on human evidence	

(Continued)

TABLE 185 (*Continued*)

Classification Based on Reproductive and Developmental Toxicity

Category	Classification	Criteria
1B	Presumed	Presumed to cause effects on human reproduction or on development based on experimental animals
2	Suspected	Suspected based on human or animal evidence possibly with other information
Additional category		Effects on or via lactation

Source: U.S. Environmental Protection Agency (2011) and United Nations (2011).

TABLE 186

Classification Based on Eye and Skin Irritation/Corrosivity, Skin Sensitization and Respiratory Sensitization

	Category	Classification	Criteria
U.S. Environmental Protection Agency Design for the Environment Program (2011)	Eye irritation/corrosivity	See Section 5, Table 90	
	Skin irritation/corrosivity	See Section 4, Table 76	
	Skin sensitization	See Section 4, Table 83	
	Respiratory sensitization	High	Occurrence in humans or evidence of sensitization in humans based on animal or other tests
		Moderate	Limited evidence including the presence of structural alerts
		Low	Adequate data available indicating lack of respiratory sensitization
United Nations Globally Harmonized System (GHS)	Eye irritation/corrosivity	See Section 5, Table 91	
	Skin irritation/corrosivity	See Section 4, Table 77	
	Skin sensitization	See Section 4, Table 84	

(Continued)

TABLE 186 (Continued)

Classification Based on Eye and Skin Irritation/Corrosivity, Skin Sensitization and Respiratory Sensitization

Category	Classification	Criteria
Respiratory sensitization		A substance is classified as a respiratory sensitizer if there is evidence in humans that the substance can lead to specific respiratory hypersensitivity and/or if there are positive results from an appropriate animal test
1A		Substances showing a high frequency of occurrence in humans, or a probability of occurrence of a high sensitization rate in humans based on animal or other tests
		Severity of reaction may also be considered.
1B		Substances showing a low-to-moderate occurrence in humans or a probability of occurrence of a low-to-moderate sensitization rate in humans based on animal or other tests. Severity of reaction may also be considered

Source: U.S. Environmental Protection Agency (2011) and United Nations (2011).

TABLE 187
EPA Categories of Toxicological Concerns[a]

Category	Concern
Acid chlorides	*Environmental toxicity.*[b] Concern is greater if the log octanol/water partition coefficient (log K_{ow}) < 8 or if molecular weight (mol. wt.) <1000
Acrylamides	*Human health and environmental toxicity.* The acrylamides of greatest concern are those with a labile substituent, for example, methylol acrylamides, that may release acrylamide *per se* under metabolic conditions
	Members of this class are considered potential carcinogens, heritable mutagens, developmental and reproductive toxicants, and are potential neurotoxins
	Structures with an acrylamide equivalent weight ≥5000 are presumed *not* to pose a hazard under *any* condition
Acrylates and methacrylates	*Environmental toxicity.* Particularly if the log of the octanol/water partition coefficient (log *P*) <5. Concerns typically confined to species with mol. wt. <1000
	Human health. Some individual compounds are irritants and sensitizers
Aldehydes	*Environmental toxicity.* Generally mol. wt. <1000, log K_{ow} < 8
Aliphatic amines	*Environmental toxicity.* Generally mol. wt. <1000
Alkoxy silanes	*Human health and environmental toxicity.* There is a concern for lung toxicity of such substances if inhaled
Aluminum compounds	*Environmental toxicity.* Soluble forms of aluminum especially with solubility >1 ppb
Anhydrides, carboxylic acid	*Human health.* Potential for pulmonary sensitization; also developmental or reproductive toxicity. Generally if mol. wt. <1000
Anilines	*Environmental toxicity.* Generally if log K_{ow} < 8 and mol. wt. <1000
Azides	*Environmental toxicity.* Generally mol. wt. <1000, log K_{ow} < ~8

(Continued)

TABLE 187 (*Continued*)

EPA Categories of Toxicological Concerns[a]

Category	Concern
Benzotriazoles	*Environmental toxicity.* Generally mol. wt. <1000, log K_{ow} < 8 are expected to manifest toxicity
Benzotriazole-hindered phenols	*Human health and environmental toxicity.* Liver, kidney, hematological and immune system effects
Boron compounds	*Human health and environmental toxicity.* Male and female reproductive toxicity, hematotoxicity, neurotoxicity
Cobalt and compounds	*Environmental toxicity.* Generally if log K_{ow} < 8.0, mol. wt. is <1000 and water solubility >1 ppb
Dianilines	*Human health and environmental toxicity.* Potential carcinogens and mutagens. Also potential retinotoxic agents, reproductive and systemic toxicants
Diazoniums (aromatic only)	*Environmental toxicity.* Those with mol. wt. <1000 are of concern
Dichlorobenzidine-based pigments	*Human health and environmental toxicity.* Concern for mutagenicity/carcinogenicity
Diisocyanates (two or more isocyanate groups)	*Human health.* Potential dermal and pulmonary sensitization and other lung effects. Some may be carcinogenic. Structures with an isocyanate equivalent weight of ≥5000 are presumed *not* to pose a hazard under any conditions
Dithiocarbamates	*Environmental toxicity.* Generally mol. wt. <1000, log K_{ow} < 19
Dyes: *acid and amphoteric*	*Environmental toxicity.* Particularly if the substance is water-soluble and mol. wt. is around 1000 or less
Dyes: cationic	*Environmental toxicity.* Any dye bearing one or more net positive charges. No mol. wt. threshold

(*Continued*)

TABLE 187 (Continued)

EPA Categories of Toxicological Concerns[a]

Category	Concern
Dyes: aminobenzothiazole (AZO)	*Human health and environmental toxicity.* There are mutagenicity/carcinogenicity concerns. There is also potential for liver, thyroid, and neurotoxicity. Ecotoxicity concerns generally relate to chronic toxicity
Epoxides	*Human health and environmental toxicity.* Concerns for cancer and reproductive effects. Structures with epoxy equivalent weights ≥1000 are presumed *not* to pose a hazard under *any* conditions
Esters	*Environmental toxicity.* Compounds with mol. wt. >1000 are not of concern
Ethylene glycol ethers	*Human health.* Irritation of skin, eyes, and mucous membranes; hemolysis, bone marrow damage, and leukopenia of both lymphocytes and granulocytes; direct and indirect kidney damage; liver damage, immunotoxicity, and central nervous system depression. Also developmental and reproductive toxicants
Hindered amines	*Human health.* May be toxic to the immune system, liver, blood, the male reproductive system, and the gastrointestinal tract
Hydrazines and related compounds	*Human health and environmental toxicity.* Concerns for carcinogenicity and chronic effects to liver, kidney, and blood
Imides	*Environmental toxicity.* Compounds with mol. wt. <1000, log K_{ow} ≤ 8 are of greater concern
Lanthanides or rare earth metals	*Environmental toxicity.* mol. wt. <1000
β-Napthylamines (monosulfonated)	*Human health.* Potential mutagens and carcinogens. Concern is restricted to those compounds where not more than two sulfonate or sulfatoethylsulfone groups are on the ring *distal* to the β-amino group
Neutral organics	*Environmental toxicity.* The molecular weights of neutral organics of concern are generally <1000 and the octanol/water partition coefficients (log *P*) are <8

(Continued)

TABLE 187 (*Continued*)

EPA Categories of Toxicological Concerns[a]

Category	Concern
Nickel compounds	*Human health and environmental toxicity.* Concern for genotoxicity/carcinogenicity, fetotoxicity, and dermatotoxicity
Nitriles, allylic/vinyl	*Environmental toxicity.* Mol. wt. <1000, log $K_{ow} \leq 8$
Organotins	*Human health and environmental toxicity.* Eye and skin irritants, systemic effects (primarily neurotoxicity), immunotoxicity. Some organotins are probable human carcinogens
Peroxides	*Human health and environmental toxicity.* Compounds assessed on a case-by-case basis
Persistent, bioaccumulative, and toxic (PBT) chemicals (e.g., DDT)	*Human health, environmental toxicity, and fate.* PBT chemical substances are chemicals that partition to water, sediment, or soil and are not removed at rates adequate to prevent their accumulation in aquatic or terrestrial species, with the potential to pose a risk via food chain toxicity. Concern for chemicals with persistence (transformation half-life) >2 months, bioaccumulation $\geq$1000 (log $K_{ow} = 4.2$) and mol. wt. <1000
Phenolphthaleins	*Human health.* Concern for carcinogenicity
Phenols	*Environmental toxicity.* Compounds of greater concern have mol. wt. <1000
Phosphates, inorganic	*Environmental toxicity.* High concern for eutrophication
Phosphinate esters	*Environmental toxicity.* Generally if log $K_{ow} \leq 8.0$, mol. wt. is $\leq$1000
Polyanionic polymers (and monomers)	*Environmental toxicity.* Compounds must be water-soluble or water self-dispersing to be in this category. Mol. wt. can be >1000, log $K_{ow} < \sim$10
Polycationic polymers	*Environmental toxicity.* The polymers must be water-soluble or water-dispersible and the molecular weights are generally >300

(*Continued*)

TABLE 187 (*Continued*)

EPA Categories of Toxicological Concerns[a]

Category	Concern
Polynitroaromatics	*Environmental toxicity.* Concern is for compounds with mol. wt. <1000
Respirable, poorly soluble particulates	*Human health.* Particles $\leq$10 μm. Effects on the lung ranging from inflammation to fibrosis and potentially cancer
Rosin	*Environmental toxicity.* Category includes rosin, abietic acid, abientinic acid and their salts, and polymeric forms with mol. wt. <1000
Stilbene, derivatives of 4,4-bis(triazin-2-ylamino)-	*Human health.* Evaluated on a case-by-case basis
Thiols/mercaptans	*Environmental toxicity.* Mol. wt. < 1000, log K_{ow} < ~9
Triarylmethane pigments/dyes with insoluble groups	*Human health and environmental toxicity.* Developmental/reproductive toxicity and carcinogenicity concerns
Substituted triazines	*Environmental toxicity.* If log $K_{ow} \leq$ 8.0, mol. wt. is <1000
Surfactants: anionic	*Environmental toxicity.* No mol. wt. boundary
Surfactants: cationic (quaternary ammonium)	*Environmental toxicity.* Little ecotoxicity is expected when the carbon chain length exceeds 22 carbons
Surfactants: nonionic	*Environmental toxicity.* Acute aquatic toxicity increases with the hydrophobic chain length when the number of ethoxy groups or the hydrophilic component is held constant. Aquatic toxicity is decreased with increasing number of ethoxylate groups when the number of carbons in the hydrophobe is constant
Vinyl esters	*Human health and environmental toxicity.* Concern for carcinogenicity, neurotoxicity, and reproductive toxicity

(*Continued*)

TABLE 187 (*Continued*)

EPA Categories of Toxicological Concerns[a]

Category	Concern
Vinyl sulfones	*Human health and environmental toxicity.* Concern for mutagenicity and carcinogenicity
Soluble complexes of zinc	Environmental toxicity
Zirconium compounds	*Environmental toxicity.* Concern for mol. wt. <1000

Source: U.S. Environmental Protection Agency (2010).

[a] This table is an overview of toxicological concerns for various chemical classes. The reader is referred to the reference which is available on the EPA website for detailed information on hazard concerns and physical/chemical boundaries for these concerns.

[b] Environmental toxicity usually indicates toxicity to algae, daphnids, and/or fish although concerns for other environmental effects could exist.

TABLE 188

Criteria Defining "High-Exposure" Chemicals

- Production greater than 100,000 kg
- More than 1000 workers exposed
- More than 100 workers exposed by inhalation to greater than 10 mg/kg/day
- More than 100 workers exposed by inhalation to 1–10 mg/day for more than 100 days/year
- More than 250 workers exposed by routine dermal contact for more than 100 days/year
- Presence of the chemical in any consumer product in which the physical state of the chemical in the product and the manner of use would make exposure likely
- More than 70 mg/year of exposure via surface water
- More than 70 mg/year of exposure via air
- More than 70 mg/year of exposure via groundwater
- More than 10,000 kg/year release to environmental media
- More than 1000 kg/year total release to surface water after calculated estimates of treatment

Source: From U.S. EPA (1988).

TABLE 189

Selected OECD Guidelines for Testing of Chemicals

Mammalian

#401	Acute oral toxicity, LD_{50}
#402	Acute dermal toxicity
#403	Acute inhalation toxicity
#404	Acute dermal irritation/corrosion
#405	Acute eye irritation
#406	Contact sensitization
#407	Repeated dose 28 day oral toxicity (rodent)
#408	Repeated dose 90 day oral toxicity (rodent)
#409	Repeated dose 90 day oral toxicity (nonrodent)
#410	Repeated dose 28 day dermal toxicity
#411	Repeated dose 90 day dermal toxicity
#412	Repeated dose 28 day inhalation toxicity
#413	Repeated dose 90 day inhalation toxicity
#414	Developmental toxicity study
#415	One generation reproduction
#416	Two generation reproduction
#417	Toxicoxinetics
#418	Acute exposure delayed neurotoxicity of organophosphorus substances
#419	Repeated exposure delayed neurotoxicity of organophosphorus substances
#420	Acute oral toxicity, fixed dose procedure
#421	Reproductive/developmental toxicity screen
#422	Combined repeated dose 28 day oral toxicity/developmental toxicity screen
#423	Acute oral toxicity, acute toxic class
#424	Neurotoxicity screening battery
#425	Acute oral toxicity, up and down method
#426	Developmental Neurotoxicity Study
#429	Skin sensitization (local lymph node assay)
#430	*In Vitro* Skin Corrosion: Transcutaneous Electrical Resistance Text Method (TER)
#431	*In vitro* skin Corrosion: reconstructed human epidermis (RHE) test method
#436	Acute inhalation toxicity–acute toxic class method
#437	Bovine Corneal Opacity and Permeability Test Method for Identifying Eye Damage
#439	*In Vitro* Skin Irritation
#440	Uterotrophic Bioassay in Rodents
#441	Hershberger Bioassay in Rats
#442A	Local Lymph Node Assay: DA
#442B	Local Lymph Node Assay: BrdU-Elisa
#442C	In Chemico Skin Sensitization

(Continued)

TABLE 189 (*Continued*)

Selected OECD Guidelines for Testing of Chemicals

#442D	*In Vitro* Skin Sensitization: ARE-Nrf2 luciferase test method
#442E	*In Vitro* Skin Sensitization: Human cell Line Activation Test h-CLAT method
#443	Extended One-Generation Reproductive Toxicity Study
#451	Carcinogenicity
#452	Chronic toxicity
#453	Combined chronic toxicity/carcinogenicity
#455	Performance-Based Text Guideline for Stably Transfected Transactivation *In Vitro* Assays to Detect Estrogen Receptor Agonists and Antagonists
#456	H295R Steroidogenesis Assay
#458	Stably Transfected Human Androgen Receptor Transcriptional Activation Assay
#491	Short Time Exposure *In Vitro* Test Method for Eye Damage
#492	Reconstructed human Cornea-like Epithelium (RhCE) test method for identifying chemicals not requiring classification and labelling for eye irritation or serious eye damage
#493	Performance-Based Test Guideling for Human Recombinant Estrogen Receptor (hrER) *In Vitro* Assays

Genetox

#471	Reverse mutation assay/*Salmonella* (Ames test)
#473	*In vitro* mammalian chromosome aberration test
#474	Micronucleus test
#475	*In vivo* bone marrow mammalian chromosome aberration test
#476	*In vitro* mammalian cell gene mutation test (mouse lymphoma)
#477	Sex-linked recessive lethal test (*Drosophila*)
#478	Rodent dominant lethal test (mouse)
#479	*In vitro* sister chromatid exchange (SCE) assay
#480	Gene mutation assay/*Saccharomyces*
#481	Mitotic recombination assay/*Saccharomyces*
#482	*In vitro* unscheduled DNA synthesis (UDS)
#483	Mammalian spermatogonial chromosome aberration assay
#484	Mouse spot test
#485	Mouse heritable translocation assay
#486	Unscheduled DNA Synthesis (UDS) Test with Mammalian Liver Cells *in vivo*
#487	*In Vitro* Mammalian Cell Micronucleus Test
#489	*In Vitro* Mammalian Alkaline Comet Assay
#490	*In Vitro* Mammalian Cell Gene Mutation Tests Using the Thymidine Kinase Gene

(*Continued*)

TABLE 189 (*Continued*)

Selected OECD Guidelines for Testing of Chemicals

Ecotox/Aquatic

#106	Absorption/desorption
#201	Algal growth inhibition
#202	Acute toxicology/*Daphnia*
#203	Acute toxicity/fish
#204	14 day prolonged toxicity/fish
#205	Avian dietary toxicity
#206	Avian reproduction
#207	Acute toxicity/earthworm
#208	Terrestrial plant growth test
#209	Activated sludge respiration inhibition
#210	Fish early life stage toxicity
#211	21 day *Daphnia* reproduction
#213	Honeybees, Acute Oral Toxicity Test
#214	Honeybees, Acute Contact Toxicity Test
#215	Fish juvenile growth test
#222	Earthworm reproduction test
#223	Avian Acute Oral Toxicity Test
#229	Fish short-term reproduction assay
#230	21 day fish assay
#231	Amphibian Metamorphosis Assay
#301	Ready biodegradability
#302	Inherent biodegradability
#305	Bioaccumulation/fish

Note: Study guidelines occasionally fall out of favor and are deleted. Some of these guidelines appear in this list for historic purposes as they may appear in older documents. The reader is cautioned to always confirm that any guideline is still acceptable for regulatory purposes.

TABLE 190

Risk (R) Phrases Previously Used in the European Community (EU)

R1	Explosive when dry
R2	Risk of explosion by shock, friction, fire, or other sources of ignition
R3	Extreme risk of explosion by shock, friction, fire, or other sources of ignition
R4	Forms very sensitive explosive metallic compounds
R5	Heating may cause an explosion
R6	Explosive with or without contact with air
R7	May cause fire
R8	Contact with combustible material may cause fire
R9	Explosive when mixed with combustible material
R10	Flammable
R11	Highly flammable
R12	Extremely flammable
R14	Reacts violently with water
R15	Contact with water liberates extremely flammable gases
R16	Explosive when mixed with oxidizing substances
R17	Spontaneously flammable in air
R18	In use may form flammable/explosive vapor–air mixture
R19	May form explosive peroxides
R20	Harmful by inhalation
R21	Harmful in contact with skin
R22	Harmful if swallowed
R23	Toxic by inhalation
R24	Toxic in contact with skin
R25	Toxic if swallowed
R26	Very toxic by inhalation
R27	Very toxic in contact with skin
R28	Very toxic if swallowed
R29	Contact with water liberates toxic gas

(Continued)

TABLE 190 (*Continued*)

Risk (R) Phrases Previously Used in the European Community (EU)

R30	Can become highly flammable in use
R31	Contact with acids liberates toxic gas
R32	Contact with acids liberates very toxic gas
R33	Danger of cumulative effects
R34	Causes burns
R35	Causes severe burns
R36	Irritating to the eyes
R37	Irritating to the respiratory system
R38	Irritating to the skin
R39	Danger of very serious irreversible effects
R40	Limited evidence of a carcinogenic effect
R41	Risk of serious damage to the eyes
R42	May cause sensitization by inhalation
R43	May cause sensitization by skin contact
R44	Risk explosion if heated under confinement
R45	May cause cancer
R46	May cause heritable genetic damage
R48	Danger of serious damage to health by prolonged exposure
R49	May cause cancer by inhalation
R50	Very toxic to aquatic organisms
R51	Toxic to aquatic organisms
R52	Harmful to aquatic organisms
R53	May cause long-term adverse effects in the aquatic environment
R54	Toxic to flora
R55	Toxic to fauna
R56	Toxic to soil organisms
R57	Toxic to bees

(*Continued*)

TABLE 190 (*Continued*)

Risk (R) Phrases Previously Used in the European Community (EU)

R58	May cause long-term adverse effects to the environment
R59	Dangerous for the ozone layer
R60	May impair fertility
R61	May cause harm to the unborn child
R62	Possible risk of impaired fertility
R63	Possible risk of harm to the unborn child
R64	May cause harm to breast-fed babies
R68	Possible risk of irreversible effects

Combination of Particular Risks

R14/15	Reacts violently with water, liberating extremely flammable gases
R15/29	Contact with water liberates toxic, extremely flammable gas
R20/21	Harmful by inhalation and in contact with skin
R20/21/22	Harmful by inhalation, in contact with skin and if swallowed
R20/22	Harmful by inhalation and if swallowed
R21/22	Harmful in contact with skin and if swallowed
R23/24	Toxic by inhalation and in contact with skin
R23/24/25	Toxic by inhalation, in contact with skin, and if swallowed
R23/25	Toxic by inhalation and if swallowed
R24/25	Toxic in contact with skin and if swallowed
R26/27	Very toxic by inhalation and in contact with skin
R26/27/28	Very toxic by inhalation, in contact with skin and if swallowed
R26/28	Very toxic by inhalation and if swallowed
R27/28	Very toxic in contact with skin and if swallowed
R36/37	Irritating to eyes, respiratory system
R36/37/38	Irritating to eyes, respiratory system, and skin
R36/38	Irritating to eyes and skin
R37/38	Irritating to respiratory system and skin

(*Continued*)

TABLE 190 (*Continued*)

Risk (R) Phrases Previously Used in the European Community (EU)

R39/23	Toxic: danger of very serious irreversible effects through inhalation
R39/23/24	Toxic: danger of very serious irreversible effects through inhalation and in contact with skin
R39/23/24/25	Toxic: danger of very serious irreversible effects through inhalation, in contact with skin and if swallowed
R39/23/25	Toxic: danger of very serious irreversible effects through inhalation and if swallowed
R39/24	Toxic: danger of very serious irreversible effects in contact with skin
R39/24/25	Toxic: danger of very serious irreversible effects in contact with skin and if swallowed
R39/25	Toxic: danger of very serious irreversible effects if swallowed
R39/26	Very toxic: danger of very serious irreversible effects through inhalation
R39/26/27	Very toxic: danger of very serious irreversible effects through inhalation and in contact with skin
R39/26/27/28	Very toxic: danger of very serious irreversible effects through inhalation, in contact with skin and if swallowed
R39/26/28	Very toxic: danger of very serious irreversible effects through inhalation and if swallowed
R39/27	Very toxic: danger of very serious irreversible effects in contact with skin
R39/27/28	Very toxic: danger of very serious irreversible effects in contact with skin and if swallowed
R39/28	Very toxic: danger of very serious irreversible effects if swallowed
R68/20	Harmful: possible risk of irreversible effects through inhalation
R68/20/21	Harmful: possible risk of irreversible effects through inhalation and in contact with skin
R68/20/21/22	Harmful: possible risk of irreversible effects through inhalation, in contact with skin, and if swallowed
R68/20/22	Harmful: possible risk of irreversible effects through inhalation and if swallowed
R68/22	Harmful: possible risk of irreversible effects if swallowed
R68/21	Harmful: possible risk of irreversible effects in contact with skin
R68/21/22	Harmful: possible risk of irreversible effects in contact with skin and if swallowed
R42/43	May cause sensitization by inhalation and skin contact
R48/20	Harmful: danger of serious damage to health by prolonged exposure through inhalation

(*Continued*)

TABLE 190 (*Continued*)

Risk (R) Phrases Previously Used in the European Community (EU)

R48/20/21	Harmful: danger of serious damage to health by prolonged exposure through inhalation and in contact with skin
R48/20/21/22	Harmful: danger of serious damage to health by prolonged exposure through inhalation, in contact with skin, and if swallowed
R48/20/22	Harmful: danger of serious damage to health by prolonged exposure through inhalation and if swallowed
R48/21	Harmful: danger of serious damage to health by prolonged exposure in contact with skin
R48/21/22	Harmful: danger of serious damage to health by prolonged exposure in contact with skin and if swallowed
R48/22	Harmful: danger of serious damage to health by prolonged exposure if swallowed
R48/23	Toxic: danger of serious damage to health by prolonged exposure through inhalation
R48/23/24	Toxic: danger of serious damage to health by prolonged exposure through inhalation and in contact with skin
R48/23/24/25	Toxic: danger of serious damage to health by prolonged exposure through inhalation, in contact with skin, and if swallowed
R48/23/25	Toxic: danger of serious damage to health by prolonged exposure through inhalation and if swallowed
R48/24	Toxic: danger of serious damage to health by prolonged exposure in contact with skin
R48/24/25	Toxic: danger of serious damage to health by prolonged exposure in contact with skin and if swallowed
R48/25	Toxic: danger of serious damage to health by prolonged exposure if swallowed
R50/53	Very toxic to aquatic organisms, may cause long-term adverse effects in the aquatic environment
R51/53	Toxic to aquatic organisms, may cause long-term adverse effects in the aquatic environment
R52/53	Harmful to aquatic organisms, may cause long-term adverse effects in the aquatic environment

TABLE 191

GHS Hazard (H) Statements

Physical Hazards

H200	Unstable explosive
H201	Explosive; mass explosion hazard
H202	Explosive; severe projection hazard
H203	Explosive; fire, blast, or projection hazard
H204	Fire or projection hazard
H205	May mass explode in fire
H220	Extremely flammable gas
H221	Flammable gas
H222	Extremely flammable aerosol
H223	Flammable aerosol
H224	Extremely flammable liquid and vapor
H225	Highly flammable liquid and vapor
H226	Flammable liquid and vapor
H227	Combustible liquid
H228	Flammable solid
H240	Heating may cause an explosion
H241	Heating may cause a fire or explosion
H242	Heating may cause a fire
H250	Catches fire spontaneously if exposed to air
H251	Self-heating; may catch fire
H252	Self-heating in large quantities; may catch fire
H260	In contact with water releases flammable gases which may ignite spontaneously
H261	In contact with water releases flammable gas
H270	May cause or intensify fire; oxidizer
H271	May cause fire or explosion; strong oxidizer
H272	May intensify fire; oxidizer
H280	Contains gas under pressure; may explode
H281	Contains refrigerated gas; may cause cryogenic burns or injury
H290	May be corrosive to metals

Health Hazards

H300	Fatal if swallowed
H301	Toxic if swallowed
H302	Harmful if swallowed
H303	May be harmful if swallowed
H304	May be fatal if swallowed and enters airways
H305	May be harmful if swallowed and enters airways
H310	Fatal in contact with skin

(Continued)

TABLE 191 (*Continued*)

GHS Hazard (H) Statements

H311	Toxic in contact with skin
H312	Harmful in contact with skin
H313	May be harmful in contact with skin
H314	Causes severe burns and eye damage
H315	Causes skin irritation
H316	Causes mild skin irritation
H317	May cause an allergic skin reaction
H318	Causes serious eye damage
H319	Causes serious eye irritation
H320	Causes eye irritation
H330	Fatal if inhaled
H331	Toxic if inhaled
H332	Harmful if inhaled
H333	May be harmful if inhaled
H334	May cause allergy or asthma symptoms or breathing difficulties if inhaled
H335	May cause respiratory irritation
H336	May cause drowsiness or dizziness
H340	May cause genetic defects
H341	Suspected of causing genetic defects
H350	May cause cancer
H351	Suspected of causing cancer
H360	May damage fertility or the unborn child
H361	Suspected of damaging fertility or the unborn child
H362	May cause harm to breast-fed children
H370	Causes damage to organs
H371	May cause damage to organs
H372	Causes damage to organs through prolonged or repeated exposure
H373	May cause damage to organs through prolonged or repeated exposure

Environmental Hazards

H400	Very toxic to aquatic life
H401	Toxic to aquatic life
H402	Harmful to aquatic life
H410	Very toxic to aquatic life and long-lasting effects
H411	Toxic to aquatic life with long-lasting effects
H412	Harmful to aquatic life with long-lasting effects
H413	May cause long-lasting harmful effects to aquatic life
H420	Harms public health and the environment by destroying ozone in the upper atmosphere

TABLE 192

Information Disclosed on a Safety Data Sheet (SDS)[a]

Section 1	*Chemical Product and Company Identification*
	Product Name
	Generic Names/Synonyms
	Product Use
	Manufacturer's Name and Address
	Name and Phone Number of the Person/Group Who Prepared the SDS
	Date SDS was Prepared
	Emergency Phone Number
Section 2	*Hazards Identification*
	Potential Human Health Hazards
	To Skin (irritancy, sensitization)
	To Eyes (irritancy)
	via Inhalation (acute effects)
	via Ingestion (acute effects)
	Delayed Effects (chronic effects)
	Carcinogenicity, reproductive and developmental effects, mutagenicity, other
Section 3	*Composition/Information on Ingredients*
	Ingredient Name(s)
	CAS Number(s)
	Percent by Weight
Section 4	*First Aid Measures*
	Specific First Aid Measures for Various Routes of Exposure
	Notes to Physician Including Antidotes and Medical Conditions Affected by the Product
Section 5	*Fire Fighting Measures*
	Flammable Properties (flashpoint, autoignition temperature, etc.)
	Extinguishing Media
	Hazardous Combustion Products
	Explosion Hazards
	Firefighting Precautions and Instructions
Section 6	*Accidental Release Measures*
	Procedures to be Followed in Case of Spill or Other Release

(Continued)

TABLE 192 (*Continued*)

Information Disclosed on a Safety Data Sheet (SDS)[a]

Section 7	*Handling and Storage*	
	Normal Handling Procedures	
	Storage Recommendations	
Section 8	*Exposure Controls/Personal Protection*	
	Engineering Controls	
	Personal Protective Equipment	
	Exposure Guidelines (TLV, PEL, other)	
Section 9	*Physical and Chemical Properties*	
	Appearance	Boiling Point
	Physical State	Melting Point
	Odor	Vapor Pressure
	Specific Gravity	Vapor Density
	Solubility	Evaporation Rate
	pH	% Volatiles
Section 10	*Stability and Reactivity*	
	Stability Conditions	
	Incompatibilities	
	Hazardous Decomposition Products	
	Hazardous Polymerization	
Section 11	*Toxicity Information*	
	Acute Effects (LD_{50}, LC_{50})	
	Subchronic and Chronic Effects	
	Irritancy	
	Sensitization	
	Neurotoxicity	
	Reprotoxicity	
	Developmental Toxicity	
	Mutagenicity	
Section 12	*Ecological Information*	
	Aquatic Toxicity	
	Terrestrial Toxicity	

(*Continued*)

TABLE 192 (*Continued*)

Information Disclosed on a Safety Data Sheet (SDS)[a]

	Bioaccumulation Potential
	Biodegradability
	Microbial Toxicity
Section 13	*Disposal Information*
Section 14	*Shipping Information*
	D.O.T. Hazard Class
	D.O.T. I.D. Number
Section 15	*Regulatory Information*
	TSCA Inventory Status
	Other Federal, State, Local
	Foreign Regulatory Information
Section 16	*Information not covered in the other 15 sections*

[a] Formerly known as Material Safety Data Sheet (MSDS)

References

Ecobichon, D.J. (1992), *The Basis of Toxicity Testing*, CRC Press, Boca Raton, FL, chap. 2.

European Economic Community (EEC) (1993), 18th Adaptation to technical progress, Directive 93/21/EEC, *Off. J. Eur. Econ. Commun.* 36, No. L110A/61, May 5.

Hodge, H.C. and Sterner, J.H. (1949), *Am. Ind. Hyg. Assoc. Q.*, 10, 4.

Schurger, M.G. and McConnell, F. (1989), Eastman Chemicals, Kingsport, TN.

U.S. Environmental Protection Agency (1988), Reported in *Pesticide and Toxic Chemical News*, 19, 34, October.

U.S. Environmental Protection Agency (2007), Evaluating Pesticides for Carcinogenic Potential, U.S. Environmental Protection Agency, www.epa.gov/pesticides/health/cancerfs.htm.

U.S. Environmental Protection Agency (2010), TSCA New Chemical Program (NCP) Chemical Categories, Office of Pollution Prevention and Toxics, August.

U.S. Environmental Protection Agency (2011), *Design for the Environment Program, Alternatives Assessment Criteria for Hazard Evaluation*, Version 2.0, Office of Pollution Prevention and Toxics, August.

United Nations (2011), *Globally Harmonized System of Classification and Labelling of Chemicals (GHS)*, fourth revision edition, New York and Geneva.

14

Pharmaceutical and Related Toxicology

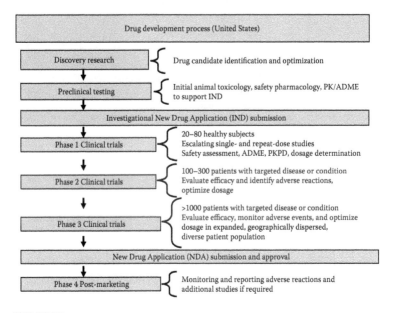

FIGURE 20
An overview of the drug development process in the United States.

TABLE 193

Typical Timing of Nonclinical Toxicology and Safety Pharmacology Studies in Drug Development

Development Phase	Study Type
Discovery	- hERG/other cardiac ion channel assays
	- *In vitro* bacterial mutagenicity screening assay
	- *In vitro* chromosome aberration screening assay
	- *In vitro* cellular toxicity
For IND and prior to phase I clinical studies	- Single-dose acute studies in rodent and nonrodent
	- 7-day dose range-finding studies in rodent and nonrodent
	- 2-week/4-week repeat-dose studies in rodent and nonrodent
	- Bacterial mutagenicity
	- *In vitro* chromosome aberrations
	- *In vivo* micronucleus
	- hERG IC$_{50}$
	- CNS, respiratory, and cardiovascular safety pharmacology (renal and GI can be addressed in repeated dose studies)
	- Intravenous/perivascular irritation, rat or rabbit (if appropriate).
Prior to phase II clinical studies	- Range finding for developmental and reproductive studies
	- Female fertility, rat
	- Embryo-fetal development, rabbit and rat
	- 1- or 3-mo repeat-dose studies in rodent and nonrodent[a]
	- ADME studies
	- Occupational studies (skin, eye irritation, and contact sensitization)
Prior to phase III clinical studies	- 6-mo rodent and 6- or 9-mo nonrodent repeat-dose studies[a]
	- Male fertility, rat
	- Prenatal/perinatal toxicity, rat

(Continued)

TABLE 193 (*Continued*)

Typical Timing of Nonclinical Toxicology and Safety Pharmacology Studies in Drug Development

Development Phase	Study Type
For NDA	- Metabolite studies as appropriate.
	- Excipient/impurity qualification studies as appropriate
	- 3-mo mouse dose range-finding study to support carcinogenicity study (for chronic indication[b])
	- Rat and mouse carcinogenicity (for chronic indication[b])
	- Juvenile rat studies to support pediatric indication (if applicable)
	- Additional studies to address special issues (e.g., immunotoxicology)

[a] Refer to Table 195 for study length.
[b] Chronic indication is commonly defined as continuous treatment for 6 mo or longer, or discontinuous/intermittent treatment for a cumulative total of 6 mo or more.

TABLE 194

Biopharmaceuticals versus Small Molecules: Major Differences That Affect Testing Strategy

	Small Molecule	Biologics
Molecular weight	Low	High
Manufacture	Chemical synthesis	Biosynthetic (cell-based)
Structure/purity of API	Pure single entity	Heterogeneous mixture
Usual route of administration	Oral, dermal, ocular, other	Intravenous, subcutaneous
Target distribution	Intra- and extracellular	Typically extracellular
Elimination	Metabolism (species differences)	Catabolism to amino acids
Toxicity	Off-target (parent or metabolite)	Exaggerated pharmacology
Species specificity	−/+	++++
Immunogenicity	−/+	++++

TABLE 195

Minimal Duration for Repeated Dose Toxicology Studies to Support Drug Clinical Trials and Marketing

Clinical Trial Duration	Rodent		Nonrodent	
	Clinical Trial	Marketing	Clinical Trial	Marketing
Up to 2 wk	2 wk[a]	1 mo	2 wk[a]	1 mo
>2 wk to 1 mo	SACT[b]	3 mo	SACT[b]	3 mo
>1 mo to 3 mo	SACT[b]	6 mo	SACT[b]	6 mo
>3 mo to 6 mo	SACT[b]	6 mo	SACT[b]	9 mo[c]
>6 mo	6 mo	6 mo	9 mo[c]	9 mo[c]

Source: ICH (2010).

Note: As a general rule, the duration of the animal studies should be equal to or exceed the human clinical trials.

[a] 2-wk studies are the minimal duration although a single-dose study with extended examinations can support single-dose human trials in the United States.

[b] SACT = Same as clinical trial.

[c] Nonrodent chronic study of 9 mo in duration is generally acceptable in all regions. A study of 6 mo duration is acceptable in the EU and in some special cases in the United States and Japan.

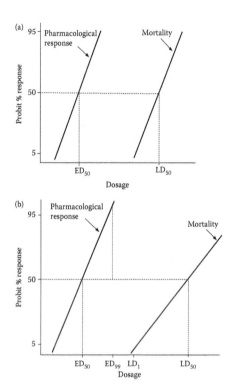

FIGURE 21

A simplistic method for assessing "safety ratios" for drugs is by comparing the ratio of the therapeutically effective dose (ED_{50}) and lethal dose (LD_{50}); this ratio of LD_{50}/ED_{50} is referred to as the therapeutic index (TI_{50}). For parallel pharmacological effect and lethality dose-response lines, the therapeutic index will be similar over a wide range of doses (upper graph). The therapeutic index may be misleading if the dose-response lines for pharmacological and lethal effects are not parallel (lower graph). As this graph shows, the margin based on LD_{50} and ED_{50} may be reasonable. However, due to the shallow slope of the mortality dose-response line, the therapeutic index will be significantly lower at the 1% and 5% level, thus the hyperreactive group may be at greater risk. In such a case, a better index of safety will be the ratio of the LD_1/ED_{99}, which is referred to as the margin of safety. (From Ballatyne, B. (1992). With permission.)

TABLE 196

Content of an Investigational New Drug (IND) Application

IND Section No.	Section Title	Description
1	IND cover sheets (Form FDA 1571)	FDA Form 1571 allows the FDA to quickly identify what is being submitted.
2	Table of contents	
3	Introductory statement and general (clinical) investigation plan	A general description of the drug, formulation to be used in the clinical trial, a description of the anticipated pharmaceutical claim and the clinical hypothesis, and description of the clinical study to be undertaken.
4	Reserved	This section can be used by the FDA for any other consideration for the IND.
5	Clinical Investigator's Brochure (IB)	Written for the clinical investigator, the IB provides an overall description of the clinical and nonclinical safety and efficacy of the drug substance and a description of the chemical and physical properties of the drug substance.
6	Proposed clinical protocol(s)	A detailed description of the proposed clinical study.
7	Chemistry, manufacturing, and control information	Detailed description of chemical and physical properties of the drug substance and drug product formulation, impurity and degradant information, manufacturing information, container closure data, etc.
8	Pharmacology and toxicology information	Description of all toxicology, pharmacology, and pharmacokinetic data. Impurities, degradants, and metabolites are also included.
9	Previous human experience with the investigational drug	Any clinical information that has been generated for the drug substance in studies outside the United States or conducted under a Screening IND (e.g., Phase 0 studies).
10	Additional information	These sections can describe the use of radiolabeled drugs in the clinical trial and a description of the safety of the radiation-absorbed dose, drug dependency, pediatric studies, and an assessment for the use of the drug in pediatric populations, etc. These sections permit the sponsor to tell the Agency about special circumstances that potentially impacts the approval of the IND.
11	Other relevant information	

TABLE 197

Order for Reporting Toxicology Data for an IND/NDA

Study Type	Species[a]	Route of Administration	Dose/group (present in order of increasing dose)	Study Results
Acute studies	Mouse	Intended route for human use followed by:	Untreated control	Mortality
Repeat-dose studies (*in order of increasing duration*)	Rat		Vehicle control	Body weight
	Hamster	Oral	Low dose	Food/water consumption
-Subchronic	Other rodents	Intravenous	Middle doses	Physical examinations (*including clinical observations, EEG, ophthalmic exam, etc.*)
-Chronic	Rabbit	Intramuscular	High dose	
Genotoxicity studies	Dog	Intraperitoneal	Positive or comparative controls	
-*In vitro*	Monkey	Subcutaneous		Hematology
-*In vivo*	Other nonrodent mammals	Inhalation		Serum chemistry
Carcinogenicity studies	Nonmammals	Topical		Urinalysis
Reproduction studies		Other *in vivo*		Organ weights
- Fertility and early embryonic development		*In vitro*		Gross pathology
				Histopathology

(Continued)

TABLE 197 (*Continued*)

Order for Reporting Toxicology Data for an IND/NDA

Study Type	Species[a]	Route of Administration	Dose/group (present in order of increasing dose)	Study Results
- Embryofetal development				
- Prenatal and postnatal development				
Local tolerance studies				
Other toxicity studies (*e.g.,immunogenicity and studies on metabolites and impurities*)				

[a] Males preceding females for each species. Data for adult animals should precede that for infant, geriatric, or disease model animals (if applicable).

TABLE 198

Content of the Common Technical Document (CTD)

Module Number	Description
1	This module contains documents specific to each region; for example, application forms or the proposed label for use in the region. The content and format of this module can be modified to be consistent with the needs of the regulatory authority.
2	CTD Summaries: Contain the bulk of the nonclinical and clinical information, provided in tabular and narrative form. It should begin with a general introduction to the pharmaceutical, including its pharmacological class, mode of action, and a thorough description of all nonclinical data.
3	Information on the drug product, physiochemical properties, drug formulation information, etc.
4	Nonclinical Study Reports: The nonclinical study reports are to be provided in this section.
5	Clinical Study Reports: If clinical data are available, the study reports and related information are to be included in this section.

TABLE 199

Pharmacokinetics: Basic Overview

Pharmacokinetics/toxicokinetics: *What the body does to a drug/chemical.*
Pharmacodynamics/toxicodynamics: *What a drug/chemical does to a body.*

Cmax: *Maximum drug plasma concentration.*
Tmax: *Time for drug to reach the maximum plasma concentration.*
AUC: *Area under the plasma concentration curve for a defined period beginning at time zero.*
T1/2: *Plasma half-life of drug.*
  The preceding four parameters can be determined directly from data generated from bioan alytical analysis of drug in plasma at various times after dose administration.

Total clearance (Cl): *The overall rate at which drug is cleared from the plasma.*

$$Cl = dose/AUC$$

Volume of distribution (Vd): *The ratio of the amount of drug in the body at any time and the corresponding plasma concentration.*

$$Vd = Cl/(0.693/T1/2)$$

Bioavailability (F): *Fraction of administered dose that reaches systemic circulation.*

$$F = AUC_{ev}/AUC_{iv}$$

where ev = extravascular administration
iv = intravenous administration

TABLE 200

Comparative Physiological Values for Frequently Used Pharmacokinetic Determinations

	Mouse (20 g)	Rat (250 g)	Rabbit (2.5 kg)	Monkey (5 kg)	Dog (10 kg)	Human (70 kg)
Total body water (mL)	14.5	167	1,790	3,465	6,036	42,000
Intracellular fluid (mL)	—	92.8	1,165	2,425	3,276	23,800
Extracellular fluid (mL)	—	74.2	625	1,040	2,760	18,200
Plasma volume (mL)	1.0	7.8	110	224	515	3,000
Urine flow (mL/d)	1.0	50.0	150	375	300	1,400
Bile flow (mL/d)	2.0	22.5	300	125	120	350
GFR (mL/min)	0.28	1.31	7.8	10.4	61.3	125
Cardiac blood flow (mL/min)	8.0	74.0	530	1,086	1,200	5,600
Liver blood flow (mL/min)	1.8	13.8	177	218	309	1,450
Renal blood flow (mL/min)	1.3	9.2	80	138	216	1,240

Source: Adapted from Davies, B. and Morris, T. (1993).

TABLE 201

Table for Predicting Human Half-Life of Xenobiotics from Rat Half-Life

Rat Half-Life (hours)	Lower			Human Half-Life Estimate (hours)	Upper		
	95%	90%	80%		80%	90%	95%
0.01	0.019	0.025	0.034	0.106	0.327	0.451	0.598
0.02	0.034	0.045	0.062	0.189	0.574	0.790	1.045
0.03	0.048	0.064	0.087	0.264	0.799	1.098	1.450
0.04	0.062	0.081	0.111	0.335	1.011	1.387	1.830
0.05	0.074	0.098	0.134	0.404	1.213	1.664	2.193
0.06	0.087	0.114	0.156	0.469	1.408	1.930	2.543
0.07	0.099	0.130	0.178	0.533	1.598	2.189	2.882
0.08	0.110	0.146	0.199	0.596	1.782	2.441	3.213
0.09	0.122	0.161	0.220	0.657	1.963	2.687	3.536
0.1	0.13	0.18	0.24	0.72	2.14	2.93	3.85
0.2	0.24	0.31	0.43	1.27	3.78	5.17	6.79
0.3	0.34	0.44	0.60	1.78	5.28	7.21	9.47
0.4	0.43	0.56	0.77	2.26	6.70	9.14	12.00
0.5	0.51	0.68	0.92	2.72	8.05	10.99	14.42
0.6	0.60	0.79	1.07	3.17	9.36	12.77	16.76
0.7	0.68	0.89	1.22	3.60	10.63	14.51	19.04
0.8	0.76	1.00	1.36	4.02	11.88	16.20	21.26
0.9	0.84	1.10	1.50	4.44	13.09	17.86	23.44
1	0.92	1.20	1.64	4.84	14.29	19.49	25.57

(Continued)

TABLE 201 (Continued)

Table for Predicting Human Half-Life of Xenobiotics from Rat Half-Life

Rat Half-Life (hours)	Lower			Human Half-Life Estimate (hours)	Upper		
	95%	90%	80%		80%	90%	95%
2	1.63	2.13	2.91	8.60	25.40	34.66	45.47
3	2.27	2.98	4.07	12.04	35.59	48.58	63.76
4	2.88	3.78	5.16	15.28	45.24	61.76	81.09
5	3.46	4.54	6.20	18.39	54.49	74.42	97.74
6	4.02	5.28	7.21	21.39	63.46	86.69	113.88
7	4.56	5.99	8.19	24.31	72.18	98.64	129.61
8	5.08	6.68	9.14	27.15	80.71	110.32	144.99
9	5.60	7.36	10.06	29.94	89.07	121.77	160.08
10	6.1	8.0	11.0	32.7	97.3	133.0	174.9
20	10.7	14.1	19.4	58.1	174.0	238.3	313.9
30	14.9	19.7	27.0	81.3	244.6	335.6	442.4
40	18.8	24.9	34.1	103.2	311.6	427.9	564.7
50	22.6	29.8	41.0	124.1	376.1	516.9	682.7
60	26.2	34.6	47.5	144.4	438.6	603.3	797.2
70	29.6	39.2	53.9	164.1	499.5	687.5	909.1
80	33.0	43.6	60.1	183.3	559.2	770.0	1018.7
90	36.3	48.0	66.1	202.1	617.7	851.1	1126.4
100	39.5	52.3	72.0	220.6	675.2	930.8	1232.4
200	68.9	91.4	126.5	391.9	1214.2	1679.5	2230.5

(Continued)

TABLE 201 (Continued)

Table for Predicting Human Half-Life of Xenobiotics from Rat Half-Life

Rat Half-Life (hours)	Lower			Human Half-Life Estimate (hours)	Upper		
	95%	90%	80%		80%	90%	95%
300	95.2	126.7	175.7	548.6	1712.7	2374.3	3159.3
400	119.8	159.7	221.7	696.4	2186.8	3036.7	4046.6
500	143.2	191.0	265.6	837.9	2643.8	3676.1	4904.4
600	165.5	221.1	307.7	974.7	3087.5	4297.9	5739.6
700	187.1	250.1	348.5	1107.6	3520.5	4905.5	6556.7
800	208.1	278.3	388.1	1237.3	3944.6	5501.3	7358.6
900	228.5	305.8	426.8	1364.3	4361.2	6086.9	8147.4
1000	248.4	332.7	464.6	1488.9	4771.1	6663.7	8925.0
1100	267.9	359.0	501.7	1611.3	5175.1	7232.7	9692.5

Source: From Bachmann, K.M. et al. (1996). With permission.

Note: The following examples indicate how this table is used. For a xenobiotic with a rat half-life of 0.8 h, the prediction or best guess of the human half-life is 4.02 h. The table indicates that the actual half-life would fall between 1.0 and 16.2 h with a confidence of 90%. Values falling between those indicated in the table can be linearly interpolated, for example, a rat half-life of 2.7 h gives a human half-life of 11.01 h.

TABLE 202

Table for Predicting Human Volume of Distribution from Rat Volume of Distribution

Rat Volume (L/kg)	Lower			Human Volume Estimate (L/kg)	Upper		
	95%	90%	80%		80%	90%	95%
0.01	0.002	0.003	0.004	0.011	0.031	0.041	0.054
0.02	0.004	0.005	0.007	0.020	0.057	0.076	0.099
0.03	0.006	0.008	0.011	0.029	0.082	0.109	0.141
0.04	0.008	0.010	0.014	0.038	0.105	0.141	0.182
0.05	0.010	0.013	0.017	0.046	0.128	0.172	0.222
0.06	0.012	0.015	0.020	0.055	0.151	0.202	0.261
0.07	0.013	0.017	0.023	0.063	0.174	0.232	0.299
0.08	0.015	0.019	0.026	0.071	0.196	0.261	0.337
0.09	0.017	0.022	0.029	0.079	0.217	0.290	0.374
0.1	0.019	0.024	0.032	0.087	0.239	0.319	0.411
0.2	0.035	0.045	0.060	0.164	0.445	0.593	0.762
0.3	0.051	0.065	0.087	0.236	0.641	0.853	1.096
0.4	0.066	0.085	0.113	0.307	0.831	1.105	1.419
0.5	0.081	0.104	0.139	0.376	1.016	1.352	1.735
0.6	0.096	0.123	0.164	0.443	1.198	1.593	2.045
0.7	0.111	0.142	0.189	0.510	1.377	1.832	2.350
0.8	0.125	0.160	0.213	0.575	1.554	2.067	2.652
0.9	0.139	0.178	0.237	0.640	1.729	2.299	2.950

(Continued)

TABLE 202 (Continued)

Table for Predicting Human Volume of Distribution from Rat Volume of Distribution

Rat Volume (L/kg)	Lower			Human Volume Estimate (L/kg)	Upper		
	95%	90%	80%		80%	90%	95%
1	0.15	0.20	0.26	0.70	1.90	2.53	3.25
2	0.29	0.37	0.49	1.32	3.57	4.75	6.09
3	0.41	0.53	0.70	1.91	5.16	6.87	8.82
4	0.53	0.69	0.91	2.48	6.71	8.93	11.47
5	0.65	0.84	1.18	3.03	8.23	10.96	14.08
6	0.77	0.99	1.32	3.58	9.72	12.95	16.64
7	0.88	1.14	1.51	4.11	11.19	14.92	19.17
8	0.99	1.28	1.71	4.64	12.65	16.86	21.68
9	1.11	1.42	1.90	5.17	14.09	18.79	24.17
10	1.21	1.56	2.08	5.69	15.52	20.70	26.63
20	2.25	2.90	3.88	10.66	29.32	39.20	50.54
30	3.22	4.16	5.57	15.40	42.60	57.04	73.63
40	4.15	5.37	7.20	19.99	55.54	74.46	96.23
50	5.06	6.54	8.78	24.48	68.24	91.59	118.48
60	5.94	7.69	10.33	28.88	80.77	108.50	140.45
70	6.80	8.81	11.84	33.22	93.14	125.22	162.21
80	7.65	9.91	13.34	37.49	105.39	141.79	183.79
90	8.48	11.00	14.81	41.72	117.54	158.23	205.22

Source: From Bachmann, K.M. et al. (1996). With permission.
Note: See note of Table 201 for examples of how this table is used.

TABLE 203

Animal/Human Dose Conversions

Species	To Convert Animal Dose (mg/kg) to Dose in mg/m², Multiply by K_m shown below	To Convert Animal Dose (mg/kg) to HED[a] (mg/kg), do either:	
		Divide Animal Dose by	Multiply Animal Dose by
Human			
Adult (60 kg)	37	—	—
Child (20 kg)[b]	25	—	—
Mouse	3	12.3	0.08
Hamster	5	7.4	0.13
Rat	6	6.2	0.16
Ferret	7	5.3	0.19
Guinea pig	8	4.6	0.22
Rabbit	12	3.1	0.32
Dog	20	1.8	0.54
Primates			
Monkey[c]	12	3.1	0.32
Marmoset	6	6.2	0.16
Squirrel monkey	7	5.3	0.19
Baboon	20	1.8	0.54
Micropig	27	1.4	0.73
Minipig	35	1.1	0.95

Source: U.S. Food and Drug Administration (2005).

[a] Assumes a 60 kg human. For species not listed or for weights outside the standard ranges, HED (human equivalent dose) can be calculated from the following formula: HED = animal dose (mg/kg) × (animal weight (kg)/human weight (kg))$^{0.33}$

[b] This km is provided for reference only because healthy children will rarely be volunteers in phase I trials.

[c] Includes cynomolgus, rhesus, and stumptail.

TABLE 204

FDA Pregnancy Categories

Category A:	Adequate and well-controlled studies in pregnant women have failed to demonstrate a risk to the fetus in the first trimester of pregnancy (and there is no evidence of a risk in later trimesters).
Category B:	Animal reproduction studies have failed to demonstrate a risk to the fetus, and there are no adequate and well-controlled studies in pregnant women.
	Or
	Animal reproduction studies have shown an adverse effect (other than decrease in fertility), but adequate and well-controlled studies in pregnant women have failed to demonstrate a risk to the fetus during the first trimester of pregnancy (and there is no evidence of a risk in later trimesters).
Category C:	Animal reproduction studies have shown an adverse effect on the fetus, there are no adequate and well-controlled studies in humans, but the benefits from the use of the drug in pregnant women may be acceptable despite its potential risks.
	Or
	There are no animal reproduction studies, and no adequate and well-controlled studies in humans.
Category D:	There is positive evidence of human fetal risk based on adverse reaction data from investigational or marketing experience or studies in humans, but the potential benefits from the use of the drug in pregnant women may be acceptable despite its potential risks (e.g., the drug is needed in a life-threatening situation or serious disease for which safer drugs cannot be used or are ineffective).
Category X:	Studies in animals or humans have demonstrated fetal abnormalities or there is positive evidence of fetal risk based on adverse reaction reports from investigational or marketing experience, or both, and the risk of the use of the drug in pregnant women clearly outweighs any possible benefit (e.g., safer drugs or other forms of therapy are available). This drug is contraindicated in women who are or may become pregnant.

Source: U.S. CFR (2006a).

TABLE 205

DEA Controlled Substances Schedules

Schedule I	Drug or substance has a high potential for abuse, has no currently accepted medical use in treatment in the United States, and there is a lack of accepted safety for its use under medical supervision (e.g., heroin, lysergic acid diethylamide/LSD, methaqualone).
Schedule II	Drug or substance has a high potential for abuse and has a currently accepted medical use in treatment in the United States with or without severe restrictions. Abuse of the drug or substance may lead to severe psychological or physical dependence (e.g., morphine, cocaine, methadone, methamphetamine).
Schedule III	The drug or substance has less potential for abuse than drugs or other substances in Schedules I and II and has a currently accepted medical use in treatment in the United States. Abuse of the drug or substance may lead to moderate or low physical dependence or high psychological dependence (e.g., codeine, hydrocodone with acetaminophen).
Schedule IV	The drug or substance has a low potential for abuse relative to the drugs or other substances in Schedule III and has a currently accepted medical use in treatment in the United States. Abuse of the drug or substance may lead to limited physical dependence or psychological dependence relative to the drugs or other substances in Schedule III (e.g., Darvon®, Valium®).
Schedule V	The drug or substance has a low potential for abuse relative to other drugs or other substances in Schedule IV and has a currently accepted medical use in treatment in the United States. Abuse of the drug or substance may lead to limited physical dependence or psychological dependence relative to drugs or other substances in Schedule IV. (e.g., cough medicines with codeine).

Source: U.S. Drug Enforcement Agency (2007).

TABLE 206

Respiratory Function Parameters Collected by Whole-Body Plethysmograph in the Sprague-Dawley Rat

End Point	Number of Animals	Mean	Standard Deviation	Minimum	Maximum
Male					
Respiratory rate (breaths/min)	621	140	45	54	430
Tidal volume (mL)	621	1.1	0.31	0.34	3.1
Minute volume (mL/min)	621	140	36	47	530
Female					
Respiratory rate (breaths/min)	80	130	51	64	360
Tidal volume (mL)	80	1.1	0.24	0.41	2.2
Minute volume (mL/min)	80	120	30	32	310

Source: Baird, T.J. and Gauvin, D.V. (2014).
Note: Statistics were generated from recording intervals of at least 1 h duration in male and female rats between 7 and 10 weeks of age.

TABLE 207

Respiratory Function Parameters Collected by Whole-Body Plethysmograph in the CD-1 Mouse

End Point	Number of Animals	Mean	Standard Deviation	Minimum	Maximum
Respiratory rate (breaths/min)	56	310	110	140	570
Tidal volume (mL)	56	0.21	0.039	0.11	0.37
Minute volume (mL/min)	56	52	22	21	120

Source: Baird, T.J. and Gauvin, D.V. (2014).
Note: Statistics were generated from recording intervals of at least 1 h duration in male and female mice between 7 and 10 weeks of age.

TABLE 208

Respiratory Function Parameters Collected by Whole-Body Plethysmograph in the Guinea Pig

End Point	Number of Animals	Mean	Standard Deviation	Minimum	Maximum
Respiratory rate (breaths/min)	8	96	21	63	190
Tidal volume (mL)	8	3.9	0.45	2.3	5.3
Minute volume (mL/min)	8	350	54	220	550

Source: Baird, T.J. and Gauvin, D.V. (2014).
Note: Statistics were generated from recording intervals of at least 1 h duration in male Crl:HA (Albino Hartley) Guinea pigs between 1 and 3 months of age.

TABLE 209

Respiratory Function Parameters Collected by Head Plethysmograph in the Beagle Dog

End Point	Number of Animals	Mean	Standard Deviation	Minimum	Maximum
Male					
Respiratory rate (breaths/min)	42	73	66	11	290
Tidal volume (mL)	42	160	110	19	490
Minute volume (mL/min)	42	7900	5500	1700	25,000
Female					
Respiratory rate (breaths/min)	28	58	56	11	270
Tidal volume (mL)	28	110	41	27	250
Minute volume (mL/min)	28	4000	2000	1000	8,500

Source: Baird, T.J. and Gauvin, D.V. (2014).
Note: Statistics were generated from recording intervals of at least 1 h duration in male and female beagle dogs between 6 months and 4 years of age.

TABLE 210

Respiratory Function Parameters Collected by Head Plethysmograph in the Nonhuman Primate (Cynomolgus Macaque)

End Point	Number of Animals	Mean	Standard Deviation	Minimum	Maximum
Male					
Respiratory rate (breaths/min)	58	42	9.4	24	71
Tidal volume (mL)	58	28	8.8	8.6	65
Minute volume (mL/min)	58	1200	510	220	2800
Female					
Respiratory rate (breaths/min)	40	38	9	23	67
Tidal volume (mL)	40	24	6.2	11	39
Minute volume (mL/min)	40	920	380	360	2400

Source: Baird, T.J. and Gauvin, D.V. (2014).

Note: Statistics were generated from recording intervals of at least 1 h duration in male and female monkeys between 2 and 5 years of age.

TABLE 211

Cardiovascular Parameters Collected by Implanted Telemetry in the Beagle Dog

End Point	Number of Animals	Mean	Standard Deviation	Minimum	Maximum
Male					
Systolic arterial pressure (mm Hg)	174	140	17	86	210
Diastolic arterial pressure (mm Hg)	174	74	12	30	140
Mean arterial pressure (mm Hg)	174	98	14	51	160
Heart rate (beats/min)	174	84	19	38	180
PR interval (ms)	174	100	15	67	160
QRS interval (ms)	174	43	4.6	18	64
QT interval (ms)	174	230	18	150	310
QT interval—corrected (ms)	174	230	14	170	300
Body temperature (°C)	174	37	1.1	33	40
Female					
Systolic arterial pressure (mm Hg)	40	130	15	89	190
Diastolic arterial pressure (mm Hg)	40	73	11	38	110
Mean arterial pressure (mm Hg)	40	96	12	59	140
Heart rate (beats/min)	40	87	19	47	170
PR interval (ms)	40	110	13	71	150
QRS interval (ms)	40	45	4.6	33	68
QT interval (ms)	40	220	14	170	280
QT interval—corrected (ms)	40	220	8.6	190	260
Body temperature (°C)	40	36	1.3	33	39

Source: Baird, T.J. and Gauvin, D.V. (2014).

Note: Statistics were generated from ~24 h recording intervals in male and female beagle dogs between 6 months and 4 years of age.

TABLE 212

Cardiovascular Parameters Collected by Implanted Telemetry in the Nonhuman Primate (Cynomolgus Macaque)

End Point	Number of Animals	Mean	Standard Deviation	Minimum	Maximum
Male					
Systolic arterial pressure (mm Hg)	91	94	14	54	150
Diastolic arterial pressure (mm Hg)	91	63	11	26	100
Mean arterial pressure (mm Hg)	91	79	12	42	120
Heart rate (beats/min)	91	120	33	56	250
PR interval (ms)	91	86	14	48	150
QRS interval (ms)	91	38	4.6	20	50
QT interval (ms)	91	250	45	120	380
QT interval—corrected (ms)	91	250	32	150	340
Body temperature (°C)	91	38	0.79	35	40
Female					
Systolic arterial pressure (mm Hg)	41	97	12	70	160
Diastolic arterial pressure (mm Hg)	41	65	11	35	110
Mean arterial pressure (mm Hg)	41	81	11	53	130
Heart rate (beats/min)	41	140	34	72	280
PR interval (ms)	41	82	8.3	59	110
QRS interval (ms)	41	37	3.9	25	51
QT interval (ms)	41	240	44	130	350
QT interval—corrected (ms)	41	240	29	150	330
Body temperature (°C)	41	38	0.89	35	40

Source: Baird, T.J. and Gauvin, D.V. (2014).

Note: Statistics were generated from ~24 h recording intervals in male and female cynomolgus monkeys between

TABLE 213

Cardiovascular Parameters Collected by Implanted Telemetry in the Gottingen Minipig

End Point	Number of Animals	Mean	Standard Deviation	Minimum	Maximum
Male					
Systolic arterial pressure (mm Hg)	18	140	14	110	180
Diastolic arterial pressure (mm Hg)	18	95	14	67	140
Mean arterial pressure (mm Hg)	18	120	13	88	160
Heart rate (beats/min)	18	93	15	56	180
PR interval (ms)	18	110	12	75	150
QRS interval (ms)	18	40	7.1	28	80
QT interval (ms)	18	300	27	220	380
QT interval—corrected (ms)	18	300	20	250	360
Body temperature (°C)	18	38	0.80	35	40
Female					
Systolic arterial pressure (mm Hg)	24	140	17	94	190
Diastolic arterial pressure (mm Hg)	24	96	14	45	140
Mean arterial pressure (mm Hg)	24	120	15	77	170
Heart rate (beats/min)	24	96	16	51	170
PR interval (ms)	24	100	9.3	78	140
QRS interval (ms)	24	38	5.6	25	55
QT interval (ms)	24	290	26	210	360
QT interval—corrected (ms)	24	290	24	230	360
Body temperature (°C)	24	38	0.80	35	40

Source: Baird, T.J. and Gauvin, D.V. (2014).

Note: Statistics were generated from ~24 h recording intervals in male and female minipigs between 6 months and 2 years of age.

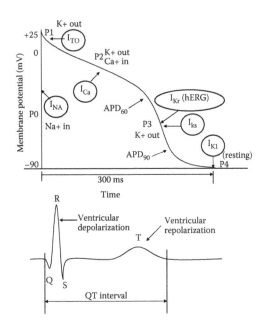

FIGURE 22

Correlation between the cardiac action potential and the QT interval of the surface electrocardiogram (ECG) is depicted. The temporal contributions of the various ion channel currents (I) to the action potential are shown. The action potential consists of five distinct phases (P0 to P4). Depolarization occurs in Phase 0 related to an influx of sodium ions and corresponding to the QRS complex of the ECG. Repolarization occurs in phases 1 to 3, restoring the resting membrane potential. The whole process lasts approximately 300 ms. The T wave of the ECG corresponds to ventricular repolarization. The duration of the QT interval on the ECG is defined as the duration between the beginning of the QRS complex and the end of the T wave. The Human Ether-a-go-go Related Gene Product (hERG) channel is the potassium channel that allows a potassium current to pass that corresponds to the rapidly activating delayed rectifier K^+ current. Significant inhibition of the hERG channel results in prolongation of the QT interval, which is believed to be associated with an increased propensity to develop a fatal ventricular tachyarrhythmia called *Torsades de Pointes*. APD_{60} and APD_{90} values, corresponding to the time where 60% and 90% of the resting membrane potential is restored, are typically reported in the Purkinje Fiber Assay.

TABLE 214

Substances Generally Recognized as Safe (GRAS)

Multiple Purpose Food Substances
Glutamic acid
Glutamic acid hydrochloride
Hydrochloric acid
Phosphoric acid
Sodium acid pyrophosphate
Aluminum sulfate
Aluminum ammonium sulfate
Aluminum potassium sulfate
Aluminum sodium sulfate
Caffeine
Calcium phosphate
Caramel
Glycerin
Methylcellulose
Monoammonium glutamate
Monopotassium glutamate
Silica aerogel
Sodium carboxymethylcellulose
Sodium caseinate
Sodium citrate
Sodium phosphate
Sodium aluminum phosphate
Sodium tripolyphosphate

Sequestrants
Sodium acid phosphate
Calcium diacetate
Calcium hexametaphosphate
Monobasic calcium phosphate
Dipotassium phosphate

(Continued)

TABLE 214 (*Continued*)

Substances Generally Recognized as Safe (GRAS)

Disodium phosphate
Sodium gluconate
Sodium hexametaphosphate
Sodium metaphosphate
Sodium phosphate
Sodium pyrophosphate
Tetra sodium pyrophosphate
Sodium tripolyphosphate

Stabilizers
Chondrus extract

Anticaking Agents
Aluminum calcium silicate
Calcium silicate
Magnesium silicate
Sodium aluminosilicate
Sodium calcium aluminosilicate hydrated
Tricalcium silicate

Chemical Preservatives
Ascorbic acid
Erythorbic acid
Sorbic acid
Thiodiproprionic acid
Ascorbyl palmitate
Butylated hydroxyanisole
Butylated hydroxytoluene
Calcium ascorbate
Calcium sorbate

(Continued)

TABLE 214 (*Continued*)

Substances Generally Recognized as Safe (GRAS)

Dilauryl thiopropionate
Potassium bisulfite
Potassium metabisulfite
Potassium sorbate
Sodium ascorbate
Sodium bisulfite
Sodium metabisulfite
Sodium sorbate
Sodium sulfite
Sulfur dioxide
Tocopherols

Nutrients
Ascorbic acid
Biotin
Calcium phosphate
Calcium pyrophosphate
Choline bitartrate
Choline chloride
Sodium phosphate
Tocopherols
α-Tocopherol acetate
Zinc chloride
Zinc gluconate
Zinc oxide
Zinc stearate
Zinc sulfate

Source: U.S. CFR (2006b).

TABLE 215

The ISO Standard 10993-1 Guidance for Selection of Biocompatibility Tests as Modified by the FDA

Device Categories		Contact Duration[a]	Biological Effect[b]									
Body Contact			Cytotoxicity	Sensitization	Irritation or Intracutaneous	Systemic Toxicity (Acute)	Subchronic Toxicity	Genotoxicity	Implantation	Hemocompatibility	Chronic Toxicity	Carcinogenicity
Surface devices	Skin	A	×	×	×							
		B	×	×	×							
		C	×	×	×							
	Mucosal membrane	A	×	×	×							
		B	×	×	×	o	o		o		o	
		C	×	×	×	o	×	×	o			
	Breached or compromised surfaces	A	×	×	×	o						
		B	×	×	×	o	o		o		o	
		C	×	×	×	o	×	×	o			
External communicating devices	Blood path, indirect	A	×	×	×	×				×		
		B	×	×	×	×	o		o	×	×	
		C	×	×	o	×	×	×		×		
	Tissue/bone/dentin communicating +	A	×	×	×	o						
		B	×	×	o	o	o	×	×			×
		C	×	×	o	o	o	×	×		o	×

(Continued)

TABLE 215 (*Continued*)

The ISO Standard 10993-1 Guidance for Selection of Biocompatibility Tests as Modified by the FDA

Device Categories		Contact Duration[a]	Cytotoxicity	Sensitization	Irritation or Intracutaneous	Systemic Toxicity (Acute)	Subchronic Toxicity	Genotoxicity	Implantation	Hemocompatibility	Chronic Toxicity	Carcinogenicity
Body Contact												
Circulating blood		A	x	x	x	x		o		x		
		B	x	x	x	x	o	x	o	x		
		C	x	x	x	x	x	x	o	x	x	x
Implant devices — Tissue/bone		A	x	x	x	o						
		B	x	x	o	o	o	x	x			
		C	x	x	o	o	o	x	x		x	x
Implant devices — Blood		A	x	x	x	x			x	x		
		B	x	x	x	x	o	x	x	x		
		C	x	x	x	x	x	x	x	x	x	x

Source: Adapted from U.S. Food and Drug Administration (1995).

Note: +, Tissue includes tissue fluids and subcutaneous spaces; ^, for all devices used in extracorporeal circuits.

[a] A, limited (24 h); B, prolonged (24 h to 30 days); C, permanent (>30 days). See text.

[b] x, ISO Evaluation Tests for Consideration; o, additional tests that may be applicable.

References

Bachmann, K.M., Pardoe, D., and White, D. (1996), Scaling basic toxicokinetic parameters from rat to man, *Environ. Health Perspect.* 104, 400–407.

Baird, T.J., Dalton, J.A., and Gauvin, D.V. (2014), Safety pharmacology, in *Handbook of Toxicology, 3rd edn.*, Derelanko, M.J. and Auletta, C.S., Eds., CRC Press, Boca Raton, chap 21.

Ballantyne, B. (1992), Exposure-dose-response relationships, in *Hazardous Materials Toxicology: Clinical Principles of Environmental Health*, Sullivan, J.B. and Krieger, G.R., Eds., Williams and Wilkins, Baltimore, MD, chap 3.

Davies, B. and Morris, T. (1993), Physiologic parameters in laboratory animals and humans, *Pharmacol. Res.* 10, 1093–1095.

ICH (2010), Guidance on the non-clinical safety studies for the conduct of human clinical trials and marketing authorization for pharmaceuticals, M3(R2). International Conference on Harmonization (ICH).

U.S. CFR (2006a), U.S. Code of Federal Regulations (CFR), title 21, Vol. 4, revised as of April 1, 2006.

U.S. CFR (2006b), U.S. Code of Federal Regulations (CFR), title 21, part 182, chap. 1, 2006 update.

U.S. Drug Enforcement Agency (2007), www.usdoj.gov/dea/pubs/abuse/1-csa.htm.

U.S. Food and Drug Administration (1995), *General Program Memorandum G95-1*.

U.S. Food and Drug Administration (2005), *Guidance for Industry, Estimating the Maximum Safe Starting Dose in Initial Clinical Trials for Therapeutics in Adult Healthy Volunteers*, July.

15

Miscellaneous Information

TABLE 216

Comparison of Physiological Parameters for Different Human Body Organs

Organ	Weight (kg)	Percent of Body Volume	Percent Water	Blood Flow (mL/min)	Plasma Flow (mL/min)	Blood Flow (mL/kg)	Blood Flow Fraction
Adrenal glands		0.03		25	15		
Blood	5.4	7	83	5000			
Bone	10	16	22	250	150		
Brain	1.5	2	75	700	420	780	
Fat	10	10	10	200	120		0.05
Heart	0.3	0.5	79	200	120	250	
Kidneys	0.3	0.4	83	1100	660	1200	
Liver	1.5	2.3	68	1350	810	1500	0.25
Portal				1050	630		
Arterial				300	180		
Lungs	1.0	0.7	79	5000	3000		
Muscle	30	42	76	750	450	900	0.19
Skin	5	18	72	300	180	250	
Thyroid gland	0.03	0.03		50	30		
Total body		100	60	5000	3000		

Source: Adapted from Illing (1989).
Note: Data are for hypothetical 70 kg human.

TABLE 217

Comparison of the Blood Flow/Perfusion and Oxygen Consumption of Liver, Lung, Intestine, and Kidney of the Rat *In Vivo* and in Organ Perfusion[a]

Parameter (Unit)	Liver	Lung	Intestine	Kidney
In vivo				
Blood flow (mL/min)	13–20	55–70	5–8	4–6
Blood pressure S/D (torr)	150/100	25/10	150/100	150/100
pO_2 (arterial) (torr)	95	40	95	95
pO_2 (venous) (torr)	40	100	50	70
O_2 consumption (μL/min)	500–800	From air	40–160	100–200
In perfusion				
Perfusion flow (mL/min)	30–50	50	6	20–35
Perfusion pressure (torr)	100–120	10–20	100–120	100–120
pO_2 (arterial) (torr)	600	600	400	600
pO_2 (venous) (torr)	200	?	180	400
Max. O_2 supply[b] (μL/min)	380–630	?	120[c]	120–220

Source: From Niesink, R.J.M. et al. (1996), *Toxicology: Principles and Applications*, CRC Press, Boca Raton, FL.

Note: S = systolic; D = diastolic.

[a] These values are indications of the most common values measured for the various organs in a rat of 250–300 g. The figures provided for the kidney apply to a single kidney. The values measured in organ perfusions may differ greatly, depending on the setup, method of gassing, etc.

[b] Calculated from pO_2 (arterial), pO_2 (venous), and perfusion flow.

[c] With 20% FC-43 emulsion in KRB; other figures apply to KRB buffer without erythrocytes or oxygen carrier (KRB = Krebs–Ringer buffer).

TABLE 218

Comparison of Physiological Characteristics of Experimental Animals and Humans

Species	Body wt. (kg)	Surface Area (m²)	Energy Metabolism[a] (cal/kg/day)	(cal/m²/day)	Heart wt. (g/100 g)	Heart Rate (beats/min)	Stroke Vol (mL/beat)	Cardiac Output (L/min)	Cardiac Index (L/m²/m)	Systolic	Diastolic
Rat	0.1–0.5	0.03–0.06	120–140 (B)	760–905 (B)	0.24–0.58	250–400	1.3–2.0	0.015–0.079	1.6	88–184	58–145
Rabbit	1–4	0.23	47	810	0.19–0.36	123–330	1.3–3.8	0.25–0.75	1.7	95–130	60–90
Monkey	2–4	0.31	49 (B)	675	0.34–0.39	165–240	8.8	1.06	—	137–188	112–152
Dog	5–31	0.39–0.78	34–39 (B)	770–800 (B)	0.65–0.96	72–130	14–22	0.65–1.57	2.9	95–136	43–66
Human	54–94	1.65–1.83	23–26 (B)	790–910 (B)	0.45–0.65	41–108	62.8	5.6	3.3	92–150	53–90
Pig	100–250	2.9–3.2	14–17 (B)	1100–1360 (B)	0.25–0.40	55–86	39–43	5.4	4.8	144–185	98–120
Ox	500–800	4.2–8.0	15 (B)	1635 (B)	0.31–0.53	40–58	244	146	—	121–166	80–120
Horse	650–800	5.8–8.0	25 (R)	2710–2770 (R)	0.39–0.94	23–70	852	188	4.4	86–104	43–86

Source: From Mitruka, B.M. and Rawnsley, H.M. (1977). With permission.

[a] B = basal; R = resting.

TABLE 219

Comparison of Certain Physiological Values of Experimental Animals and Humans

Species	Body Temperature (°C)	Whole Blood Volume (mL/kg body wt)	Plasma Volume (mL/kg body wt)	Plasma pH	Plasma CO_2 Content (mM/L)	CO_2 Pressure (mm Hg)
Mouse	36.5 ± 0.70	74.5 ± 17.0	48.8 ± 17.0	7.40 ± 0.06	22.5 ± 4.50	40.0 ± 5.40
Rat	37.3 ± 1.40	58.0 ± 14.0	31.3 ± 12.0	7.35 ± 0.09	24.0 ± 4.70	42.0 ± 5.70
Hamster	36.0 ± 0.50	72.0 ± 15.0	45.5 ± 7.50	7.39 ± 0.08	37.3 ± 2.50	59.0 ± 5.00
Guinea pig	37.9 ± 0.95	74.0 ± 7.00	38.8 ± 4.50	7.35 ± 0.09	22.0 ± 6.60	40.0 ± 9.80
Rabbit	38.8 ± 0.65	69.4 ± 12.0	43.5 ± 9.10	7.32 ± 0.03	22.8 ± 8.60	40.0 ± 11.5
Chicken	41.4 ± 0.25	95.5 ± 24.0	65.6 ± 12.5	7.52 ± 0.04	23.0 ± 2.50	26.0 ± 4.50
Cat	38.6 ± 0.70	84.6 ± 14.5	47.7 ± 12.0	7.43 ± 0.03	20.4 ± 3.50	36.0 ± 4.60
Dog	38.9 ± 0.65	92.6 ± 29.5	53.8 ± 20.1	7.42 ± 0.04	21.4 ± 3.90	38.0 ± 5.50
Monkey	38.8 ± 0.80	75.0 ± 14.0	44.7 ± 13.0	7.46 ± 0.06	29.3 ± 3.8	44.0 ± 4.8
Pig	39.3 ± 0.30	69.4 ± 11.5	41.9 ± 8.90	7.40 ± 0.08	30.2 ± 2.5	43.0 ± 5.60
Goat	39.5 ± 0.60	71.0 ± 14.0	55.5 ± 13.0	7.41 ± 0.09	25.2 ± 2.8	50.0 ± 9.40
Sheep	38.8 ± 0.80	58.0 ± 8.50	41.9 ± 12.0	7.48 ± 0.06	26.2 ± 5.00	38.0 ± 8.50
Cattle	38.6 ± 0.30	57.4 ± 5.00	38.8 ± 2.50	7.38 ± 0.05	31.0 ± 3.0	48.0 ± 4.80
Horse	37.8 ± 0.25	72.0 ± 15.0	51.5 ± 12.0	7.42 ± 0.03	28.0 ± 4.00	47.0 ± 8.50
Man	36.9 ± 0.35	77.8 ± 15.0	47.9 ± 8.70	7.39 ± 0.06	27.0 ± 2.00	42.0 ± 5.00

Source: From Mitruka, B.M. and Rawnsley, H.M. (1977). With permission.

TABLE 220

Overview of Major Mammalian Hormones

Source	Hormone	Target	Major Effect
Anterior pituitary	Growth hormone (GH)	Multiple sites	Stimulates bone and muscle growth and metabolic functions
	Adrenocorticotropic hormone (ACTH)	Adrenal cortex	Stimulates secretion of adrenal cortex hormones
	Thyroid stimulating hormone (TSH)	Thyroid	Stimulates secretion of thyroid hormone
	Follicle-stimulating hormone (FSH)	Ovaries Testes	Stimulates production of ova and sperm
	Luteinizing hormone (LH)	Ovaries Testes	Stimulates ovulation and production of estrogen, progesterone, and testosterone
	Prolactin (LTH)	Mammary Ovary	Stimulates milk production and maintains estrogen and progesterone secretion
	Melanocyte-stimulating hormone (MSH)	Melanocyte	Stimulates dispersal of pigment
Hypothalamus/ posterior pituitary	Oxytocin	Uterus Mammary	Stimulates contraction of uterus and secretion of milk
	Antidiuretic hormone (ADH)	Kidney	Promotes water retention
Thyroid	Triiodothyronine (T_3) and thyroxin (T_4)	Multiple	Stimulates and maintains metabolism
	Calcitonin	Bone	Lowers blood calcium
Parathyroid	Parathyroid hormone	Bone Kidney Digestive tract	Raises blood calcium

(Continued)

TABLE 220 (*Continued***)**

Overview of Major Mammalian Hormones

Source	Hormone	Target	Major Effect
Ovary	Estrogens (estradiol)	Uterus Multiple sites	Stimulates growth of uterine lining, promotes development and maintenance of secondary sex characteristics
Ovary/placenta	Progesterone	Uterus Breast	Promotes growth of uterine lining
Placenta	Chrorionic gonadotropin	Anterior pituitary	Stimulates release of FSH and LH
Testis	Androgens (testosterone)	Multiple sites	Supports spermatogenesis, promotes development and maintenance of secondary sex characteristics
Adrenal cortex	Glucocorticoids (cortisol)	Multiple sites	Raises blood glucose
	Mineralocorticoids (aldosterone)	Kidney	Maintains sodium and phosphate balance
Adrenal medulla	Epinephrine (adrenalin) Norepinephrine	Muscle Liver Blood vessels	Raises blood glucose, increases metabolism, and constricts certain blood vessels
Pineal gland	Melatonin	Multiple sites	Regulates biorhythms, influences reproduction in some species
Pancreas	Insulin	Multiple sites	Lowers blood glucose
	Glucagon	Liver Fatty tissue	Raises blood glucose
Thymus	Thymosin		Stimulates T-lymphocytes
Gastrointestinal Tract			
Duodenum	Secretin	Pancreas	Stimulates secretion of pancreatic enzymes
	Cholecystokinin	Gallbladder	Stimulates release of bile
Stomach	Gastrin	Stomach Intestinal tract	Stimulates acid secretion and contraction of intestinal tract

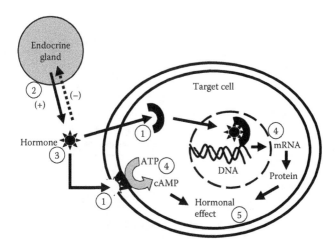

FIGURE 23

A generalized schematic of mammalian hormonal regulation showing major points for potential disruption/modulation. Endocrine disrupters/modulators are defined as exogenous substances that can alter or modulate endocrine function resulting in adverse effects at the level of the organism, its progeny, and/or populations of organisms. The site of action most focused on for these substances is at the target cell hormone receptor (1). A substance that has an affinity for binding to either a peptide-hormone membrane receptor or a steroid-hormone cytoplasmic receptor might act either as an agonist inducing the hormonal action or as an antagonist, preventing the natural hormone from inducing its effect. However, endocrine disrupters/modulators need not interact with a receptor to affect hormonal regulation. They may affect the synthesis and secretion of the hormone or its regulatory control at the endocrine gland (2) or the transport or elimination of the hormone (3) resulting in increased or decreased levels of hormone reaching the target cell. Endocrine disrupters/ modulators might interfere with or alter the cellular mechanisms through which the hormone exerts its effect (4). For steroid hormones, this involves gene activation and synthesis of specific proteins or for peptide hormones, activation of a second-messenger sequence producing cellular effects such as enzyme activation and alteration of cell membrane permeability. Endocrine disrupters/modulators may alter the temporal expression of the hormonal effect (5) such as causing the premature expression of the hormonal effect at a critically sensitive period during sexual development.

TABLE 221

Tissue Localization of Xenobiotic-Metabolizing Enzymes

Relative Amount	Tissue
High	Liver
Medium	Lung, kidney, intestine
Low	Skin, testes, placenta, adrenals
Very low	Nervous system tissues

TABLE 222

Metabolic Phase I and Phase II Reactions

Phase I	Phase II
Oxidation	Glucuronidation
Reduction	Glucosidation
Hydrolysis	Ethereal sulfation
Isomerization	Methylation
Others	Acetylation
	Amino acid conjugation
	Glutathione conjugation
	Fatty acid conjugation
	Condensation

TABLE 223

Major Cytochrome P450 Enzymes

	Substrates	Substrate Examples	Comments
Subfamily CYP1A			
13% of human liver cytochrome P450 enzymes metabolizes ~4% of drugs			
CYP1A1	Polycyclic aromatic hydrocarbons	7-ethoxyresorufin, interferon	Expressed only after exposure to inducers (human), induced in placenta and lung after exposure to cigarette smoke
CYP1A2	Aromatic and heterocyclic amines, neutral or basic lypophylic planar drugs	Phenacetin, caffeine, estradiol	Present in most human livers, induced by cigarette smoke
Subfamily CYP2A			
CYP2A1		Testosterone	Expressed in liver
CYP2A6		Coumarin	Expressed in liver. Important in precarcinogen activation. Exhibits significant ethnic-related genotypic or phenotypic deficiencies
Subfamily CYP1B			
CYP1B1		7-ethoxyresorufin	TCDD inducible in keratinocytes, present in other tissues after exposure to inducers

(Continued)

TABLE 223 (Continued)

Major Cytochrome P450 Enzymes

	Substrates	Substrate Examples	Comments
Subfamily CYP2B			
CYP2B1		Testosterone	Expressed in liver and extrahepatic tissues
CYP2B6		Coumarin, cyclophosphamide	Expressed in human liver. Activates aflatoxin B1. Induced by phenobarbital
Subfamily CYP2C			
20% of human liver cytochrome P450 enzymes metabolizes ~10% of drugs			
CYP2C8		Paclitaxel	Expressed in human liver and at lower levels in intestine. CYP2C in rats
CYP2C9	Neutral or acidic, often amphipathic drugs	Diclofenac, ibuprofen	Expressed in human liver and at lower levels in intestine
CYP2C19	Limited range of drugs, primarily psychotherapeutic agents	(S)-mephenytoin phenobarbitone	Expressed in human liver and at lower levels in intestine. Polymorphic in humans
Subfamily CYP2D			
2% of human liver cytochrome P450 enzymes metabolizes ~30% of drugs			
CYP2D1		Bufuralol, debrisoquine	Rat and human
CYP2D6	Lipophylic aryl-alkyl-amines	Bufuralol, amitriptyline, haloperidol, chlorpromazine, lidocaine, oxycodone	Responsible for a common genetic defect in oxidation of many drugs in humans. Not inducible

(Continued)

TABLE 223 (*Continued*)

Major Cytochrome P450 Enzymes

	Substrates	Substrate Examples	Comments
Subfamily CYP2E			
7% of human liver cytochrome P450 enzymes metabolizes ~2% of drugs			
CYP2E1	Drugs with mol. wt. < 200, volatile anesthetics	p-Nitrophenol, benzene, ethanol, aniline, chlorzoxazone, acetaminophen, and halothane	Expressed in human liver and inducible by a variety of mechanisms
Subfamily CYP3A			
30% of human liver cytochrome P450 enzymes metabolizes ~50% of drugs			
CYP3A3			Expressed in liver of most adults
CYP3A4	A wide range of lipophyilic, neutral, or basic compounds.	6β-testosterone, erythromycin, cyclosporin, terfenidine, verapamil, and hydorcortisone	Most abundant P450 in human liver. Plays a major role in the metabolism of drugs
CYP3A5		See CYP3A4	Expressed in about 20% of human liver samples
CYP3A7		See CYP3A4	Expressed in human fetal liver and kidney
Subfamily CYP4A			
CYP4A11		Lauric acid	Not extensively studied in humans

Note: Bold = key enzyme in drug metabolism.

TABLE 224

Xenobiotic Steady State and Half-Life

Number of Half-Life	Xenobiotic Steady State (%)	Xenobiotic Left in Body (%)
1	50.00	50.00
2	75.00	25.00
3	87.50	12.50
4	93.75	6.25
5	96.87	3.13

TABLE 225

Greek Alphabet

Greek Letter		Greek Name	English Equivalent	Greek Letter		Greek Name	English Equivalent
A	α	Alpha	a	N	ν	Nu	n
B	β	Beta	b	Ξ	ξ	Xi	x
Γ	γ	Gamma	g	O	o	Omicron	o
Δ	δ	Delta	d	Π	π	Pi	p
E	ε	Epsilon	ĕ	P	ρ	Rho	r
Z	ζ	Zeta	z	Σ	σ ς	Sigma	s
H	ή	Eta	ē	T	τ	Tau	t
Θ	θ	Theta	th	Y	υ	Upsilon	u
I	ι	Iota	i	Φ	φ φ	Phi	ph
K	κ	Kappa	k	X	χ	Chi	ch
Λ	λ	Lambda	l	Ψ	ψ	Psi	ps
M	μ	Mu	m	Ω	ω	Omega	o

Source: From Beyer, W.H. (1991).

TABLE 226

Prefixes and Symbols for Decimal
Multiples and Submultiples

Factor	Prefix	Symbol
10^{18}	Exa	E
10^{15}	Peta	P
10^{12}	Tera	T
10^{9}	Giga	G
10^{6}	Mega	M
10^{3}	Kilo	k
10^{2}*	Hecto	h
10^{1}*	Deka	da
10^{-1}*	Deci	d
10^{-2}*	Centi	c
10^{-3}	Milli	m
10^{-6}	Micro	μ
10^{-9}	Nano	n
10^{-12}	Pico	p
10^{-15}	Femto	f
10^{-18}	Atto	a

Source: From Beyer, W.H. (1991).

Note: The preferred multiples and submultiples listed above change the quantity by increments of 10^3 or 10^{-3}. The exceptions to these recommended factors are indicated by the *.

TABLE 227

Table of Equivalents

kg	= 1000 g, 1 million mg, 2.2 lbs
g	= 1000 mg, 1 million µg, approx. 0.035 oz
mg	= 1000 µg, 1 million ng
µg	= 1000 ng
cm^2	= 0.0001 m^2
m^2	= 10,000 cm^2
L	= approx. 1 quart, approx. 33 oz
lb	= 16 oz, 454.5 g, 0.45 kg
oz	= 28.4 g
acre	= 4047 m^2
hectare	= 2.5 acres

When referring to the concentration of a chemical in food or other medium:

mg/kg	= ppm, µg/g
mg/L	= ppm = 0.0001%
µg/kg	= ppb, ng/g
ng/kg	= ppt
ppm	= mg/kg, µg/g
ppb	= µg/kg, ng/g
ppt	= ng/kg

Source: From Beyer, W.H. (1991); Lide, D.R. (1992).

TABLE 228

Approximate Metric and Apothecary Weight Equivalents

Metric	Apothecary	Metric	Apothecary
1 gram (g)	= 15 grains	0.05 g (50 mg)	= 3/4 grain
0.6 g (600 mg)	= 10 grains	0.03 g (30 mg)	= 1/2 grain
0.5 g (500 mg)	= 7½ grains	0.015 g (15 mg)	= 1/4 grain
0.3 g (300 mg)	= 5 grains	0.001 g (1 mg)	= 1/80 grain
0.2 g (200 mg)	= 3 grains	0.6 mg	= 1/100 grain
0.1 g (100 mg)	= 1½ grains	0.5 mg	= 1/120 grain
0.06 g (60 mg)	= 1 grain	0.4 mg	= 1/150 grain

Approximate Household, Apothecary, and Metric Volume Equivalents

Household	Apothecary	Metric
1 teaspoon (t or tsp)	= 1 fluidram	= 5 mL[a,b.]
1 tablespoon (T or tbs)	= ½ fluidounce	= 15 mL
2 tablespoons	= 1 fluidounce	= 30 mL
1 measuring cupful	= 8 fluidounces	= 240 mL
1 pint (pt)	= 16 fluidounces	= 473 mL
1 quart (qt)	= 32 fluidounces	= 946 mL
1 gallon (gal)	= 128 fluidounces	= 3785 mL

Source: From Beyer, W.H. (1991).

[a] 1 mL = 1 cubic centimeter (cc); however, mL is the preferred measurement term today.

[b] 1 teaspoon (metric) = 5 mL, 1 teaspoon (United States) = 4.9 mL, 1 teaspoon (United Kingdom) = 3.6 mL.

TABLE 229

Conversion Factors: Metric to English

To Obtain	Multiply	By
Inches	Centimeters	0.3937007874
Feet	Meters	3.280839895
Yards	Meters	1.093613298
Miles	Kilometers	0.6213711922
Ounces	Grams	$3.527396195 \times 10^{-2}$
Pounds	Kilograms	2.204622622
Gallons (US liquid)	Liters	0.2641720524
Fluid ounces	Milliliters (cc)	$3.381402270 \times 10^{-2}$
Square inches	Square centimeters	0.1550003100
Square feet	Square meters	10.76391042
Square yards	Square meters	1.195990046
Cubic inches	Milliliters (cc)	$6.102374409 \times 10^{-2}$
Cubic feet	Cubic meters	35.31466672
Cubic yards	Cubic meters	1.307950619

Source: From Beyer, W.H. (1991).

TABLE 230

Conversion Factors: English to Metric[a]

To Obtain	Multiply	By
Centimeters	Inches	**2.54**
Meters	Feet	**0.3048**
Meters	Yards	**0.9144**
Kilometers	Miles	**1.609344**
Grains	Ounces	28.34952313
Kilograms	Pounds	**0.45359237**
Liters	Gallons (US liquid)	**3.785411784**
Milliliters (cc)	Fluid ounces	29.57352956
Square centimeters	Square inches	**6.4516**
Square meters	Square feet	**0.09290304**
Square meters	Square yards	**0.83612736**
Milliliters (cc)	Cubic inches	**16.387064**
Cubic meters	Cubic feet	$2.831684659 \times 10^{-2}$
Cubic meters	Cubic yards	0.764554858

Source: From Beyer, W.H. (1991).

[a] Boldface numbers are exact; others are given to 10 significant figures where so indicated by the multiplier factor. See end of section for source.

TABLE 231

Temperature Conversion Factors[a]

F = 9/5 (°C) + 32

Fahrenheit temperature = 1.8 (temperature in kelvins) − 459.67
  °C = 5/9 [(°F) − 32]

Celsius temperature = temperature in kelvins − 273.15

Fahrenheit temperature = 1.8 (Celsius temperature) + 32

Conversion of Temperatures

From	To	
°Celsius	°Fahrenheit	$t_F = (t_c \times 1.8) + 32$
	Kelvin	$T_K = t_c + 273.15$
	°Rankine	$T_R = (t_c + 273.15) \times 18$
°Fahrenheit	°Celsius	$T_Y = \dfrac{t_F - 32}{\mathbf{1.8}}$
	Kelvin	$T_K = \dfrac{t_F - \mathbf{32}}{\mathbf{1.8}} + 273.15$
	°Rankine	$T_R = t_F + 459.67$
Kelvin	°Celsius	$t_c = T_K - 273.15$
	°Rankine	$T_R = T_K \times 1.8$
°Rankine	°Fahrenheit	$t_F = T_R - 459.67$
	Kelvin	$T_K = \dfrac{t_R}{\mathbf{1.8}}$

Source: From Beyer, W.H. (1991); Lide, D.R. (1992).

[a] Boldface numbers are exact; others are given to 10 significant figures where so indicated by the multiplier factor.

TABLE 232

Temperature Conversions

°F	°C	°F	°C	°F	°C	°F	°C	°F	°C
−10	−23.3	35	+1.6	85	29.4	135	57.2	185	85.0
−5	−20.5	40	4.4	90	32.2	140	60.0	190	87.8
0	−17.8	45	7.2	95	35.0	145	62.8	195	90.5
+5	−15.0	50	10.0	100	37.8	150	65.5	200	93.3
10	−12.2	55	12.8	105	40.5	155	68.3	205	96.1
15	−9.4	60	15.5	110	43.3	160	71.1	210	98.9
20	−6.6	65	18.3	115	46.1	165	73.9	212	100
25	−3.9	70	21.1	120	48.9	170	76.6		
30	−1.1	75	23.9	125	51.6	175	79.4		
32	0	80	26.6	130	54.4	180	82.2		

TABLE 233

Conversion of Hematological Values from Conventional Units into SI Units

Constituent	Conventional Units	Conversion Factor	SI Units
Clotting time	minutes	1	minutes
Prothrombin time	seconds	1	seconds
Hematocrit	%	0.01	fraction of 1
Hemoglobin	g/dL	10	g/L
Leukocyte (WBC) count	$\times 10^3/\mu L$	1	$\times 10^9/L$
Erythrocyte (RBC) count	$\times 10^6/\mu L$	1	$\times 10^{12}/L$
Mean corpuscular volume (MCV)	μm^3	1	fL
Mean corpuscular hemoglobin (MCH)	pg/cell	1	pg/cell
Mean corpuscular hemoglobin concentration (MCHC)	g/dL	10	g/L
Erythrocyte sedimentation rate	mm/h	1	mm/h
Platelet count	$\times 10^3\ \mu L$	1	$\times 10^9 L$
Reticulocyte count	$\times 10^3\ \mu L$	1	$\times 10^9 L$

Source: American Medical Association (2017) http://www.amamanualofstyle.com/page/si-conversion-calculator; Quest Diagnostics (2017) https://www.questdiagnostics.com/dms/Documents/test-center/si_units.pdf; Access Medicine (2017) http://accessmedicine.mhmedical.com/content.aspx?bookid=1069&sectionid=60775149.

Note: Multiply the constituent value in conventional units by the conversion factor to obtain the SI units; divide the value in SI units by the conversion factor to obtain the conventional units.

TABLE 234

Conversion of Clinical Laboratory Values from Conventional Units into SI Units

Constituent	Conventional Units	Conversion Factor	SI Units
Alkaline phosphatase	U/L	0.0167	µkat/L
Alanine aminotransferase (ALT)	U/L	0.0167	µkat/L
Aspartate aminotransferase (AST)	U/L	0.0167	µkat/L
Gamma-glutamyltransferase (GGT)	U/L	0.0167	µkat/L
Glucose	mg/dL	0.0555	mmol/L
Bilirubin: total and direct (conjugated)	mg/dL	17.104	µmol/L
Bile acids	µg/mL	2.448	µmol/L
Creatinine	mg/dL	88.4	µmol/L
Urea	mg/dL	0.357	mmol/L
Total protein	g/dL	10	g/L
Albumin	g/dL	10	g/L
Sodium	mEq/L	1	mmol/L
Potassium	mEq/L	1	mmol/L
Chloride	mEq/L	1	mmol/L
Calcium	mg/dL	0.25	mmol/L
Phosphorous	mg/dL	0.323	mmol/L
Cholesterol	mg/dL	0.0259	mmol/L
Triglycerides	mg/dL	0.0113	mmol/L
Lactate dehydrogenase (LDH)	U/L	0.0167	µkat/L
Creatine kinase	U/L	0.0167	µkat/L
Lipids (total)	mg/dL	0.01	g/L
Uric acid	mg/dL	0.0595	mmol/L

Source: American Medical Association (2017) http://www.amamanualofstyle.com/page/si-conversion-calculator; Quest Diagnostics (2017) https://www.questdiagnostics.com/dms/Documents/test-center/si_units.pdf; Access Medicine (2017) http://accessmedicine.mhmedical.com/content.aspx?bookid=1069&sectionid=60775149.

Note: Multiply the constituent value in conventional units by the conversion factor to obtain the SI units; divide the value in SI units by the conversion factor to obtain the conventional units.

The SI unit katal is the amount of enzyme generating 1 mol of product per second. It is provisionally recommended as the SI unit for enzymatic activity.

TABLE 235

Transformation of Percentages into Logits

Percentage	0	1	2	3	4	5	6	7	8	9
50	0	0.04	0.08	0.12	0.16	0.20	0.24	0.28	0.32	0.36
60	0.41	0.45	0.49	0.53	0.58	0.62	0.66	0.71	0.75	0.80
70	0.85	0.90	0.94	0.99	1.05	1.10	1.15	1.21	1.27	1.32
80	1.38	1.45	1.52	1.59	1.66	1.73	1.82	1.90	1.99	2.09
90	2.20	2.31	2.44	2.59	2.75	2.94	3.18	3.48	3.89	4.60
99	4.60	4.70	4.82	4.95	5.11	5.29	5.52	5.81	6.21	6.91

Source: From Tallarida, R.J. (1992).

TABLE 236

Transformation of Percentages into Probits

Percentage	0	1	2	3	4	5	6	7	8	9
0	[–]	2.67	2.95	3.12	3.25	3.36	3.45	3.52	3.59	3.66
10	3.72	3.77	3.82	3.87	3.92	3.96	4.01	4.05	4.08	4.12
20	4.16	4.19	4.23	4.26	4.29	4.33	4.36	4.39	4.42	4.45
30	4.48	4.50	4.53	4.56	4.59	4.61	4.64	4.67	4.69	4.72
40	4.75	4.77	4.80	4.82	4.85	4.87	4.90	4.92	4.95	4.97
50	5.00	5.03	5.05	5.08	5.10	5.13	5.15	5.18	5.20	5.23
60	5.25	5.28	5.31	5.33	5.36	5.39	5.41	5.44	5.47	5.50
70	5.52	5.55	5.58	5.61	5.64	5.67	5.71	5.74	5.77	5.81
80	5.84	5.88	5.92	5.95	5.99	6.04	6.08	6.13	6.18	6.23
90	6.28	6.34	6.41	6.48	6.55	6.64	6.75	6.88	7.05	7.33
99	7.33	7.37	7.41	7.46	7.51	7.58	7.65	7.75	7.88	8.07

Source: From Tallarida, R.J. (1992).

TABLE 237

Molarity, Molality, Normality, Osmolarity
Calculations

1. Molarity (M) $= \dfrac{\text{Number of moles of solute}}{\text{Liter of solution}}$

   where: Number of moles $= \dfrac{\text{Grams of chemical}}{\text{Molecular weight}}$

2. Molality (m) $= \dfrac{\text{Number of moles of solute}}{\text{Kilogram of solution}}$

3. Normality (N) $= \dfrac{\text{Number of equivalents of solute}}{\text{Liter of solution}}$

   where: Number of equivalents $= \dfrac{\text{Grams of chemical}}{\text{Equivalent weight}}$

   Equivalent weight $= \dfrac{\text{Molecular weight}}{n}$

   For acids and bases, n is the number of replaceable H^+ or OH^- ions per molecule.

4. Normality $= n$ Molarity

   where: n is the number of replaceable H^+ or OH^- ions per molecule.

5. Osmolarity $= n$ Molarity

   where: n is the number of dissociable ions per molecule.

TABLE 238

Solution Calculations

1. Volume percent (% v/v) $= \dfrac{\text{Volume of solute}}{\text{Volume of solution}} \times 100$

2. Weight percent (% w/w) $= \dfrac{\text{Weight of solute}}{\text{Weight of solution}} \times 100$

3. Weight/volume percent (% w/v) $= \dfrac{\text{Weight of solute (g)}}{\text{Volume of solution (mL)}} \times 100$

4. Milligram percent (mg%) $= \dfrac{\text{Weight of solute (mg)}}{100 \text{ mL of solution}} \times 100$

5. Parts per million (ppm) $= \dfrac{\text{Weight of solute}}{\text{Weight of solution}} \times 10^6$

6. Parts per million (for gases)

   $\text{ppm} = \dfrac{(\text{mg/m}^3)(R)}{\text{Molecular weight}}$

   where: $R = 24.5$ at 25°C.

   $(\text{volume}_C)(\text{concentration}_C) = (\text{volume}_D)(\text{concentration}_D)$

   where

       C is the concentrated solution

       D is the dilute solution

   The aforementioned relationship is useful in preparing dilute solutions from concentrated solutions.

TABLE 239

pH Calculations

1. $\mathrm{pH} = -\log[\mathrm{H}^+] = \log \dfrac{1}{[\mathrm{H}^+]}$

2. $\mathrm{pH} = \mathrm{pK_a} + \log \dfrac{[A^-]}{[HA]}$

   where: $\mathrm{HA} \leftrightarrow \mathrm{H}^+ + A^-$
   (weak acid) (conjugate base)
   $\mathrm{p}K_a = -\log Ka$
   (equilibrium constant)

TABLE 240

Mammalian Toxicology Tests—Cost and Material Requirements

Study Type	Typical Costs[d]($)	Estimated Material Requirements[e]
Acute Oral Toxicity in Rats, Limit Test	4000	25 g
Acute Oral Toxicity in Rats, LD50 (4 levels)	14,000	50 g
Acute Dermal Toxicity in Rabbits, Limit Test	5000	50 g
Acute Dermal Toxicity in Rabbits, LD50 (4 levels)	16,000	100 g
Acute Whole-Body Inhalation Toxicity in Rats, Limit Test	20,000	100–5000 g
Acute Whole-Body Inhalation Toxicity in Rats, LC50	50,000	500–50,000 g
Primary Eye Irritation in Rabbits[a]	3800	7.5 g
Primary Skin Irritation in Rabbits[a]	3800	7.5 g
Dermal Sensitization in G. Pigs, Maximization[b]	12,000	40–80 g
Dermal Sensitization in G. Pigs, Buehler Type[b,c]	10,500	40–80 g
1-Month Oral Toxicity in Rats, Gavage	200,000	250 g
1-Month Whole-Body Inhalation Toxicity in Rats	325,000	1–200 kg
1-Month Intravenous Toxicity in Rats	240,000	20 g
1-Month Dermal Toxicity in Rats	220,000	250 g
1-Month Oral Toxicity in Dogs, Capsule	290,000	2.5 kg
1-Month Intravenous Toxicity in Dogs	300,000	200 g
3-Month Oral Toxicity in Rats, Gavage	325,000	750 g
3-Month Dietary Toxicity Study in Rats	300,000	800 g
3-Month Whole-Body Inhalation Toxicity in Rat	550,000	3–600 kg
3-Month Dermal Toxicity in Rats	350,000	750 g
3-Month Oral Toxicity in Dogs, Capsule	350,000	8 kg
3-Month Dietary Toxicity Study in Dogs	375,000	10 kg
6-Month Oral Toxicity in Rats, Gavage	500,000	1 kg
9-Month Toxicity in Dogs, Capsule	800,000	15 kg
24-Month Oncogenicity in Mice, Gavage	1,750,000	1.5 kg
24-Month Oncogenicity in Rats, Gavage	1,750,000	18 kg

(Continued)

TABLE 240 *(Continued)*

Mammalian Toxicology Tests—Cost and Material Requirements

Study Type	Typical Costs[d] ($)	Estimated Material Requirements[e]
24-Month Whole-Body Inhalation in Rats	3,750,000	20–4000 kg
General Fertility and Reproductive Performance (Segment 1) in Rats	200,000	750 g
Range-Finding Developmental Toxicity Study in Rats	50,000	90 g
Developmental Toxicity Study (Segment II) in Rats	140,000	200 g
Range-Finding Developmental Toxicity Study in Rabbits	66,000	400 g
Developmental Toxicity Study (Segment II) in Rabbits	180,000	1 kg
Perinatal and Postnatal Study (Segment III) in Rats	325,000	750 g
2-Generation Reproduction Study in Rats	325,000	8 kg

[a] Additional cost if extended observation periods are required.
[b] Additional cost for positive control.
[c] Number of induction times may vary.
[d] Based on 2012 costs.
[e] Material requirements, especially for longer-term studies by whole-body inhalation, can vary considerably depending on toxic potency of the chemical substance as well as its physical properties (e.g., dust versus gas for inhalation studies).

TABLE 241

Genetic Toxicology Tests—Cost and Material Requirements

Study Type	Typical Costs[a]($)	Estimated Material Requirements (g)
Ames Assay	7000	1.5–5
Mouse Lymphoma Assay	33,000	5
In Vitro Chromosome Aberrations (CHO)	31,000	5
In Vitro Chromosome Aberrations (Human Lymphocytes)	38,000	5
In Vitro Chromosome Aberrations (Rat Lymphocytes)	36,000	5
In Vivo Chromosome Aberrations (Mouse Bone Marrow)	40,000	10–15
In Vivo Chromosome Aberrations (Rat Bone Marrow)	41,000	60
In Vitro Unscheduled DNA Synthesis (UDS)	19,000	5
In Vivo/In Vitro UDS	50,000	25–50
In Vitro SHE Cell Transformation (Syrian Hamster Embryo Cells)	26,000	25–30
Mouse Micronucleus	14,000 (males) 27,000 (males and females)	10–15
In Vitro Sister Chromatid Exchange (SCE)	16,000	5
In Vivo SCE (Mouse)	36,000	15
In Vivo Rodent Comet Assay (Rat)	28,000	30 g
In Vivo Rodent Comet and Micronucleus Combination Assay (Rat)	48,000	30 g

[a] Based on 2012 costs.

TABLE 242

Aquatic/Ecotoxicology Tests—Cost and Material Requirements

Study Type	Typical Costs[a]($)	Estimated Material Requirements (g)
Fish Static Acute (96 hr)	4500	10
Fish Early Life Stage	44,000	350
Daphnid Static Acute (48 hr)	4200	5
Daphnid 21-Day Chronic Reproduction	20,000	100
Algal Static Acute (96 hr)	5000	5
Algal Static 14-Day	13,000	5
Fish Bioconcentration	78,000	200
Earthworm (48 hr — Filter Paper)	5000	5
Earthworm (14-Day — Soil)	6500	30

[a] Based on 2012 costs.

TABLE 243

Chemical Functional Groups

Acetamido (acetylamino)	CH_3CONH-
Acetimido (acetylimino)	$CH_3C(=NH)-$
Acetoacetamido	CH_3COCH_2CONH-
Acetoacetyl	CH_3COCH_2CO-
Acetonyl	CH_3COCH_2-
Acetonylidene	$CH_3COCH=$
Acetyl	CH_3CO-
Acrylyl	$CH_2=CHCO-$
Adipyl (from adipic acid)	$-OC(CH_2)_4CO-$
Alanyl (from alanine)	$CH_3CH(NH_2)CO-$
β-alanyl	$HN(CH_2)_2CO-$
Allophanoyl	$H_2NCONHCO-$
Allyl (2-propenyl)	$CH_2=CHCH_2-$
Allylidene (2-propenylidene)	$CH_2=CHCH=$
Amidino (aminoiminomethyl)	$H_2NC(=NH)-$
Amino	H_2N-
Amyl (pentyl)	$CH_3(CH_2)_4-$
Anilino (phenylamino)	C_6H_5NH-
Anisidino	$CH_3OC_6H_4NH-$
Anisyl (from anisic acid)	$CH_3OC_6H_4CO-$
Anthranoyl (2-aminobenzoyl)	$2-H_2NC_6H_4CO-$
Arsino	AsH_2-
Azelaoyl (from azelaic acid)	$-OC(CH_2)_7CO-$
Azido	N_3-
Azino	$=NN=$
Azo	$-N=N-$
Azoxy	$-N(O)N-$
Benzal	$C_6H_5CH=$
Benzamido (benzylamino)	C_6H_5CONH-
Benzhydryl (diphenylmethyl)	$(C_6H_5)_2CH-$

(Continued)

TABLE 243 (*Continued*)

Chemical Functional Groups

Benzimido (benzylimino)	$C_6H_5C(=NH)-$
Benzoxy (benzoyloxy)	C_6H_5COO-
Benzoyl	C_6H_5CO-
Benzyl	$C_6H_5CH_2-$
Benzylidine	$C_6H_5CH=$
Benzylidyne	$C_6H_5C\equiv$
Biphenylyl	$C_6H_5C_6H_5-$
Biphenylene	$-C_6H_4C_6H_4-$
Butoxy	C_4H_9O-
sec-butoxy	$C_2H_5CH(CH_3)O-$
tert-butoxy	$(CH_3)_3CO-$
Butyl	$CH_3(CH_2)_3-$
iso-butyl (3-methylpropyl)	$(CH_3)_2(CH_2)_2-$
sec-butyl (1-methylpropyl)	$C_2H_5CH(CH_3)-$
tert-butyl (1,1, dimethylethyl)	$(CH_3)_3C-$
Butyryl	C_3H_7CO-
Caproyl (from caproic acid)	$CH_3(CH_2)_4CO-$
Capryl (from capric acid)	$CH_3(CH_2)_8CO-$
Capryloyl (from caprylic acid)	$CH_3(CH_2)_6CO-$
Carbamido	$H_2NCONH-$
Carbamoyl (aminocarbonyl)	H_2NCO-
Carbamyl (aminocarbonyl)	H_2NCO-
Carbazoyl (hydrazinocarbonyl)	$H_2NNHCO-$
Carbethoxy	$C_2H_5O_2C-$
Carbobenzoxy	$C_6H_5CH_2O_2C-$
Carbonyl	$-C=O-$
Carboxy	$HOOC-$
Cetyl	$CH_3(CH_2)_{15}-$
Chloroformyl (chlorocarbonyl)	$ClCO-$
Cinnamyl (3-phenyl-2-propenyl)	$C_6H_5CH=CHCH_2-$

(*Continued*)

TABLE 243 (*Continued*)

Chemical Functional Groups

Cinnamoyl	$C_6H_5CH=CHCO-$
Cinnamylidene	$C_6H_5CH=CHCH=$
Cresyl (hydroxymethylphenyl)	$HO(CH_3)C_6H_4-$
Crotoxyl	$CH_3CH=CHCO-$
Crotyl (2-butenyl)	$CH_3CH=CHCH_2-$
Cyanamido (cyanoamino)	$NCNH-$
Cyanato	$NCO-$
Cyano	$NC-$
Decanedioyl	$-OC(CH_2)_6CO-$
Decanoyl	$CH_3(CH_2)_6CO-$
Diazo	$N_2=$
Diazoamino	$-NHN=N-$
Disilanyl	H_2SiSiH_2-
Disiloxanoxy	$H_3SiOSiH_2O-$
Disulfinyl	$-S(O)S(O)-$
Dithio	$-SS-$
Enanthyl	$CH_3(CH_2)_5CO-$
Epoxy	$-O-$
Ethenyl (vinyl)	$CH_2=CH-$
Ethinyl	$HC\equiv C-$
Ethoxy	C_2H_5O-
Ethyl	CH_3CH_2-
Ethylthio	C_2H_5S-
Formamido (formylamino)	$HCONH-$
Formyl	$HCO-$
Fumaroyl (from fumaric acid)	$-OCCH=CHCO-$
Furfuryl (2-furanylmethyl)	$OC_4H_3CH_2-$
Furfurylidene (2-furanylmethylene)	$OC_4H_3CH=$
Furyl (furanyl)	OC_4H_3-

(*Continued*)

TABLE 243 (*Continued*)

Chemical Functional Groups

Glutamyl (from glutamic acid)	$-OC(CH_2)_2CH(NH_2)CO-$
Glutaryl (from glutaric acid)	$-OC(CH_2)CO-$
Glycidyl (oxiranylmethyl)	CH_2-CHCH_2-
Glycinamido	H_2NCH_2CONH-
Glycolyl (hydroxyacetyl)	$HOCH_2CO-$
Glycyl (aminoacetyl)	H_2NCH_2CO-
Glyoxylyl (oxoacetyl)	$HCOCO-$
Guanidino	$H_2NC(=NH)NH-$
Guanyl	$H_2NC(=NH)-$
Heptadecanoyl	$CH_3(CH_2)_{15}CO-$
Heptanamido	$CH_3(CH_2)_{15}CONH-$
Heptanedioyl	$-OC(CH_2)_5CO-$
Heptanoyl	$CH_3(CH_2)_5CO-$
Hexadecanoyl	$CH_3(CH_2)_4CO-$
Hexamethylene	$-(CH_2)_6-$
Hexanedioyl	$-OC(CH_2)_4CO-$
Hippuryl (N-benzoylglycyl)	$C_6H_5CONHCH_2CO-$
Hydantoyl	$H_2NCONHCH_2CO-$
Hydrazino	N_2NNH-
Hydrazo	$-HNNH-$
Hydrocinnamoyl	$C_6H_5(CH_2)_2CO-$
Hydroperoxy	$HOO-$
Hydroxyamino	$HONH-$
Hydroxy	$HO-$
Imino	$HN=$
Iodoso	$OI-$
Isoamyl (isopentyl)	$(CH_3)_2CH(CH_2)_2-$
Isobutenyl (2-methyl-1-propenyl)	$(CH_3)_2C=CH-$
Isobutoxy	$(CH_3)_2CHCH\,O-$
Isobutyl	$(CH_3)_2CHCH_2-$

(Continued)

TABLE 243 (*Continued*)

Chemical Functional Groups

Isobutylidene	$(CH_3)_2CHCH=$
Isobutyryl	$(CH_3)_2CHCO-$
Isocyanato	$OCN-$
Isocyano	$CN-$
Isohexyl	$(CH_3)_2CH(CH_2)_3-$
Isoleucyl (from isoleucine)	$C_2H_3CH(CH_3)CH(NH_4)CO-$
Isonitroso	$HON=$
Isopentyl	$(CH_3)_2CH(CH_2)_2-$
Isopentylidene	$(CH_3)_2CHCH_2CH=$
Isopropenyl	$H_2C=C(CH_3)-$
Isopropoxy	$(CH_3)_2CHO-$
Isopropyl	$(CH_3)_2CH-$
Isopropylidene	$(CH_3)_2C=$
Isothiocyanato (isothiocyano)	$SCN-$
Isovaleryl (from isovaleric acid)	$(CH_3)_2CHCH_2CO-$
Keto (oxo)	$O=$
Lactyl (from lactic acid)	$CH_3CH(OH)CO-$
Lauroyl (from lauric acid)	$CH_3(CH_2)_{10}CO-$
Leucyl (from leucine)	$(CH_3)_2CHCH_2CH(NH_2)CO-$
Levulinyl (From levulinic acid)	$CH_3CO(CH_2)_2CO-$
Malonyl (from malonic acid)	$-OCCH_2CO-$
Mandelyl (from mandelic acid)	$C_6H_5CH(OH)CO-$
Mercapto	$HS-$
Methacrylyl (from methacrylic acid)	$CH_2=C(CH_3)CO-$
Methallyl	$CH_2=C(CH_3)CH_2-$
Methionyl (from methionine)	$CH_3SCH_2CH_2CH(NH_2)CO-$
Methoxy	CH_3O-
Methyl	H_3C-
Methylene	$H_2C=$

(*Continued*)

TABLE 243 (*Continued*)

Chemical Functional Groups

Methylenedioxy	$-OCH_2O-$
Methylenedisulfonyl	$-O_2SCH_2SO_2-$
Methylol	$HOCH_2-$
Methylthio	CH_2S-
Myristyl (from myristic acid)	$CH_3(CH_2)_{12}CO-$
Naphthol	$(C_{10}H_7)CH=$
Naphthobenzyl	$(C_{10}H_7)CH_2-$
Naphthoxy	$(C_{10}H_7)O-$
Naphthyl	$(C_{10}H_7)-$
Naphthylidene	$(C_{10}H_6)=$
Neopentyl	$(CH_3)_3CCH_2-$
Nitramino	O_2NNH-
Nitro	O_2N-
Nitrosamino	$ONNH-$
Nitrosimino	$ONN=$
Nitroso	$ON-$
Nonanoyl (from nonanoic acid)	$CH_3(CH_2)_7CO-$
Oleyl (from oleic acid)	$CH_3(CH_2)_7CH=CH(CH_2)_7CO-$
Oxalyl (from oxalic acid)	$-OCCO-$
Oxamido	$H_2NCOCONH-$
Oxo (keto)	$O=$
Palmityl (from palmitic acid)	$CH_3(CH_2)_{14}CO-$
Pelargonyl (from pelargonic acid)	$CH_3(CH_2)_7CO-$
Pentamethylene	$-(CH_2)_5-$
Pentyl	$CH_3(CH_2)_4-$
Phenacyl	$C_6H_5COCH_2-$
Phenacylidene	$C_6H_5COCH=$
Phenanthryl	$(C_{14}H_9)-$
Phenethyl	$C_6H_5CH_2CH_2-$

(*Continued*)

TABLE 243 (*Continued*)

Chemical Functional Groups

Phenoxy	C_6H_5O-
Phenyl	C_6H_5-
Phenylene	$-C_6H_4-$
Phenylenedioxy	$-OC_6H_4O-$
Phosphino	H_2P-
Phosphinyl	$H_2P(O)-$
Phospho	O_2P-
Phosphono	$(HO)_2P(O)-$
Phthalyl (from phthalic acid)	$1,2-C_6H_4(CO-)_2$
Picryl (2,4,6-trinitrophenyl)	$2,4,6-(NO_2)_2C_6H_2-$
Pimelyl (from pimelic acid)	$-OC(CH_2)_5CO-$
Piperidino	$C_5H_{10}N-$
Piperidyl (piperidinyl)	$(C_5H_{10}N)-$
Piperonyl	$3,4-(CH_2O_2)C_6H_3CH_2-$
Pivalyl (from pivalic acid)	$(CH_3)_3CCO-$
Prenyl (3-methyl-2-butenyl)	$(CH_3)_2C=CHCH_2-$
Propargyl (2-propynyl)	$HC\equiv CCH_2-$
Propenyl	$CH_2=CHCH_2-$
iso-Propenyl	$(CH_3)_2C=$
Propionyl	CH_3CH_2CO-
Propoxy	$CH_3CH_2CH_2O-$
Propyl	$CH_3CH_2CH_2-$
iso-Propyl	$(CH_3)_2CH-$
Propylidene	$CH_3CH_2CH=$
Pyridino	C_5H_5N-
Pyridyl (pyridinyl)	$(C_5H_4N)-$
Pyrryl (pyrrolyl)	$(C_3H_4N)-$
Salicyl (2-hydroxybenzoyl)	$2-HOC_6H_4CO-$
Selenyl	$HSe-$
Seryl (from serine)	$HOCH_2CH(NH_2)CO-$

(Continued)

TABLE 243 (*Continued*)

Chemical Functional Groups

Siloxy	H_3SiO-
Silyl	H_3Si-
Silylene	$H_2Si=$
Sorbyl (from sorbic acid)	$CH_3CH=CHCH=CHCO-$
Stearyl (from stearic acid)	$CH_3(CH_2)_{16}CO-$
Styryl	$C_6H_5CH=CH-$
Suberyl (from suberic acid)	$-OC(CH_2)_6CO-$
Succinamyl	$H_2NCOCH_2CH_2CO-$
Succinyl (from succinic acid)	$-OCCH_2CH_2CO-$
Sulfamino	$HOSO_2NH-$
Sulfamyl	H_2NSO-
Sulfanilyl	$4-H_2NC_6H_4SO_2-$
Sulfeno	$HOS-$
Sulfhydryl (mercapto)	$HS-$
Sulfinyl	$OS=$
Sulfo	HO_3S-
Sulfonyl	$-SO_2-$
Terephthalyl	$1,4-C_6H_4(CO-)_2$
Tetramethylene	$-(CH_2)_4-$
Thenyl	$(C_4H_3S)CH-$
Thienyl	$(C_4H_3S)-$
Thiobenzoyl	C_6H_5CS-
Thiocarbamyl	H_2NCS-
Thiocarbonyl	$-CS-$
Thiocarboxy	$HOSC-$
Thiocyanato	$NCS-$
Thionyl (sulfinyl)	$-SO-$
Thiophenacyl	$C_6H_5CSCH_2-$
Thiurain(aminothioxomethyl)	H_2NCS-
Threonyl (from threonine)	$CH_3CH(OH)CH(NH_2)CO-$

(*Continued*)

TABLE 243 (*Continued*)

Chemical Functional Groups

Toluidino	$CH_3C_6H_4NH-$
Toluyl	$CH_3C_6H_4CO-$
Tolyl (methylphenyl)	$CH_3C_6H_4-$
α-tolyl	$C_6H_5CH_2-$
Tolylene (methylphenylene)	$(CH_3C_6H_3)=$
α-tolylene	$C_6H_5CH=$
Tosyl [(4-methylphenyl) sulfonyl]	$4-CH_3C_6H_4SO_2-$
Triazano	$H_2NNHNH-$
Trimethylene	$-(CH_2)_3-$
Triphenylmethyl (trityl)	$(C_6H_5)_3C-$
Tyrosyl (from tyrosine)	$4-HOC_6H_4CH_2CH(NH_2)CO-$
Ureido	$H_2NCONH-$
Valeryl (from valeric acid)	C_4H_9CO
Valyl (from valine)	$(CH_3)_2CHCH(NH_2)CO-$
Vinyl	$CH_2=CH-$
Vinylidene	$CH_2=C=$
Xenyl (biphenylyl)	$C_6H_5C_6H_4-$
Xylidino	$(CH_3)_2C_6H_3NH-$
Xylyl (dimethylphenyl)	$(CH_3)_2C_6H_3-$
Xylylene	$-CH_2C_6H_4CH_2-$

Source: From Lide, D.R. (1992).

TABLE 244

Standard Atomic Weights

Name	Symbol	Atomic Number	Atomic Weight
Actinium[a]	Ac	89	[227]
Aluminum	Al	13	26.981539(5)
Americium[a]	Am	95	[243]
Antimony (Stibium)	Sb	51	121.75(3)
Argon	Ar	18	39.948(1)
Arsenic	As	33	74.92159(2)
Astatine[a]	At	85	[210]
Barium	Ba	56	137.327(7)
Berkelium[a]	Bk	97	[247]
Beryllium	Be	4	9.012182(3)
Bismuth	Bi	83	208.98037(3)
Boron	B	5	10.811(5)
Bromine	Br	35	79.904(l)
Cadmium	Cd	48	112.411(8)
Cesium	Cs	55	132.90543(5)
Calcium	Ca	20	40.078(4)
Californium[a]	Cf	98	[251]
Carbon	C	6	12.011(1)
Cerium	Ce	58	140.115 (4)
Chlorine	Cl	17	35.4527(9)
Chromium	Cr	24	51.9961(6)
Cobalt	Co	27	58.93320(1)
Copper	Cu	29	63.546(3)
Curium[a]	Cm	96	[247]
Dysprosium	Dy	66	162.50(3)
Einsteinium[a]	Es	99	[252]
Erbium	Er	68	167.26(3)

(*Continued*)

TABLE 244 (*Continued*)

Standard Atomic Weights

Name	Symbol	Atomic Number	Atomic Weight
Europium	Eu	63	151.965(9)
Fermium[a]	Fm	100	[257]
Fluorine	F	9	18.9984032(9)
Francium[a]	Fr	87	[223]
Gadolinium	Gd	64	157.25(3)
Gallium	Ga	31	69.723(1)
Germanium	Ge	32	72.61(2)
Gold	Au	79	196.96654(3)
Hafnium	Hf	72	178.49(2)
Helium	He	2	4.002602(2)
Holmium	Ho	67	164.93032(3)
Hydrogen	H	1	1.00794(7)
Indium	In	49	114.82(1)
Iodine	I	53	126.90447(3)
Iridium	Ir	77	192.22(3)
Iron	Fe	26	55.847(3)
Krypton	Kr	36	83.80(1)
Lanthanum	La	57	138.9055(2)
Lawrencium[a]	Lr	103	[262]
Lead	Pb	82	207.2(1)
Lithium	Li	3	6.941(2)
Lutetium	Lu	71	174.967(1)
Magnesium	Mg	12	24.3050(6)
Manganese	Mn	25	54.93805(1)
Mendelevium[a]	Md	101	[258]
Mercury	Hg	80	200.59(3)
Molybdenum	Mo	42	95.94(1)
Neodymium	Nd	60	144.24(3)

(Continued)

TABLE 244 (*Continued*)

Standard Atomic Weights

Name	Symbol	Atomic Number	Atomic Weight
Neon	Ne	10	20.1797(6)
Neptunium[a]	Np	93	[237]
Nickel	Ni	28	58.69(1)
Niobium	Nb	41	92.90638(2)
Nitrogen	N	7	14.00674(7)
Nobelium[a]	No	102	[259]
Osmium	Os	76	190.2(1)
Oxygen	O	8	15.9994(3)
Palladium	Pd	46	105.42(1)
Phosphorus	P	15	30.973762(4)
Platinum	Pt	78	195.08(3)
Plutonium[a]	Pu	94	[244]
Polonium[a]	Po	84	[209]
Potassium (Kalium)	K	19	39.0983(1)
Praseodymium	Pr	59	140.90765(3)
Promethium[a]	Pm	61	[145]
Protactinium	Pa	91	231.03588(2)
Radium[a]	Ra	88	[226]
Radon[a]	Rn	86	[222]
Rhenium	Re	75	186.207(1)
Rhodium	Rh	45	102.90550(3)
Rubidium	Rb	37	85.4678(3)
Ruthenium	Ru	44	101.07(2)
Samarium	Sm	62	150.36(3)
Scandium	Sc	21	44.955910(9)
Selenium	Se	34	78.96(3)
Silicon	Si	14	28.0855(3)
Silver	Ag	47	107.8682(2)

(*Continued*)

TABLE 244 *(Continued)*

Standard Atomic Weights

Name	Symbol	Atomic Number	Atomic Weight
Sodium (Natrium)	Na	11	22.989768(6)
Strontium	Sr	38	87.62(l)
Sulfur	S	16	32.066(6)
Tantalum	Ta	73	180.9479(l)
Technetium[a]	Tc	43	[98]
Tellurium	Te	52	127.60(3)
Terbium	Tb	65	158.92534(3)
Thallium	Tl	81	204.3833(2)
Thorium[a]	Th	90	232.0381(1)
Thulium	Tm	69	168.93421(3)
Tin	Sn	50	118.710(7)
Titanium	Ti	22	47.88(3)
Tungsten (Wolfram)	W	74	183.85(3)
Uranium[a]	U	92	238.0289(1)
Vanadium	V	23	50.9415(1)
Xenon	Xe	54	131.29(2)
Ytterbium	Yb	70	173.04(3)
Yttrium	Y	39	88.90585(2)
Zinc	Zn	30	65.39(2)
Zirconium	Zr	40	91.224(2)

Source: Lide, D.R. (1992).

Note: The values of atomic weight given here apply to elements as they exist naturally on earth. The number in parentheses following atomic weight value gives uncertainty in the last digit. An entry in brackets indicates the mass number of the longest-lived isotope of an element that has no stable isotopes and for which s standard atomic weight cannot be defined.

[a]   Element has no stable nuclides.

TABLE 245

Frequently Encountered Acronyms

AAALAC	Association for Assessment and Accreditation of Laboratory Animal Care (International)
AADA	Abbreviated Antibiotic Drug Application
AALAS	American Association for Laboratory Animal Science
AAPCO	Association of American Pesticide Control Officials
ACB	Analytical Chemistry Branch (re: OPP)
ACC	American Chemistry Council
ACP	Associates of Clinical Pharmacology
ACS	American Chemical Society
ACT	American College of Toxicology
ACUP	Animal care and use procedure
ADE	Adverse drug experience/effect/event
ADI	Acceptable daily intake (see also RfD)
ADME	Absorption distribution, metabolism, and excretion
ADR	Adverse drug reaction
AE	Adverse experience/event
AERS	Adverse Event Reporting System
AHI	Animal Health Institute
ai/A	Active ingredient per acre
AI	Active ingredient
ALISS	"A-List" Inventory Support System (re: SRRD)
ALJ	Administrative law judge
ANADA	Abbreviated new animal drug application
ANDA	Abbreviated new drug application
ANSI	American National Standards Institute
AOAC	Association of Official Analytical Chemists
ARAR	Applicable or relevant and appropriate requirements (re: Superfund)
APB	Antimicrobial Program Branch
ARB	Accelerated Reregistration Branch (re: SRRD of OPP)
ARS	Agricultural Research Service (re: USDA)
ARTS	Accelerated Reregistration Tracking System (re: SRRD of OPP)

(Continued)

TABLE 245 (*Continued*)

Frequently Encountered Acronyms

ASAP	Administrative System Automations Project
ASQC	American Society of Quality Control
ASR	Analytical Summary Report
ASTHO	Association of State and Territorial Health Officials
ASTM	Association of Standard Test Methods
AWA	Animal Welfare Act
BAB	Biological Analysis Branch (re: BEAD of OPP)
BARQA	The British Association of Research Quality Assurance
BDAT	Best demonstrated available technology
BEAD	Biological and Economic Analysis Division (re: OPP)
CANDAs	Computer-assisted new drug applications
CANADA	Computer-assisted new animal drug application
CAP	Compliance audit program
CAPER	Computer-assisted preclinical electronic review
CAPLA	Computer-assisted product license application (Re: Biologics)
CB	Communications branch (re: OPP)
CB I & II	Chemistry Branch I and II (re: OPP)
CBER	Center for Biologics Evaluation and Research (re: FDA)
CBI	Confidential business information
CDC	Centers for Disease Control (see also USCDC)
CDER	Center for Drug Evaluation and Research (re: FDA)
CDRH	Center for Devices and Radiological Health (re: FDA)
CEO	Council on Environmental Quality
CERCLA	Comprehensive, Environmental, Response, Compensation, and Liability Act
CFD	Call for data
CFR	Code of Federal Regulations
CGMP	Current good manufacturing practices (see also GMP)
CLIA	Clinical Laboratory Improvement Act
CMA	Chemical Manufacturers Association

(*Continued*)

TABLE 245 (*Continued*)

Frequently Encountered Acronyms

CMC	Chemistry, Manufacturing, and Controls
CNAEL	Committee on National Accreditation for Environmental Laboratories
CORT	Toxicology studies set: *c*hronic feeding; *o*ncogenicity; *r*eproduction; *t*eratology
CPDA	Chemical Producers and Distributors Association
CPG	Compliance Policy Guide
CPGM	Compliance program guidance manuals (re: bioresearch monitoring program)
CPSC	Consumer Product Safety Commission
CRA	Clinical research associate
CRADA	Cooperative research and development agreement
CRF	Case report form
CRO	Contract research organization
CRP	Child-resistant packaging
CSA	Clinical & Scientific Affairs
CSF	Confidential Statement of Formula
CSMA	Chemical Specialities Manufacturers' Association
CSO	Consumer safety officer (re: FDA)
CSRS	Cooperative State Research Service
CTB	Certification and Training Branch (re: FOD of OPP)
CV	Curriculum vitae
CVM	Center for Veterinary Medicine (re: FDA)
CWA	Clean Water Act
DAMOS	Drug Application Methodology with Optical Storage (re: EC)
DCI	Data callin notice (re: RD or SRRD of OPP)
DEA	Drug Enforcement Agency
DEB	Dietary exposure branch (see also CB I and II)
DFE	Design for the Environment
DI	Department of Interior (see also USDI)
DIA	Drug Information Association
DIS	Drug information system

(*Continued*)

TABLE 245 (*Continued*)

Frequently Encountered Acronyms

DISLODG	Dislodgeable foliar residue (re: EPA)
DMF	Drug master file
DOE	Department of Energy (see also USDOE)
DOT	Department of Transportation
DQOs	Data quality objectives (re: EPA work)
DRES	Dietary risk evaluation system (re: OPP)
EAB	Economic Analysis Branch (re: OPP)
EC	Emulsifiable concentrate
EC	European Community (see also EEC)
EDF	Environmental Defense Fund
EEB	Ecological Effects Branch (re: OPP)
EEC	European Economic Community (see also EC)
EFED	Environmental Fate and Effects Division (re: OPP)
EFGWB	Environmental Fate and Groundwater Branch (re: OPP)
EIR	Establishment inspection report
ELA	Establishment license report (re: Biologics)
ELGIN	Environmental Liaison Group International
ELI	Environmental Law Institute
EMO	Experimental manufacturing order
EP	End-use product
EPA	Environmental Protection Agency (see also USEPA)
EPCRA	Emergency Planning and Community Right-to-Know Act
EPRS	Establishment/product registration system
ESA	Entomological Society of America
EUP	Experimental use permit (re: EPA FIFRA)
FACTS	Field Accomplishments and Compliance Tracking System
FDA	Food and Drug Administration (see also USFDA)
FDB	Field data book
FD&C	Federal Food, Drug, and Cosmetic Act (see also FFDCA, FDCL)
FDCL	Food, Drug, and Cosmetic Law (see also FD&C, FFDCA)

(*Continued*)

TABLE 245 (*Continued*)

Frequently Encountered Acronyms

FDLI	Food and Drug Law Institute
FFDCA	Federal Food, Drug, and Cosmetic Act (see also FD&C, FDCL)
FHB	Fungicide–Herbicide Branch (re: RD of OPP)
FHSA	Federal Health and Safety Act
FIFRA	Federal Insecticide, Fungicide, and Rodenticide Act
FOD	Field Operations Division (re: OPP)
FOI	Freedom of information (see also FOIA)
FOIA	Freedom of Information Act (see also FOI)
FPLA	Fair Packaging and Labeling Act
FR	Federal Register
FRD	Field research director
FTC	Federal Trade Commission
FWS	Fish and Wildlife Service (see also USFWS)
GALP	Good automated laboratory practices
GARPs	Good academic research practices (draft 1992)
GAO	General Accounting Office
GATT	General Agreement on Tariffs and Trade
GCPs	Good Clinical Practices
GLPs	Good Laboratory Practices
GLPSs	Good Laboratory Practice Standards
GMPs	Good manufacturing practices
GH_2O	Groundwater studies (re: EPA)
GRAE	Generally recognized as effective
GRAS	Generally recognized as safe
HDT	Highest dose tested (re: EPA)
HED	Health Effects Division (re: OPP)
HEI	Health Effects Institute
HES	Health and environmental safety
HHS	Health and human services
HIMA	Health Industry Manufacturers Association

(Continued)

TABLE 245 (*Continued*)

Frequently Encountered Acronyms

HPB	Health Protection Branch (re: Canada)
HPVSIDS	High Production Volume Screening Info Data Set (re: OECD)
HRS	Hazard Ranking System (re: Superfund)
IACUC	Institutional Animal Care and Use Committee
IB	Investigator's brochure
ICH	International Conference on Harmonisation
ICR	Information collection request
IDB	Investigational drug brochure
IDE	Investigational device exemption
IG	Inspector general (see also OIG)
INAD	Investigational new animal drug
IND	Investigational new drug
IPM	Integrated pest management (re: OPP)
IR-4	Interregional Research Project #4 for Minor Crops (re: USDA)
IRB	Insecticide and Rodenticide Branch (re: RD of OPP)
IRB	Institutional review board
IS	Information standards
ISA	Information systems architecture
ISB	Information Services Branch (re: PMSD of OPP)
ISQA	International Society of Quality Assurance
ITC	Interagency Testing Committee (re: TSCA)
LAC	Laboratory Accreditation Committee
LADD	Lifetime average daily dose (re: OPP)
LADD	Lowest acceptable daily dose (re: OPP)
LC_{50}	Lethal concentration for 50% of test population
LD_{50}	Lethal dose for 50% of test population
LDT	Lowest dose tested
LEL	Lowest effective level
LRD	Laboratory research director
LUIS	Label Use Information System (re: OPP)

(*Continued*)

TABLE 245 (*Continued*)

Frequently Encountered Acronyms

MARSQA	Mid-Atlantic Region Society of Quality Assurance
MCA	Medicines Control Agency (United Kingdom equivalent to FDA)
MCL	Maximum contaminant level
MCLG	Maximum contaminant level goal
MDDI	Medical devices, diagnostics and instrumentation
MNVP	Medically necessary veterinary product (re: FDACVM)
MOE	Margin of exposure
MOS	Margin of safety
MOU	Memorandum of Understanding
MP	Manufacturing use product (re: OPP, see also MUP)
MPI	Maximum permitted intake
MRID#	Master record identification number
MS	Master schedule
MSS	Master schedule sheet
MSDS	Material safety data sheet
MTL	Master testing list
MUP	Manufacturing use product (re: OPP, see also MP)
MURS	Multiuser regulatory submission (International Multiagency Project)
NACA	National Agricultural Chemical Association
NADA	New animal drug application
NADE	New animal drug evaluation, office of (re: FDACVM)
NAF	Notice of adverse findings
NAI	No action indicated
NAICC	National Alliance of Independent Crop Consultants
NARA	National Agrichemical Retailers Association
NAS	National Academy of Sciences
NASDA	National Association of State Departments of Agriculture
NCAMP	National Coalition Against the Misuse of Pesticides
NCE	New chemical entity
NCP	National contingency plan (re: Superfund)

(*Continued*)

TABLE 245 (*Continued*)

Frequently Encountered Acronyms

NBS	National Bureau of Standards (now called NIST)
NDA	New drug application
NDS	New drug submission
NEIC	National Enforcement Investigation Center (re: USEPA)
NIEHS	National Institute of Environmental Health Sciences
NIH	National Institutes of Health
NIOSH	National Institute for Occupational Safety and Health
NIST	National Institute of Standards and Technology (formerly NBS)
NPCA	National Pest Control Association
NPDES	National Pollutant Discharge Elimination System
NPIRS	National Pesticide Information Retrieval System
NPL	National Priority List (re: Superfund)
NPTN	National Pesticide Telecommunications Network
NRDC	Natural Resources Defense Council
NTIS	National Technical Information Service
OAI	Official action indicated
OASIS	Operational and Administrative System for Import Support
OCSPP	Office of Chemical Safety and Pollution Prevention
ODW	Office of drinking water
OECD	Organisation for Economic Co-operation and Development
OECA	Office of Enforcement and Compliance Assurance
OES	Office of Endangered Species (re: FWS of DI)
OGD	Office of Generic Drugs (re: FDA)
OIG	Office of the Inspector General (see also IG)
OLTS	Online tracking system
OMB	Office of Management and Budget
OPM	Office of Personnel Management
OPP	Office of Pesticide Programs (re: EPA)
OPPT	Office of Pollution Prevention and Toxics (formerly OTS; re. TSCA)
OPPTS	Office of Prevention, Pesticides, and Toxic Substances (now OCSPP)

TABLE 245 (*Continued*)

Frequently Encountered Acronyms

OREB	Occupational and Residential Exposure Branch (re: HED of OPP)
OSB	Occupational Safety Branch (re: FOD of OPP)
OSHA	Occupational Safety and Health Administration
OSW	Office of Solid Waste (re: EPA)
OSWER	Office of Solid Waste and Emergency Response
OTA	Office of Technology Assessment (re: Congress)
OTC	Over the counter
OTS	Office of Toxic Substances (now called OPPT)
OWPE	Office of Waste Programs Enforcement
PAG	Pesticide Assessment Guidelines
PAI	Pure active ingredient
PBA	Preliminary benefit analysis
PCO	Pest control operator
PDA	Parenteral Drug Association
PDMS	Pesticide document management system
PDR	Physician's Desk Reference
PDUFA	Prescription Drug User Fee Act
PES	Planning and evaluation staff
PHED	Pesticide handlers exposure database (re: EPA)
PHI	Preharvest interval
PHI	Postharvest interval
PIMS	Pesticide Incident Monitoring System
PLA	Product license application (re: Biologics)
PM	Product manager
PMA	Pharmaceutical Manufacturers Association
PMN	Premanufacture notification (re: TSCA)
PMSD	Program management and support division (re: OPP)
P&P guide	Policy and procedures guide (re: FDACVM)
PPIS	Pesticide Product Information Systems
PR#	Pesticide clearance request number

(*Continued*)

TABLE 245 (*Continued*)

Frequently Encountered Acronyms

PR notice	Pesticide registration notice
PRATS	Pesticide Registration Activity Tracking System
PRCSQA	Pacific Regional Chapter of the Society of Quality Assurance
PRP	Potentially responsible party (re: Superfund)
PSA	Product safety assurance
QA	Quality assurance
QAAS	Quality assurance advisory subcommittee
QAO	Quality assurance officer
QAP	Quality assurance project plan
QAU	Quality assurance unit
QC	Quality control
QMP	Quality management plan
QUA	Qualitative use assessment
R&D	Research and development
R&E	Research and experimental
RAC	Raw agricultural commodity (see also RACPC)
RAC	Risk Assessment Council
RACPC	Raw agricultural commodity processing (re: EPA)
RAF	Risk Assessment Forum
RAPS	Regulatory affairs professional society
RB	Reregistration Branch
RCFs	Refractory ceramic fibers
RCRA	Resource Conservation and Recovery Act
RD	Registration Division (re: OPP)
RDRA	Remedial design/remedial action (re: Superfund)
RED	Reregistration Eligibility Document
REI	Reentry interval (re: OPP)
RFC	Regional Field Coordinator
RfD	Reference dose
RFD	Recommended for development

(*Continued*)

TABLE 245 (*Continued*)

Frequently Encountered Acronyms

RI/FS	Remedial investigation/feasibility study (re: Superfund)
RLC	Regional laboratory coordinator
RM	Review Manager (re: OPP)
RM1	Risk management–1 (re: EPA)
RM2	Risk management–2 (re: EPA)
RMEB	Resource Management and Evaluation Branch (re: PMSD of OPP)
ROD	Record of decision (re: Superfund)
RPAR	Rebuttable presumption against registration (re: OPP; see also SR)
RRC	Regulatory Review Committee of the Society of Quality Assurance
RRD	Residue Research Director
RS	Registration standard (re: OPP)
RSB	Registration Support Branch
RTECS	Registry of Toxic Effects of Chemical Substances
RUP	Restricted use pesticide
SAB	Scientific advisory board (re: EPA)
SACB	Science Analysis and Coordination Branch (re: HED of OPP)
SACS	Science Analysis and Coordination Staff (re: EFED of OPP)
SAES	State agricultural experiment stations
SAP	Scientific advisory panel (re: FIFRA)
SARA	Superfund Amendments and Reauthorization Act
SB	Systems Branch (re: PMSD of OPP)
SD	Study director
SDLC	Software development life cycles
SDWA	Safe Drinking Water Act
SETAC	Society of Environmental Toxicology & Chemistry
SFIREG	State FIFRA Issues, Research and Evaluation Group
SGML	Standard general markup language (computer language)
SITE	Superfund Innovative Technology Evaluation Program
SMART	Submission management and review tracking
SMARTS	Simple Maintenance of ARTS (see also ARTS)

(*Continued*)

TABLE 245 (*Continued*)

Frequently Encountered Acronyms

SOP	Standard operating procedure
SOT	Society of Toxicology
SPI	Standard practice instructions
SQA	Society of Quality Assurance
SR	Special review (re: OPP; formerly RPAR)
SRB	Special Review Branch (re: SRRD of OPP)
SRRD	Special Review and Reregistration Division (re: OPP)
STARS	Submission Tracking And Reporting System
STP	Society of Toxicologic Pathologists
TCLP	Toxicity characteristic leaching procedure, RCRA
TEP	Typical end-use product
TFM	Testing facility management
TGAI	Technical-grade active ingredient
TMRC	Theoretical maximum residue contribution (re: OPP)
TOSCA	Toxic Substances Control Act (see also TSCA)
TQ	Total quality
TQM	Total quality management
TQSS	Total Quality Specialty Section of the Society of Quality Assurance
TRI	Toxics Release Inventory (re: EPCRA)
TSCA	Toxic Substances Control Act (see also TOSCA)
TSCATS	TSCA test submissions
TVOCs	Total volatile organic compounds
UN	United Nations
USC	United States Code
USCDC	United States Centers for Disease Control (see also CDC)
USDA	United States Department of Agriculture
USDOE	United States Department of Energy (see also DOE)
USDI	United States Department of Interior (see also DI)
USEPA	United States Environmental Protection Agency (see also EPA)
USFDA	United States Food and Drug Administration (see also FDA)

(*Continued*)

TABLE 245 (*Continued*)

Frequently Encountered Acronyms

USFWS	United States Fish and Wildlife Service (see also FWS)
VAI	Voluntary action indicated
WEXWPS	Worker Exposure Studies—Worker Protection Standards (re: EPA)
WHO	World Health Organization
WP	Wettable powders
WSS	Weed Science Society

Source: From Society of Quality Assurance. With permission.

References

Beyer, W.H., Ed. (1991), *CRC Standard Mathematical Tables and Formulae*, 29th ed., CRC Press, Boca Raton, FL.

Illing, H.P.A. (1989), *Xenobiotic Metabolism and Disposition: The Design of Studies on Novel Compounds*, CRC Press, Boca Raton, FL.

Lide, D.R., Ed. (1992), *CRC Handbook of Chemistry and Physics*, 73rd ed., CRC Press, Boca Raton, FL.

Mitruka, B.M. and Rawnsley, H.M. (1977), *Clinical Biochemical and Hematological Reference Values in Normal Experimental Animals*, Masson Publishing, New York.

Niesink, R.J.M., deVries, J., and Hollinger, M.A. (1996), *Toxicology: Principles and Applications*, CRC Press, Boca Raton, FL.

Shackelford, J. and Alexander, W., Eds. (1992), *CRC Materials Science and Engineering Handbook*, CRC Press, Boca Raton, FL.

Tallarida, R.J. (1992), *Pocket Book of Integrals and Mathematical Formulas*, 2nd ed., CRC Press, Boca Raton, FL.

16

Glossary—by Subject

Carcinogenesis*

Adduct: The covalent linkage or addiction product between an alkylating agent and cellular macromolecules such as protein, RNA, and DNA.

Alkylating agent: A chemical compound that has positively charged (electron-deficient) groups that can form covalent linkages with negatively charged portions of biological molecules such as DNA. The covalent linkage is referred to as an adduct and may have mutagenic or carcinogenic effects on the organism. The alkyl species is the radical that results when an aliphatic hydrocarbon loses one hydrogen atom to become electron deficient. Alkylating agents react primarily with guanine, adding their alkyl group to N7 of the purine ring.

Altered focus: A histologically identifiable clone of cells within an organ that differs phenotypically from the normal parenchyma. Foci of altered cells usually result from increased cellular proliferation, represent clonal expansions of initiated cells, and are frequently observed in multistage animal models of carcinogenesis. Foci of cellular alteration are most commonly observed in the liver

---

* Maronpot (1991). Reprinted in part with permission.

of carcinogen-treated rodents and are believed by some to represent preneoplastic lesions.

Benign: A classification of anticipated biological behavior of neoplasms in which the prognosis for survival is good. Benign neoplasms grow slowly, remain localized, and usually cause little harm to the patient.

Choristoma: A mass of well-differentiated cells from one organ included within another organ, for example, adrenal tissue present in the lung.

Chromosomal aberration: A numerical or structural chromosomal abnormality.

Cocarcinogen: An agent not carcinogenic alone but that potentiates the effect of a known carcinogen.

Cocarcinogenesis: The augmentation of neoplasm formation by simultaneous administration of a genotoxic carcinogen and an additional agent (cocarcinogen) that has no inherent carcinogenic activity by itself.

Direct carcinogen: Carcinogens that have the necessary structure to directly interact with cellular constituents and cause neoplasia. Direct acting carcinogens do not require metabolic conversion by the host to be active. They are considered genotoxic because they typically undergo covalent binding to DNA.

Dysplasia: Disordered tissue formation characterized by changes in size, shape, and orientational relationships of adult types of cells. Primarily seen in epithelial cells.

Epigenetic: Change in phenotype without a change in DNA structure. One of two main mechanisms of carcinogen action, epigenetic carcinogens are nongenotoxic, that is, they do not form reactive intermediates that interact with genetic material in the process of producing or enhancing neoplasm formation.

Genotoxic carcinogen: An agent that interacts with cellular DNA either directly in its parent form (direct carcinogen) or after metabolic biotransformation.

Hyperplasia: A numerical increase in the number of phenotypically normal cells within a tissue or organ.

Hypertrophy: Increase in the size of an organelle, cell, tissue, or organ within a living organism. To be distinguished from hyperplasia, hypertrophy refers to an increase in size rather than an increase in number. Excessive hyperplasia in a tissue may produce hypertrophy of the organ in which that tissue occurs.

Initiation: The first step in carcinogenesis whereby limited exposure to a carcinogenic agent produces a latent but heritable alteration in a cell, permitting its subsequent proliferation and development into a neoplasm after exposure to a promoter.

Initiator: A chemical, physical, or biological agent that is capable of irreversibly altering the genetic component (DNA) of the cell. While initiators are generally considered to be carcinogens, they are typically used at low noncarcinogenic doses in two-stage initiation–promotion animal model systems. Frequently referred to as a "tumor initiator."

In situ carcinoma: A localized intraepithelial form of epithelial cell malignancy. The cells possess morphological criteria of malignancy but have not yet gone beyond the limiting basement membrane.

Malignant: A classification of anticipated biological behavior of neoplasms in which the prognosis for survival is poor. Malignant neoplasms grow rapidly, invade, destroy, and are usually fatal.

Metaplasia: The substitution in a given area of one type of fully differentiated cell for the fully differentiated cell type normally present in that area, for example, squamous epithelium replacing ciliated epithelium in the respiratory airways.

Metastasis: The dissemination of cells from a primary neoplasm to a noncontiguous site and their growth therein. Metastases arise by dissemination of cells from the primary neoplasm via the vascular or lymphatic system and are an unequivocal hallmark of malignancy.

Mitogenesis: The generation of cell division or cell proliferation.

Maximum tolerated dose (MTD): Refers to the maximum amount of an agent that can be administered to an animal in a carcinogenicity test without adversely affecting the animal due to toxicity other than carcinogenicity. Examples of having exceeded the MTD include excessive early mortality, excessive loss of body weight, production of anemia, production of tissue necrosis, and overloading of the metabolic capacity of the organism.

Mutation: A structural alteration of DNA that is hereditary and gives rise to an abnormal phenotype. A mutation is always a change in the DNA base sequence and includes substitutions, additions, rearrangements, or deletions of one or more nucleotide bases.

Oncogene: The activated form of a proto-oncogene. Oncogenes are associated with the development of neoplasia.

Preneoplastic lesion: A lesion usually indicative that the organism has been exposed to a carcinogen. The presence of preneoplastic lesions indicates that there is enhanced probability for the development of neoplasia in the affected organ. Preneoplastic lesions are believed to have a high propensity to progress to neoplasia.

Procarcinogen: An agent that requires bioactivation in order to give rise to a direct acting carcinogen. Without metabolic activation, these agents are not carcinogenic.

Progression: Processes associated with the development of an initiated cell to a biologically malignant neoplasm. Sometimes used in a more limited sense to describe the process whereby a neoplasm develops from a benign to a malignant proliferation or from a low-grade to a high-grade malignancy. Progression is that stage of neoplastic development characterized by demonstrable changes associated with increased growth rate, increased invasiveness, metastases, and alterations in biochemical and morphological characteristics of a neoplasm.

Promoter: *Use in multistage carcinogenesis*—an agent that is not carcinogenic itself but when administered after an initiator of carcinogenesis stimulates the clonal expansion of the initiated cell to produce a neoplasm. *Use in molecular biology*—a DNA sequence that initiates the process of transcription and is located near the beginning of the first exon of a structural gene.

Promotion: The enhancement of neoplasm formation by the administration of a carcinogen followed by an additional agent (promoter) that has no intrinsic carcinogenic activity by itself.

Protooncogene: A normal cellular structural gene that, when activated by mutations, amplifications, rearrangements, or viral transduction, functions as an oncogene and is associated with the development of neoplasia. Protooncogenes regulate functions related to normal growth and differentiation of tissues.

Regulatory gene: A gene that controls the activity of a structural gene or another regulatory gene. Regulatory genes usually do not undergo transcription into messenger RNA.

Sister chromatid exchange: The morphological reflection of an interchange between DNA molecules at homologous loci within a replicating chromosome.

Somatic cell: A normal diploid cell of an organism as opposed to a germ cell, which is haploid. Most neoplasms are believed to begin when a somatic cell is mutated.

Transformation: Typically refers to tissue culture systems where there is conversion of normal cells into cells with altered phenotypes and growth properties. If such cells are shown to produce invasive neoplasms in animals, malignant transformation is considered to have occurred.

Ultimate carcinogen: That form of the carcinogen that actually interacts with cellular constituents to cause the neoplastic transformation. The final product of metabolism of the procarcinogen.

Clinical Pathology

Activated partial thromboplastin time: A measure of the relative activity of factors in the intrinsic clotting sequence and the common pathway necessary in normal blood coagulation.

Alanine aminotransferase (ALT): An enzyme, primarily of liver origin, whose blood levels can rise in response to hepatocellular toxicity. Also known as serum glutamic pyruvic transaminase (SGPT).

Albumin: The most abundant blood protein synthesized by the liver.

Alkaline phosphatase: An enzyme whose blood levels can rise in response to hepatobiliary disease or increased osteoblastic (bone cell) activity. Serum alkaline phosphatase activity can decrease in fasted rats because the intestinal isozyme is an important component of serum enzyme activity.

Anemia: Any conditions in which RBC count, hemoglobin concentration, and hematocrit are reduced.

Anisocytosis: Variations in the size of red blood cells.

Aspartate aminotransferase (AST): An enzyme whose blood levels can rise in response to hepatotoxicity, muscle damage, or hemolysis. Also known as serum glutamic oxaloacetic transaminase (SGOT).

Azotemia: An increase in serum urea nitrogen and/or creatinine levels.

Creatine kinase (CK): An enzyme that is concentrated in skeletal muscle, brain, and heart tissue.

Creatinine: The end product of creatine metabolism in muscle. Elevated blood levels can indicate renal (glomerular) injury.

Fibrinogen: A glycoprotein that is involved in the formation of fibrin.

Gamma glutamyltransferase (γGT): An enzyme of liver origin, whose blood concentration can be elevated in hepatobiliary disease.

Globulin: A group of blood proteins synthesized by lymphatic tissue in the liver.

Hemolysis: The destruction of red blood cells resulting in the liberation of hemoglobin into plasma.

Icteric: Relating to a jaundiced condition, typically as a result of elevated serum bilirubin levels.

Lactate dehydrogenase: An enzyme found in several organs, including liver, kidney, heart, and skeletal muscle.

Mean corpuscular hemoglobin (MCH): The average amount of hemoglobin per red blood cell.

Mean corpuscular hemoglobin concentration (MCHC): The average hemoglobin concentration per red blood cell.

Mean corpuscular volume (MCV): The average size of the red blood cell.

Methemoglobin: Oxidized hemoglobin incapable of carrying oxygen.

Packed cell volume: The percent of blood that contains RBC components; synonymous with hematocrit.

Poikilocytosis: Variations in the shape of red blood cells.

Polychromasia: Increased basophilic staining of erythrocytes.

Polycythemia: An increase in the number of red blood cells.

Prothrombin time: A measure of the relative activity of factors in the extrinsic clotting sequence and the common pathway necessary in normal blood coagulation.

Reticulocyte: An immature (polychromatic) erythrocyte.

Reticulocytosis: Increased numbers of reticulocytes in the circulation, typically seen in response to regenerative anemia.

Sorbitol dehydrogenase (SDH): An enzyme of liver origin, whose blood concentration rises in response to hepatocellular injury.

Triglycerides: Synthesized primarily in the liver and intestine; the major form of lipid storage.

Urea nitrogen (BUN): The end product of protein catabolism. Blood levels can rise after renal (glomerular) injury.

Dermal Toxicology

Acanthosis: Hypertrophy of the stratum spinosum and granulosum.

Blanching: To take color from, to bleach. Characterized by a white or pale discoloration of the exposure area due to decreased blood flow to the skin (ischemia).

Contact dermatitis: A delayed type of induced sensitivity (allergy) of the skin with varying degrees of erythema, edema, and vesiculation, resulting from cutaneous contact with a specific allergen.

Contact urticaria: Wheal-and-flare response elicited with 30–60 min after cutaneous exposure to test substance. May be IgE mediated or nonimmunologically mediated.

Corrosion: Direct chemical action on normal living skin that results in its disintegration and irreversible alteration at the site of contact. Corrosion is manifested by ulceration and necrosis with subsequent scar formation.

Cumulative irritation: Primary irritation resulting from repeated exposures to materials that do not in themselves cause acute primary irritation.

Dermatitis: Inflammation of the skin.

Desquamation: The shedding of the cuticle in scales or the outer layer of any surface. To shred, peel, or scale off, as the casting off of the epidermis in scales or shred or the shedding of the outer layer of any surface.

Eczema: Inflammatory condition in which the skin becomes red and small vesicles, crusts, and scales develop.

Edema: An excessive accumulation of serious fluid or water in cells, tissues, or serous cavities.

Erythema: An inflammatory redness of the skin, as caused by chemical poisoning or sunburn, usually a result of congestion of the capillaries.

Eschar: A dry scab, thick coagulated crust or slough formed on the skin as a result of a thermal burn or by the action of a corrosive or caustic substance.

Exfoliation: To remove in flakes or scales, peel. To cast off in scales, flakes, or the like. To come off or separate, as scales, flakes, sheets, or layers. Detachment and shedding of superficial cells of an epithelium or from any tissue surface. Scaling or desquamation of the horny layer of epidermis, which varies in amount from minute quantities to shedding the entire integument.

Hyperkeratosis: Hypertrophy and thickening of the stratum corneum.

Irritant: A substance that causes inflammation and other evidences of irritation, particularly of the skin, on first contact or exposure; a reaction of irritation not dependent on a mechanism of sensitization.

Irritation: A local reversible inflammatory response of normal living skin to direct injury caused by a single application of a toxic substance, without the involvement of an immunological mechanism.

Necrosis: Pathological death of one or more cells or of a portion of tissue or organ, resulting from irreversible damage.

Nonocclusive: The site of application of test substance is open to the air.

Occlusive: A bandage or dressing that covers the skin and excludes it from air. Prevents loss of a test substance by evaporation and by increasing tissue penetration.

Photoallergy: An increased reactivity of the skin to ultraviolet (UV) and/or visible radiation produced by a chemical agent on an immunological basis. Previous allergy sensitized by exposure to the chemical agent and appropriate radiation is necessary. The main role of light in photoallergy appears to be in the conversion of the hapten to a complete allergen.

Photoirritation: Irritation resulting from light-induced molecular changes in the structure of chemicals applied to the skin.

Photosensitization: Sensitization of the skin to UV light, usually due to the action of certain drugs, plants, or other substances; may occur shortly after administration of the substance or may occur only after latent period of days to months. The processes whereby foreign substances, absorbed either locally into the skin or systemically, may be subjected to photochemical reactions within the skin, either leading to chemically induced photosensitivity reactions or altering the "normal" pathologic effects of light. UV-A is usually responsible for most photosensitivity reactions.

Semiocclusive: The site of application of test substance is covered; however, movement of air through covering is not restricted.

Sensitization (allergic contact dermatitis): An immunologically mediated cutaneous reaction to a substance.

Superficial sloughing: Characterized by dead tissue separated from a living structure. Any outer layer or covering that is shed. Necrosed tissue separated from the living structure.

Ulceration: The development of an inflammatory, often suppurating lesion, on the skin or an internal mucous surface of the body caused by superficial loss of tissue, resulting in necrosis of the tissue.

Ecotoxicology*

Bioaccumulation: General term describing a process by which chemicals are taken up by aquatic organisms directly from water as well as from exposure through other routes, such as consumption of food and sediment containing chemicals.

Bioaccumulation factor (BAF): The ratio of tissue chemical residue to chemical concentration in an external environmental

---

* Rand (1995). With permission.

phase (i.e., water, sediment, or food). BAF is measured as steady state in situations where organisms are exposed from multiple sources (i.e., water, sediment, and food), unless noted otherwise.

Biochemical oxygen demand (BOD): Sometimes called *biological oxygen demand*, a measure of the rate at which molecular oxygen is consumed by microorganisms during oxidation of organic matter. The standard test is the 5-day BOD test, in which the amount of dissolved oxygen required for oxidation over a 5-day period is measured. The results are measured in mg of oxygen/L (mg/L) or parts per million (ppm).

Bioconcentration: A process by which there is a net accumulation of a chemical directly from water into aquatic organisms resulting from simultaneous uptake (e.g., by gill or epithelial tissue) and elimination.

Bioconcentration factor (BCF): A term describing the degree to which a chemical can be concentrated in the tissues of an organism in the aquatic environment as a result of exposure to waterborne chemical. At steady state during the uptake phase of a bioconcentration test, the BCF is a value that is equal to the concentration of a chemical in one or more tissues of the exposed aquatic organisms divided by the average exposure water concentration of the chemical in the test.

Biodegradation: The transformation of a material resulting from the complex enzymatic action of microorganisms (e.g., bacteria, fungi). It usually leads to disappearance of the parent chemical structure and to the formation of smaller chemical species, some of which are used for cell anabolism. Although typically used with reference to microbial activity, it may also refer to general metabolic breakdown of a substance by any living organism.

Chemical oxygen demand (COD): COD is measured instead of BOD when organic materials are not easily degraded by microorganisms. Strong oxidizing agents (e.g., potassium permanganate) are used to enhance oxidation. COD values will be larger than BOD values.

EC_{50} (median effective concentration): The concentration of chemical in water to which test organisms are exposed that is estimated to be effective in producing some sublethal response in 50% of the test organisms. The EC_{50} is usually expressed as a time-dependent value (e.g., 24 h or 96 h EC_{50}). The sublethal response elicited from the test organisms as a result of exposure to the chemical must be clearly defined (e.g., test organisms may be immobilized, lose equilibrium, or undergo physiological or behavioral changes).

Fate: Disposition of a material in various environmental compartments (e.g., soil or sediment, water, air, biota) as a result of transport, transformation, and degradation.

LC_{50} (median lethal concentration): The concentration of chemical in water to which test organisms are exposed that is estimated to be lethal to 50% of the test organisms. The LC_{50} is often expressed as a time-dependent value (e.g., 24 h or 96 h LC_{50}).

Maximal acceptable toxicant concentration (MATC): The hypothetical toxic threshold concentration lying in a range bounded at the lower end by the highest tested concentration having no observed effect (NOEC) and at the higher end by the lowest concentration having a statistically significant toxic effect (LOEC) in a life cycle (full chronic) or a partial life cycle (partial chronic) test. This can be represented by NOEC < MATC < LOEC.

Octanol–water partition coefficient (K_{ow}): The ratio of the solubility of a chemical in n-octanol and water at steady state;

also expressed as *P*. The logarithm of *P* or K_{ow} (i.e., log *P* or K_{ow}) is used as an indication of the propensity of a chemical for bioconcentration by aquatic organisms.

TLm or TL_{50} (median tolerance limit): The concentration of material in water at which 50% of the test organisms survive after a specified time of exposure. The TL_m (or TL_{50}) is usually expressed as a time-dependent value (e.g., 24 h or 96 h TL_{50}).

Genetic Toxicology

Aneuploidy: An abnormal number of chromosomes in a cell or organism that is not an exact multiple of the haploid number.

Base substitution: The substitution of one or more base(s) for another in the nucleotide sequence.

Clastogen: An agent that produces structural changes of chromosomes.

Frameshift mutation: A mutation in the genetic code in which one base or two adjacent bases are inserted or deleted to the nucleotide sequence of a gene.

Gene mutation: A detectable permanent change (point mutation, insertion, or deletion) within a single gene or its regulating sequences.

Micronucleus: A microscopically detectable particle in a cell that contains nuclear DNA, usually 1/20th to 1/5th the size of the main nucleus. It may be composed of a broken centric or acentric part of a chromosome or a whole chromosome.

Mitotic index: The ratio of the number of cells in a population in various stages of mitosis to the number of cell in the population not in mitosis.

Plasmid: An autonomously replicating DNA molecule distinct from the normal genome. A plasmid may insert into the host chromosome or form an extra chromosomal element.

Point mutation: Change in the genetic code, usually confined to a single base pair.

Unscheduled DNA synthesis (UDS): DNA synthesis that occurs at some stage in the cell cycle other than S-phase in response to DNA damage and is usually associated with DNA excision repair.

Immunotoxicology

ADCC (antibody-dependent cell-mediated cytotoxicity): Cell-mediated immunity in which a specific antibody binds to a target cell, targeting it for cytolytic activity by an effector cell (generally a macrophage or natural killer [NK] cell).

Adjuvant (immunological): A material that enhances an immune response but does not confer immunity by itself. Examples include oil emulsions, aluminum salts, and toll-like receptor agonists. Complete Freund's Adjuvant (CFA) was first developed by Jules Freund in the 1940s and contains a water-in-oil emulsion and mycobacterial cell fragments. Incomplete Freund's Adjuvant does not contain the mycobacterial cell fragments.

Allogeneic: From a different genetic background. In the context of immunotoxicology, this usually refers to the use of genetically dissimilar cells in *in vitro* assays to elicit a cell-mediated immune reaction.

Antibody complex: Macromolecules produced by plasma cells that recognize specific antigens.

Antibodies: Are also referred to as immunoglobulins (Ig). They consist of two basic units: the antigen-binding fragment (Fab), which contains variable regions coding for antigen recognition, and the constant fragment (Fc), which determines the function of the antibody. Based on the Fc region, immunoglobulins are designated IgA, IgD, IgE, IgG, or IgM. The cross-linking of antibody molecules on the surface of a cell leads to the activation of complement, resulting in the destruction of the target by lytic cells or in phagocytosis by macrophages.

Antibody-forming cell (AFC) assay: Also termed plaque-forming cell (PFC) assay. This assay measures the ability of animals to produce either IgM or IgG antibodies against a T-dependent or T-independent antigen following *in vivo* (or less frequently *in vitro*) immunization. Because of the involvement of multiple cellular and humoral elements in mounting an antibody response, the assay evaluates several immune parameters simultaneously. It is considered to be one of the most sensitive indicator systems for rodent immunotoxicology studies.

Antigen: A molecule that is the target of a specific immune reaction. Antigens are recognized in a cognate fashion by either immunoglobulins or the antigen receptor on the surface of T cells. Antigens are usually proteinaceous in nature.

Antigen-presenting cell (APC): Cells responsible for making antigens accessible to immune effector and regulatory cells. Following internalization and degradation of the antigen (e.g., by phagocytes), a fragment of the antigen molecule is presented on the APC cell surface in association with histocompatibility molecules. The resulting complex is subsequently recognized either by B cells via surface-bound Ig molecules or by T cells via the T cell antigen receptor (TCR). The induction of a specific immune response then proceeds. Representative APC includes macrophages, dendritic cells, and certain B cells.

Autoimmunity: Reaction of the immune system against the host organism. In the context of drug development, autoimmunity may take the form of escape from tolerance, as when a drug modifies a host antigen, which is subsequently seen as foreign. Drug-induced autoimmunity may also result from bystander damage to host tissues from a drug-specific immune reaction.

B cell/B lymphocyte: Lymphocytes that recognize antigen via surface-bound Ig. B cells that have been exposed to cognate antigen subsequently proliferate and differentiate into plasma cells, which are responsible for producing specific antibody. B cells differentiate in the bone marrow in mammals and in an organ known as the bursa of Fabricius in birds.

Bioassay: A functional assay that depends on living cells as an indicator system. It may be performed either *in vivo* or *in vitro/ex vivo*.

Biologics/biotherapeutics: Biotechnology-derived pharmaceuticals (biopharmaceuticals) such as monoclonal antibodies.

Cell-mediated immunity (CMI): Antigen-specific reactivity mediated primarily by T cells. CMI may take the form of immunoregulatory activity (mediated by CD4+ helper T cells) or immune effector activity (mediated by CD8+ killer T cells). Other forms of direct cellular activity in host defense (e.g., NK cells and macrophages) are not antigen specific and are more accurately referred to as innate immunity.

Chemokine/chemotactic cytokines: Small chemoattractant proteins important in the stimulation and migration of immune cells. Chemokines are divided into four groups depending on the positions and spacing of the cysteine residues.

Cluster of differentiation (CD): A series of molecules expressed by immune cells on their cell surface. These proteins serve various physiological roles *in vivo*, usually as receptors or ligands. The CD nomenclature was standardized in 1982 and has been expanded since then as new molecules are characterized. The CD nomenclature is especially useful in immunophenotyping immune cell types based on the proteins present on their surface.

Complement: A group of approximately 20 protein precursor molecules assemble into a complex that intercalates into the membrane of a cell and forms a pore resulting in osmotic lysis of the target cell.

Cytokine: Small peptides produced primarily by cells of the immune system, particularly helper T cells. Cytokines are grouped into nonexclusive categories including interleukins, tumor necrosis factors, interferons, colony-stimulating factors, and various miscellaneous cytokines. Related molecules include peptide growth factors, transforming growth factors, and chemokines. Cytokines form an interactive network with both hormones and neuropeptides. Cytokines may be referred to in the older literature as lymphokines.

Cytotoxic T lymphocyte (CTL): A subset of CD8+ T cells able to kill target cells following the induction of a specific immune response. The mechanism of lysis appears to be a combination of direct lysis by extravasation of lytic molecules (such as perforins and granzymes), as well as the induction of apoptosis in the target cell. Measurement of CTL activity is a sensitive indicator of CMI.

Delayed-type hypersensitivity (DTH): A form of cell-mediated immunity in which secondary exposure to an antigen results in an inflammatory reaction mediated by CD4$^+$ T cells.

ELISA (Enzyme-linked immunosorbent assay): A type of immunoassay in which specific antibodies are used to capture and detect molecules of interest from a fluid matrix. The most common form is a "sandwich" ELISA in which the antibodies are bound to a substrate such as a plastic culture plate, and a second labeled antibody is used to detect the bound molecules.

Hapten: Low molecular weight molecules that are not antigenic by themselves but that are recognized as antigens when bound to larger molecules, usually proteins.

Host defense: The ability of an organism to protect itself against disease associated with exposure to infectious organisms, foreign tissue, or neoplasia. Host defense assays measure protection from infectious or neoplastic disease and are mediated by an immunological effect involving all components of the immune system—innate, cell-mediated, and humoral-mediated immunity. Host resistance assays are the only way to demonstrate immunological reserve.

Humoral-mediated immunity (HMI): Specific immune responses mediated primarily by humoral factors including antibodies and complement. The induction of HMI generally, although not exclusively, requires the cooperation of cellular immune mechanisms.

Hypersensitivity: A vigorous and often inappropriate immune response to seemingly innocuous antigens. Hypersensitivity is classified into subtypes depending on the mechanisms of action and the target cells or tissues.

Immunoassay: Refers to any assay that employs specific antibodies as reagents.

Immunological reserve: The concept that the immune response exhibits multiple immunological functions acting in an orchestrated manner, such that a decrease in one function is compensated by other immune functions. Immunological reserve prevents infectious or neoplastic disease due to acute reductions in one or two immune functions. This reserve would theoretically prevent a severe reduction in host resistance following a temporary immunosuppression of selected immune functions. This concept is important in interpretation of immunotoxicology data.

Immunostimulation: Enhancement of immune function above an accepted baseline (control) level. Immunostimulation may be beneficial, for example, therapeutics designed to restore a suboptimal immune response. It may conversely be detrimental, as would be the case with autoimmunity or hypersensitivity.

Immunosuppression: Depression/reduction of immune function below an accepted baseline (control) level. Immunosuppression may result from inadvertent exposure to drugs or other chemical or physical agents, intentional modification for therapeutic reasons (e.g., organ transplantation), or following exposure to certain infectious agents (e.g., HIV). An important consideration in immunotoxicology is the ability to determine the amount of immunosuppression necessary to alter host defense. Immunosuppression may result in a state of immunodeficiency.

Immunotoxicology: The discipline of synergistically applying cardinal principles of both immunology and toxicology to study the ability of certain treatments to alter the immune response in an adverse manner.

Inflammation: A nonspecific host defense mechanism characterized by the infiltration of leukocytes into the peripheral tissue, followed by the release of various mediators eliciting nonspecific physiological defense mechanisms. A normal prelude to a specific immune response; unchecked inflammation can result in extensive tissue damage.

Innate (natural or nonspecific) immunity: Host defense mechanisms that do not require prior exposure to antigen; often are antigen nonspecific in nature. Nonspecific immunity is mediated by NK cells, macrophages, neutrophils, γδT cells, and complement.

Lymphoproliferation: Proliferation of lymphocytes in response to stimulation with cellular activators such as antigens, mitogens, or allogeneic cells. Because proliferation is one of the initial consequences of activation, lymphoproliferation is used as a nonspecific indicator of immune responsiveness. This reaction is also referred to as "blastogenesis" or "mitogenesis."

Macrophage: A bone marrow-derived cell present in the peripheral tissue that serves a wide variety of host defense functions, acting as both nonspecific phagocytes and killer cells, as well as regulators of specific immune reactions. Macrophages have different designations depending on the tissue in which they are located, such as Kupffer cells in the liver and veiled cells in the lymphatic system.

Major histocompatibility complex (MHC): Cell surface molecules that determine tissue compatibility and regulate self-recognition and tolerance. Two major classes are recognized: Class I (present on all nucleated cells) and Class II (present on B cells, T cells, and macrophages). MHC molecules direct the course of immune reactivity and are presented in association with antigens by antigen presenting cells. In humans, MHCs are specifically referred to as human leukocyte antigen (HLA).

Mitogen: Molecules capable of inducing cellular activation; these may be protein or polysaccharide in nature.

Mixed lymphocyte response/reaction (MLR): An *in vitro* assay that measures the ability of lymphocytes to respond to the presence of allogeneic cells. This proliferation represents the initial stage of the acquisition of CTL function by CD8$^+$ T cells, and thus serves as a measure of CMI. The MLR is a form of lymphoproliferation. Also referred to as mixed lymphocyte culture (MLC).

Mononuclear phagocyte system: Previously known as the reticuloendothelial system (RES), this system is composed of all phagocytic cells of the body, including monocytes/macrophages and polymorphonuclear cells (i.e., neutrophils).

Mucosa-associated lymphoid tissue (MALT): Previously known as gut-associated lymphoid tissue (GALT). Lymphoid cells and tissues lining the mucosa that serve as the first point of contact with antigen encountered via this route. MALT comprises Peyer's patches, the appendix, tonsils, and lymphoid cells in the lamina propria of the gut.

Natural (innate or nonspecific) immunity: Host defense mechanisms that do not require prior exposure to antigen; often are antigen-nonspecific in nature. Nonspecific immunity is mediated by NK cells, macrophages, neutrophils, γδT cells, and complement.

Natural killer (NK) cells: A population of lymphocytes distinct from T and B cells, also referred to as large granular lymphocytes (LGLs) because of their microscopic appearance. NK cells exhibit cytotoxicity against virally infected cells and certain tumor cells. Assessment of NK cell function provides a good measure of innate immunity.

Peripheral blood mononuclear cells (PBMCs)/peripheral blood mononuclear leukocytes (PBMLs): Leukocytes derived from the peripheral circulation. Because of their accessibility, these cells are often used in *ex vivo* immune function assessment.

Skin immune system (SIS): Cells associated with the skin that participate in immunity. Includes Langerhans cells, dendritic cells, and keratinocytes. Alternatively known as skin-associated lymphoid tissue (SALT).

T cell/T lymphocyte: Lymphocytes that recognize specific antigens via a complex of molecules known collectively as the T cell antigen receptor (TCR). T cells are primarily responsible for the induction and maintenance of CMI, although they also regulate HMI and some nonimmune effector mechanisms. A variety of T cell subpopulations exist, including helper T cells, cytotoxic T cells, inducer T cells, and regulatory T cells. T cells mature in the thymus.

Xenobiotic: Any substance that is foreign to an organism. In the context of immunotoxicology, the term generally refers to nonbiological chemicals or drugs.

Inhalation Toxicology

Dust: The airborne state of a chemical that is solid at room temperature but is dispensed into a particulate atmosphere.

Fume: The airborne state of a chemical that is liquid or solid at room temperature and pressure but is heated and allowed to condense into a particulate atmosphere.

Gas: The airborne state of a chemical that boils at or below room temperature and pressure.

Geometric standard deviation (GSD): In inhalation toxicology, the relative dispersion of the MMAD such that a value approaching 1 indicates a monodisperse atmosphere.

Head-only exposure: A system for exposing test animals via inhalation in which they are restrained in a tube in such a way that only their heads are exposed directly to the test material.

Intratracheal dosing: A method of delivering, via a syringe and blunt needle, test material directly into the trachea of a test animal.

Liquid aerosol: The airborne state of a chemical that is liquid at room temperature and pressure but is nebulized into a particulate atmosphere.

Mass median aerodynamic diameter (MMAD): In inhalation toxicology, the median-sized particle based on mass measurement relative to a unit density sphere.

Nose-only exposure: A system for exposing test animals via inhalation in which they are restrained in a tube in such a way that only their nose or snout is exposed directly to the test material.

Smoke: The airborne state of a chemical that is combusted and allowed to condense into a particulate matter.

Vapor: The airborne state of a chemical that is liquid at room temperature and pressure but is volatile.

Whole-body exposure: A system for exposing test animals via inhalation in which they are placed in a chamber in such a way that their entire bodies are exposed directly to the test material.

Medical Devices

Biomaterial: A material that has direct or indirect patient contact. A biomaterial (also termed a *biomedical material*) may be composed of any synthetic or natural rubber or fiber, polymeric or elastomeric formulation, alloy, ceramic, bonding agent, ink, or other nonviable substance, including tissue rendered nonviable, used as a device or any part thereof.

Class testing: The testing of plastics for biological reactivity according to predetermined testing requirements defined by the US Pharmacopeia (USP).

Combination (medical device) product: A product containing both a drug and a device component that are physically, chemically, or otherwise combined to result in a medical product that is used therapeutically as a single entity. The medical device component must be evaluated for safety according to device requirements, the drug component must be evaluated for safety as per drug requirements, and the safety of the finished combined product must be also be evaluated.

Direct contact: When the materials of a medical device are in direct (i.e., intimate) contact with the surface or tissues of the body (e.g., adhesive bandages, pacemaker leads).

Extract: A solution produced by the incubation of a material/medical device in an appropriate vehicle. After incubation, the vehicle contains the soluble chemicals (or leachables) that have dissolved out of, or off, the material/medical device.

Indirect contact: When materials of a medical device do not contact the surface or tissues of the body, the materials of the device may influence the body. In this case, a solution or other material that contacts the device may become contaminated with leachables from the device that in turn contacts tissues of the body (e.g., intravenous infusion bag).

ISO: International Standards Organization.

Medical device: Any instrument, apparatus, appliance, material, or other articles, including software, whether used alone or in combination, intended by the manufacturer for use by human beings solely or principally for the purpose of diagnosis, prevention, monitoring, or treatment; alleviation of disease, injury, or handicap; investigation, replacement, or modification of the anatomy or of a physiological process; control of conception; and that which does not achieve its principal intended action of the body by pharmaceutical, immunological, or metabolic means but may be assisted in its function by such means.

Predicate device: A previously marketed medical device that is substantially equivalent to a proposed device. The predicate device is used as a comparison to the proposed device to establish safety and efficacy.

Processing aid: A material that contacts a medical device product during the manufacturing process and, therefore, has a potential for affecting product quality and/or may elicit a biological response following the use of a medical device. Solvents, cleaning products, lubricants, and mold-release agents are examples of processing aids.

USP negative control plastic RS: A standardized plastic produced by the USP for use as a control material in some biocompatibility assays.

Neurotoxicology

Akinesia: Absence or the loss of power of voluntary motion; immobility.

Ataxia: Incoordination; the inability to coordinate the muscles in the execution of voluntary movement.

Catalepsy: Condition in which there is waxy rigidity of the limbs that may be placed in various positions that will be maintained for a time.

Clonic convulsion: A convulsion in which the muscles alternately relax and contract.

Clonus: A form of movement characterized by contractions and relaxations of a muscle.

Convulsion: A violent spasm of the face, trunk, or extremities.

Dysarthria: Disturbance of articulation due to emotional stress or to paralysis, incoordination, or spasticity of muscles used in vocalizing.

Dyskinesia: Difficulty in performing voluntary movements; a movement disorder characterized by insuppressible, stereotyped, and automatic movements.

Dystonia: Abnormal tonicity (hyper or hypo) in any tissues.

Fasciculations: Involuntary contractions, or twitching, of groups of muscle fibers.

Hyperkinesia: Excessive muscular activity.

Myoclonus: Brief, involuntary twitching of a muscle or a group of muscles.

Myotonia: Delayed relaxation of a muscle after an initial contraction.

Opisthotonus: A form of muscle spasm of the back and rear neck muscles producing abnormal posturing in which the head, neck, and spine are arched backwards so that the heels approximate to the head.

Paresthesia: An abnormal sensation, such as burning, prickling, tickling, or tingling.

Stereotypy: The constant repetition of gestures or movements that appear to be excessive or purposeless.

Tonic convulsion: A convulsion in which muscle contraction is sustained.

Ocular Toxicology

Anterior chamber: The aqueous-containing cavity of the eye, bounded by the cornea anteriorly, the chamber angle structures peripherally, and the iris and lens posteriorly.

Blepharitis: Inflammation of the eyelids.

Blepharospasm: Involuntary spasm of the lids.

Cataract: An opacity of the lens or its capsule.

Chemosis: Intense edema of the conjunctiva. The conjunctiva is loose fibrovascular connective tissue that is relatively rich in lymphatics and responds to noxious stimuli by swelling to the point of prolapse between the lids.

Choroid: The vascular middle coat between the retina and sclera.

Ciliary body: Portion of the uveal tract between the iris and the choroid consisting of ciliary processes and the ciliary muscle.

Conjunctiva: Mucous membrane that lines the posterior aspect of the eyelids (palpebral conjunctiva) and the anterior sclera (bulbar conjunctiva).

Conjunctivitis: Inflammation of the conjunctiva.

Cornea: Transparent portion of the outer coat of the eyeball forming the anterior wall of the anterior chamber.

Exophthalmos: Abnormal protrusion of the eyeball.

Fluorescein (fluorescein sodium): A fluorescent dye, the simplest of the fluorane dyes and the mother substance of eosin, which is commonly used intravenously to determine the state of adequacy of circulation in the retina and a lesser degree the choroid and iris. Another important use is to detect epithelial lesions of the cornea and conjunctiva. Peak excitation occurs with light at a wavelength between 485 and 500 nm, and peak emission occurs between 520 and 530 nm.

Fovea: Depression in the macula adapted for most acute vision.

Fundus: The posterior portion of the eye visible through an ophthalmoscope.

Hyperemia: Excess of blood in a part due to local or general relaxation of the arterioles. Blood vessels become congested and give the area involved a reddish or red-blue color.

Injection: Congestion of blood vessels.

Iris: The circular pigmented membrane behind the cornea and immediately in front of the lens; the most anterior portion of the vascular tunic of the eye. It is composed of the dilator and sphincter muscles and the two-layered posterior epithelium and mesodermal components that form the iris stroma.

Iritis: Inflammation of the iris, manifested by vascular congestion (hyperemia). An outpouring of serum proteins into the aqueous (flare) may accompany the inflammatory reaction.

Keratitis: Inflammation of the cornea.

Lens: A transparent biconvex structure suspended in the eyeball between the aqueous and the vitreous. Its function is to bring rays of light to a focus on retina. Accommodation is produced by variations in the magnitude of this effect.

Miotic: A drug causing pupillary constriction.

Mydriatic: A drug causing pupillary dilatation.

Nystagmus: An involuntary, rapid movement of the eyeball that may be horizontal, vertical, rotatory, or mixed.

Optic disk: Ophthalmoscopically visible portion of the optic nerve.

Palpebral: Pertaining to the eyelid.

Pannus: Vascularization and connective-tissue deposition beneath the epithelium of the cornea.

Posterior chamber: Space filled with aqueous anterior to the lens and posterior to the iris.

Ptosis: Drooping of the upper eyelid.

Pupil: The round opening at the center of the iris that allows transmission of light to the posterior of the eyeball.

Retina: The innermost or nervous tunic of the eye that is derived from the optic cup (the outer layer develops into the complex sensory layer).

Sclera: The white tough covering of the eye that, with the cornea, forms the external protective coat of the eye.

Vitreous: Transparent, colorless, mass of soft, gelatinous material filling the space in the eyeball posterior to the lens and anterior to the retina.

Pharmaceuticals

Abbreviated new drug application (ANDA): An ANDA contains data that, when submitted to FDA's Center for Drug Evaluation and Research, Office of Generic Drugs, provide for the review and ultimate approval of a generic drug product. Generic drug applications are called "abbreviated" because they are generally not required to include preclinical (animal) and clinical (human) data to establish safety and effectiveness. Instead, a generic applicant must scientifically demonstrate that its product is bioequivalent (i.e., performs in the same manner as the innovator drug). Once approved, an applicant may manufacture and market the generic drug product to provide a safe, effective, low-cost alternative.

Active ingredient: Any component that provides pharmacological activity or other direct effect in the diagnosis, cure, mitigation, treatment, or prevention of disease or affects the structure or any function of the body of humans or animals.

ADME: Absorption, distribution, metabolism, and excretion.

Biological license application (BLA): Biological products are approved for marketing under the provisions of the Public Health Service (PHS) Act. The Act requires a firm who manufactures a biologic for sale in interstate commerce to hold a license for the product. A biologics license application is a submission that contains specific information on the manufacturing processes, chemistry, pharmacology, clinical pharmacology, and the medical effects of the biological product. If the information provided meets FDA requirements, the application is approved, and a license is issued allowing the firm to market the product.

Biological product: A biological product is any virus, serum, toxin, antitoxin, vaccine, blood, blood component or derivative, allergenic product, or analogous product applicable to the prevention, treatment, or cure of diseases or injuries. Biological products are a subset of "drug products" distinguished by their manufacturing processes (biological process vs. chemical process). In general, the term "drugs" includes biological products.

CBER: FDA's Center for Biologics Evaluation and Research—the division charged with regulating biological products.

CDER: FDA's Center for Drug Evaluation and Research—the division charged with developing and enforcing policy with regard to the safety, effectiveness, and labeling of all drug products for human use.

Drug: A substance recognized by an official pharmacopoeia or formulary. A substance intended for use in the diagnosis, cure, mitigation, treatment, or prevention of disease. A substance (other than food) intended to affect the structure or any function of the body. A substance intended for use as a component of a medicine but not a device or a component, part, or accessory of a device. Biological products are included within this definition and are generally covered by the same laws and regulations, but differences exist regarding their manufacturing processes (chemical process vs. biological process).

Human equivalent dose (HED): A dose in humans anticipated to provide the same degree of effect as that observed in animals at a given dose. In drug development, the term HED is usually used to refer to the HED of the no-observed-adverse-effect level (NOAEL).

ICH: The International Council for Harmonization of Technical Requirements for Pharmaceuticals for Human Use (ICH) is a project that brings together the regulatory authorities of Europe, Japan, and the United States and experts from the pharmaceutical industry in the three regions to discuss scientific and technical aspects of pharmaceutical product registration. The ICH develops scientific guidelines that are harmonized between the three regions.

Identification threshold: A limit (%) above which an impurity or degradation product should be identified.

IND: Investigational New Drug Application—a request to initiate a clinical study of a new drug product.

Label: The FDA-approved label is the official description of a drug product that includes indication (what the drug is used for); who should take it; adverse events (side effects); instructions for uses in pregnancy, children, and other populations; and safety information for the patient. Labels are often found inside drug product packaging.

Maximum recommended starting dose (MRSD): The highest dose recommended as the initial dose of a drug in a clinical trial. The MRSD is predicted to cause no adverse reactions.

New drug application (NDA): When the sponsor of a new drug believes that enough evidence on the drug's safety and

effectiveness has been obtained to meet FDA's requirements for marketing approval, the sponsor submits to
FDA a NDA. The application must contain data from specific technical viewpoints for review, including chemistry,
pharmacology, medical, biopharmaceutics, and statistics.
If the NDA is approved, the product may be marketed
in the United States. For internal tracking purposes, all
NDAs are assigned an NDA number.

Pharmacologically active dose (PAD): The lowest dose tested
in an animal species with the intended pharmacological
activity.

PLA: Product License Application for a biologic.

Qualification: For pharmaceuticals, the process of acquiring and
evaluating data that establishes the biological safety of an
individual impurity or degradation product or a given
impurity or degradation profile at the level(s) specified.

Qualification threshold: A limit (%) above which an impurity or
degradation product should be qualified.

Reporting threshold: A limit (%) above which an impurity or
degradation product should be reported.

Reproductive/Developmental Toxicology

Aberration: A minor structural change. It may be a retardation (a provisional delay in morphogenesis), a variation
(external appearance controlled by genetic and extragenetic factors), or a deviation (resulting from altered
differentiation).

Ablepharia: Absence or reduction of the eyelid(s).

Abrachius: Without arms, forelimbs.

Acardia: Absence of the heart.

Acaudia (anury): Agenesis of the tail.

Accessory spleen: An additional spleen.

Acephaly: Congenital absence of the head.

Achondroplasia: A hereditary defect in the formation of epiphysial cartilage, resulting in a form of dwarfism with short limbs, normal trunk, small face, normal vault, etc.

Acrania: Partial or complete absence of the skull.

Acystia: Absence of the urinary bladder.

Adactyly: Absence of digits.

Agastria: Absence of the stomach.

Agenesis: Absence of an organ or part of an organ.

Agenesis of the kidney: Absence of the kidney(s).

Agenesis of the lung (lobe): Complete absence of a lobe of the lung.

Aglossia: Absence of the tongue.

Agnathia: Absence of lower jaw (mandible).

Anal atresia: Congenital absence of the anus.

Anencephaly: Congenital absence of the cranial vault with missing or small brain mass.

Anomaly or abnormality: A morphological or functional deviation from normal limit. It can be a malformation or a variation.

Anophthalmia: Absence of eye(s).

Anorchism: Congenital absence of one or both testes.

Anotia: Absence of the external ear(s).

Aphalangia: Absence of a finger or a toe; corresponding metacarpals not affected.

Aplasia: Lack of development of an organ, frequently used to designate complete suppression or failure of development of a structure from the embryonic primordium.

Aplasia of the lung: The trachea shows rudimentary bronchi, but pulmonary and vascular structures are absent.

Apodia: Absence of one or both feet.

Aproctia: Imperforation or absence of anus.

Arrhinia: Absence of nose.

Arthrogryposis: Persistent flexure or contracture of a joint; flexed paw (bent at wrist) is the most common form of arthrogryposis.

Aspermia: No ejaculate

Astomia: Absence of oral orifice.

Asthenozoospermia: More than 50% spermatozoa with poor (<2 grade) forward progression.

Azoospermia: No spermatozoa in the ejaculate.

Brachydactyly: Shortened digits.

Brachyury (short tail): Tail that is reduced in length.

Bulbous rib: Having a bulge or balloon-like enlargement some-where along its length.

Cardiomegaly: Hypertrophy (enlargement) of the heart.

Cardiovascular situs inversus: Mirror-image transposition of the heart and vessels to the other side of the body.

Cephalocele: A protrusion of a part of the cranial contents, not necessarily neural tissue.

Conceptus: The sum of derivatives of a fertilized ovum at any stage of development from fertilization until birth. Also term used when the stage of prenatal development at the time of initial insult is not known; also referred to as *embryo–fetus.*

Corpus luteum: The yellow endocrine body formed in the ovary at the site of the ruptured Graafian follicle.

Craniorachischisis: Exencephaly and holorrachischisis (fissure of the spinal cord).

Cranioschisis: Abnormal fissure of the cranium; may be associated with meningocele or encephalocele.

Cryptorchidism (undescended testes, ectopic testes): Failure of the testes to descend into the scrotum (can be unilateral).

Cyclopia: One central orbital fossa with none, one, or two globes.

Deflection: A turning, or state of being turned, aside.

Deformity: Distortion of any part or general disfigurement of the body.

Deviation: Variation from the regular standard or course.

Dextrocardia: Location of the heart in the right side of the thorax; a developmental disorder that is associated with total or partial situs inversus (transposition of the great vessels and other thoracoabdominal organs) or occurs as an isolated anomaly.

Dextrogastria: Having the stomach on the right side of the body.

Displaced rib: Out of normal position.

Dysgenesis: Defective development; malformation.

Dysmelia: Absence of a portion of one or several limbs.

Dysplasia: (a) Abnormal development of tissues; (b) alteration in size, shape, or organization of adult cells.

Dystocia: Abnormal labor.

Ectocardia: Displacement of the heart inside or outside the thorax.

Ectopic: Out of the normal place.

Ectopic esophagus: Displacement of the esophagus (description of position should be included).

Ectopic pinna: Displaced external ear.

Ectrodactyly: Absence of all or of only a part of digit (partial ectrodactyly).

Ectromelia: Aplasia or hypoplasia of one or more bones of one or more limbs (this term includes amelia, hemimelia, and phocomelia).

Encephalocele: A partial protrusion of brain through an abnormal cranial opening; not as severe as exencephaly.

Estrus: Phase of the sexual cycle of female mammals characterized by willingness to mate.

Exencephaly: Brain outside of the skull as a result of a large cranial defect.

Exomphalos: Congenital herniation of abdominal viscera into umbilical cord.

Exophthalmos: Protrusion of the eyeball ("pop" eye).

Fecundity: Ability to produce offspring rapidly and in large numbers.

Feticide: The destruction of the fetus in the uterus.

Gamete: A male (spermatozoa) or female (ovum) reproductive cell.

Gastroschisis: Fissure of the abdominal wall (median line) not involving the umbilicus, usually accompanied by protrusion of the small part of the large intestine, not covered by membranous sac.

Hemivertebra: Presence of only one-half of a vertebral body.

Hepatic lobe agenesis: Absence of a lobe of the liver.

Hepatomegaly: Abnormal enlargement of the liver.

Hydrocephaly: Enlargement of the head caused by abnormal accumulation of cerebrospinal fluid in subarachnoid cavity (external hydrocephaly) or ventricular system (internal hydrocephaly).

Hydronephrosis: Dilatation of the renal pelvis usually combined with destruction of renal parenchyma and often with dilation of the ureters (bilateral, unilateral). Note: This is a pathology term and should have histological confirmation.

Hypoplasia of the lung: Bronchial tree is poorly developed, and pulmonary tissue shows an abnormal histological picture (total or partial); incomplete development, smaller.

Hypospadias: Urethra opening on the underside of the penis or on the perineum (males) or into the vagina (females).

Imperforate: Not open; abnormally closed.

Incomplete ossification (delayed, retarded): The extent of ossification is less than what would be expected for that developmental age, not necessarily associated with reduced fetal or pup weight.

Levocardia: Displacement of the heart in the extreme left hemithorax.

Lordosis: Anterior concavity in the curvature of the cervical and lumbar spine as viewed from the side.

Macrobrachia: Abnormal size or length of the arm.

Macrodactylia: Excessive size of one or more digits.

Macroglossia: Enlarged tongue, usually protruding.

Macrophthalmia: Enlarged eye(s).

Malformation: Defective or abnormal formation; deformity. A permanent structural deviation that generally is incompatible with, or severely detrimental to, normal survival or development.

Meiosis: Cell division occurring in maturation of the sex cell (gametes) by means of which each daughter nucleus receives half the number of chromosomes characteristic of the somatic cells of the species.

Microcephaly: Small head.

Micrognathia: Shortened lower jaw (mandible); tongue may protrude.

Microphthalmia: Small eyes.

Microstomia: Small mouth opening.

Microtia: Small external ear.

Monocardium: Possessing a heart with only one atrium and one ventricle.

Multigravida: A female pregnant for the second (or more) time.

Naris (nostril) atresia: Absence or closure of nares.

Nasal agenesis: Absence of the nasal cavity and external nose.

Normozoospermia: Normal semen sample.

Nulliparous: A female that has never borne a viable offspring.

Oligodactyly: Fewer than normal number of digits.

Oligohydramnios (oligoamnios): Reduction in the amount of amniotic fluid.

Oligozoospermia: A subnormal sperm concentration in ejaculate.

Omphalocele: Midline defect in the abdominal wall at the umbilicus, through which the intestines and often other viscera (stomach, spleen, and portions of the liver) protrude. These are always covered by a membranous sac. As a rule, the umbilical cord emerges from the top of the sac.

Pachynsis: Abnormal thickening.

Patent ductus arteriosus (ductus botalli): An open channel of communication between the main pulmonary artery and the aorta; may occur as an isolated abnormality or in combination with other heart defects.

Polydactyly: Extra digits.

Polysomia: A fetal malformation consisting of two or more imperfect and partially fused bodies.

Pseudopregnancy: (a) False pregnancy: condition occurring in animals in which anatomical and physiological changes occur similar to those of pregnancy; (b) the premenstrual stage of the endometrium: so called because it resembles the endometrium just before the implantation of the blastocyst.

Rachischisis: Absence of vertebral arches in limited area (partial rachischisis) or entirely (rachischisis totalis).

Renal hypoplasia: Incomplete development of the kidney.

Resorption: A conceptus that, having implanted in the uterus, subsequently died and is being or has been resorbed.

Rhinocephaly: A developmental anomaly characterized by the presence of a proboscis-like nose above the eyes, partially or completely fused into one.

Rudimentary rib: Imperfectly developed riblike structure.

Schistoglossia: Cleft tongue.

Seminiferous epithelium: The normal cellular components within the seminiferous tubule consisting of Sertoli cells, spermatogonia, primary spermatocytes, secondary spermatocytes, and spermatids.

Septal agenesis: Absence of nasal septum.

Sertoli cells: Cells in the testicular tubules providing support, protection, and nutrition for the spermatids.

Spermatocytogenesis: The first stage of spermatogenesis in which spermatogonia develop into spermatocytes and then into spermatids.

Spermiation: The second stage of spermatogenesis in which the spermatids transform into spermatozoa.

Spina bifida: Defect in closure of bony spinal cavity.

Sympodia: Fusion of the lower extremities.

Syndactyly: Partially or entirely fused digits.

Tetralogy of Fallot: An abnormality of the heart that includes pulmonary stenosis, ventricular septal defect, dextroposition of the aorta overriding the ventricular septum and receiving blood from both ventricles, and right ventricular hypertrophy.

Teratozoospermia: Fewer than 30% spermatozoa with normal morphology.

Thoracogastroschisis: Midline fissure in the thorax and abdomen.

Totalis or partialis: Total or partial transposition of viscera (due to incomplete rotation) to the other side of the body; heart most commonly affected (dextrocardia).

Tracheal stenosis: Constriction or narrowing of the tracheal lumen.

Unilobular lung: In the rat fetus, a condition in which the right lung consists of one lobe instead of four separate lobes.

Vaginal plug: A mass of coagulated semen that forms in the vagina of animals after coitus; also called copulation plug or bouchon vaginal.

Variation: An alteration that represents a retardation in development, a transitory alteration, or a permanent alteration not believed to adversely affect survival, growth, development, or functional competence.

Risk Assessment/General Toxicology

Absorbed dose: The amount of a substance penetrating across the exchange boundaries of an organism and into body fluids and tissues after exposure.

Acceptable daily intake (ADI): A value used for noncarcinogenic effects that represents a daily dose that is very likely to be safe over an extended period of time. An ADI is similar to a reference dose (RfD) (defined in the succeeding text) but less strictly defined.

Administered dose: The amount of a substance given to a human or test animal in determining dose–response relationships, especially through ingestion or inhalation (see Applied dose). Administered dose is actually a measure of exposure, because even though the substance is "inside" the organism, once ingested or inhaled, administered dose does not account for absorption (see Absorbed dose).

Aggregate risk: The sum of individual increased risks of an adverse health effect in an exposed population.

Applied dose: The amount of a substance given to a human or test animal in determining dose–response relationships, especially through dermal contact. Applied dose is actually a measure of exposure, since it does not take absorption into account (see Absorbed dose).

Biological significant effect: A response in an organism or other biological systems that is considered to have a substantial or noteworthy effect (positive or negative) on the well-being of the biological system. Used to distinguish from statistically significant effects or changes, which may or may not be meaningful to the general state of health of the system.

Cancer potency factor (CPF): The statistical 95% upper confidence limit on the slope of the dose–response relationship at low doses for a carcinogen. Values are in units of lifetime risk per unit dose (mg/kg/day). A plausible upper bound risk is derived by multiplying the extended lifetime average daily dose (LADD) by the CPF.

Case-control study: A retrospective epidemiologic study in which individuals with the disease under study (cases) are compared with individuals without the disease (controls) in order to contrast the extent of exposure in the diseased group with the extent of exposure in the controls.

Ceiling limit: A concentration limit in the work place that should not be exceeded, even for a short time, to protect workers against frank health effects.

CFR: Code of Federal Regulations (United States).

Cohort study: A study of a group of persons sharing a common experience (e.g., exposure to a substance) within a defined time period; the experiment is used to determine if an increased risk of a health effect (disease) is associated with that exposure.

Confidence limit: The confidence interval is a range of values that has a specified probability (e.g., 95%) of containing a given parameter or characteristic. The confidence limit often refers to the upper value of the range (e.g., upper confidence limit).

Critical endpoint: A chemical may elicit more than one toxic effect (endpoint), even in one test animal, in tests of the same or different duration (acute, subchronic, and chronic exposure studies). The doses that cause these effects may differ. The critical endpoint used in the dose–response assessment is the one that occurs at the lowest dose. In the event that data from multiple species are available, it is often the most sensitive species that determines the critical endpoint. This term is applied in the derivation of risk reference doses (RfDs).

Cross-sectional study: An epidemiologic study assessing the prevalence of a disease in a population. These studies are most useful for conditions or diseases that are not expected to have a long latent period and do not cause death or withdrawal from the study population. Potential bias in case ascertainment and exposure duration must be addressed when considering cross-sectional studies.

De minimus risk: From the legal maxim, "de minimus non curat lex" or the law is not concerned with trifles. As relates to risk assessment of carcinogens, it is commonly interpreted to mean that a lifetime risk of 1×10^{-6} is a *de minimus* level of cancer risk (i.e., insignificant and therefore acceptable) and is of no public health consequence.

Dispersion model: A mathematical model or computer simulation used to predict the movement of airborne or water contaminants. Models take into account a variety of mixing mechanisms that dilute effluents and transport them away from the point of emission.

DNEL (Derived no effect level): A European occupational exposure level of chemical determined for REACH above which humans should not be exposed. A DNEL is typically derived from the No Observed Adverse Effect Level (NOAEL) for a principal toxic effect in an animal study divided by assessment (safety/uncertainty) factors. The risk for a chemical substance is considered acceptable if the exposure levels do not exceed the DNEL. DNELs are communicated in Safety Data Sheets (SDS).

Dose: The amount of substance administered to an animal or human generally expressed as the weight or volume of the substance per unit of body weight (e.g., mg/kg, mL/kg).

Dose–response relationship: A relationship between the dose, often actually based on "administered dose" (i.e., exposure) rather than absorbed dose, and the extent of toxic injury produced by that chemical. Response can be expressed as either the severity of injury or proportion of exposed subjects affected. A dose–response assessment is one of the steps in a risk assessment.

Duration of exposure: Generally referred to in toxicology as acute (one time), subacute (repeated over several weeks), subchronic (repeated for a fraction of a lifetime), and chronic (repeated for nearly a lifetime).

Endemic: Present in a community or among a group of people; said of a disease prevailing continually in a region.

Environmental fate: The destiny of a chemical or biological pollutant after release into the environment. Environmental fate involves temporal and spatial considerations of transport, transfer, storage, and transformation.

Exposure: Contact of an organism with a chemical, physical, or biological agent. Exposure is quantified as the amount of the agent available at the exchange boundaries of the organism (e.g., skin, lungs, digestive tract) and available for absorption.

Exposure frequency: The number of times an exposure occurs in a given period. The exposure(s) may be continuous, discontinuous but regular (e.g., once daily), or intermittent.

Extrapolation: An estimate of response or quantity at a point outside the range of the experimental data. Also refers to the estimation of a measured response in a different species or by a different route than that used in the experimental study of interest (i.e., species to species, route to route, acute to chronic, high to low).

Fence line concentration: Modeled or measured concentrations of pollutants found at the boundaries of a property on which a pollution source is located. Usually assumed to be the nearest location at which an exposure of the general population could occur.

Frank effect level (FEL): Related to biological responses to chemical exposures (compare with NOAEL and LOEL); the exposure level that produces an unmistakable adverse health effect (such as inflammation, severe convulsions, or death).

Hazard: The inherent ability of a substance to cause an adverse effect under defined conditions of exposure.

Hazard index: The ratio of the maximum daily dose (MDD) to the ADI used to evaluate the risk of noncarcinogens. A value of less than 1 indicates that the risk from the exposure is likely insignificant; a value greater than 1 indicates a potentially significant risk.

Human equivalent dose (HED): The human dose of an agent expected to induce the same type and severity of toxic effect that an animal dose has induced.

Immediately dangerous to life and health (IDLH): A concentration representing the maximum level of a pollutant from which an individual could escape within 30 min without escape-impairing symptoms or irreversible health effects.

Incidence: The number of new cases of a disease within a specified time period. It is frequently presented as the number of new cases per 1,000, 10,000, or 100,000. The incidence rate is a direct estimate of the probability or risk of developing a disease during a specified time period.

Involuntary risk: A risk that impinges on an individual without their awareness or consent.

IUCLID: An acronym for International Uniform Chemical Information Database software application used to capture, store, maintain, and exchange data on intrinsic and hazard properties of chemical substances. Co-developed by the European Chemicals Agency (ECHA) and the OECD, ECHA maintains and makes the software available for the submission of chemical information under REACH.

Latency: The period of time between exposure to an injurious agent and the manifestation of a response.

LC_{LO} (lethal concentration low): The lowest concentration of a chemical required to cause death in some of the population after exposure for a specified period of time and observed for a specified period of time after exposure. Refers to inhalation time exposure in the context of air toxics (may refer to water concentration for tests of aquatic organisms).

LC_{50} (median lethal concentration): The concentration of a chemical required to cause death in 50% of the exposed population when exposed for a specified time period and observed for a specified period of time after exposure. Refers to inhalation exposure concentration in the context of air toxics (may refer to water concentration for tests of aquatic organisms).

LD_{LO} (lethal dose low): The lowest dose of a chemical required to cause death in some of the population after noninhalation exposure (e.g., injection, ingestion), for a specified observation period after exposure.

LD_{50} (median lethal dose): The dose of a chemical required to cause death in 50% of the exposed population after noninhalation exposure (e.g., injection, ingestion), for a specified observation period after exposure.

Lifetime average daily dose (LADD): The total dose received over a lifetime multiplied by the fraction of a lifetime during which exposure occurs, expressed in mg/kg body weight/day.

Lifetime risk: A risk that results from lifetime exposure.

Lowest-observed-adverse-effect level (LOAEL): The lowest dose or exposure level of a chemical in a study at which there is a statistically or biologically significant increase in the frequency or severity of an *adverse* effect in the exposed population as compared with an appropriate, unexposed control group.

Lowest-observed-effect level (LOEL): In a study, the lowest dose or exposure level of a chemical at which a statistically or biologically significant effect is observed in the exposed population compared with an appropriate unexposed control group. The effect is generally considered not to have an adverse effect on the health and survival of the animal. This term is occassionally misused in place of a LOAEL.

Margin of exposure (MOE): The ratio of the NOAEL to the estimated human exposure. The MOE was formerly referred to as the margin of safety (MOS).

Maximum contaminant level (MCL): The maximum level of a contaminant permissible in water as defined by regulations promulgated under the Safe Drinking Water Act.

Maximum daily dose (MDD): Maximum dose received on any given day during a period of exposure generally expressed in mg/kg body weight/day.

Maximum tolerated dose (MTD): The highest dose of a toxicant that causes toxic effects without causing death during a chronic exposure and that does not decrease the body weight by more than 10%.

Modifying factor (MF): A factor that is greater than zero and less than or equal to 10; used in the operational derivation of an RfD. Its magnitude depends upon an assessment of the scientific uncertainties of the toxicological database not explicitly treated with standard uncertainty factor (UF) (e.g., the completeness of the overall database). The default value for the MF is 1. The use of an MF was discontinued by the EPA in 2004.

Multistage model: A mathematical function used to extrapolate the probability of incidence of disease from a bioassay in animals using high doses, to that expected to be observed at the low doses that are likely to be found in chronic human exposure. This model is commonly used in quantitative carcinogenic risk assessments where the chemical agent is assumed to be a complete carcinogen and the risk is assumed to be proportional to the dose in the low region.

Nonthreshold toxicant: An agent considered to produce a toxic effect from any dose; any level of exposure is deemed to involve some risk. Usually only in regard to carcinogenesis.

No-observed-adverse-effect level (NOAEL): The highest experimental dose at which there are not statistically or biologically significant increases in frequency or severity of *adverse* health effects, as seen in the exposed population compared with an appropriate, unexposed population. Effects may be produced at this level, but they are not considered to be adverse.

No-observed-effect level (NOEL): The highest experimental dose at which there is not statistically or biologically significant increase in the frequency or severity of effects seen in the exposed compared with an appropriate unexposed population.

Occupational exposure limit (OEL): A generic term denoting a variety of values and standards, generally time-weighted average (TWA) concentrations of airborne substances to which a worker can be safely exposed during defined work periods.

Permissible exposure limit (PEL): Similar to an OEL.

PNEC (Predicted no effect concentration): A European exposure concentration of a chemical determined for REACH, which marks the limit at which below no adverse effects in an ecosystem are expected to occur during short-term or long-term exposure. PNECs are calculated by dividing endpoints of toxicity such as mortality (LC50), or the No Observed Effect Concentration (NOAEC) for growth and reproduction by assessment (safety/uncertainty) factors.

Potency: A comparative expression of chemical or drug activity measured in terms of the relationship between the incidence or intensity of a particular effect and the associated dose of a chemical, to a given or implied standard of reference. Can be used for ranking the toxicity of chemicals.

ppb: Parts per billion.

ppm: Parts per million.

Prevalence: The percentage of a population that is affected with a particular disease at a given time.

q1*: The symbol used to denote the 95% upper bound estimate of the linearized slope of the dose–response curve in the low-dose region as determined by the multistage model.

REACH: Regulation (EC) No. 1907/2006 of the European Parliament and of the Council concerning the Registration, Evaluation, Authorization and Restriction of Chemicals, known in short as the "REACH," requires manufacturers and importers of chemicals to gather information on the properties of their chemical substances, which will allow their safe handling, and to register the information in a central database in the European Chemicals Agency (ECHA) in Helsinki. The Agency is the central point in the REACH system: it manages the databases necessary to operate the system, coordinates the in-depth evaluation of suspicious chemicals and it builds a public database for chemical hazard information.

Reference dose (RfD): An estimate (with uncertainty spanning perhaps an order of magnitude or more) of the daily exposure to the human population (including sensitive subpopulations) that is likely to be without deleterious effects during a lifetime. The RfD is reported in units of mg of substance/kg body weight/day for oral exposures or mg/substance/m^3 of air breathed for inhalation exposures (RfC).

Risk: The probability that an adverse effect will occur under a particular condition of exposure.

Risk assessment: The scientific activity of evaluating the toxic properties of a chemical and the conditions of human exposure to it in order both to ascertain the likelihood that exposed humans will be adversely affected and to characterize the nature of the effects they may experience. May contain some or all of the following four steps:

Hazard identification: The determination of whether a particular chemical is or is not causally linked to particular health effect(s)

Dose–response assessment: The determination of the relation between the magnitude of exposure and the probability of occurrence of the health effects in question

Exposure assessment: The determination of the extent of human exposure

Risk characterization: The description of the nature and often the magnitude of human risk, including attendant uncertainty

Risk management: The decision-making process that uses the results of risk assessment to produce a decision about environmental action. Risk management includes the consideration of technical, scientific, social, economic, and political information.

Short-term exposure limit (STEL): A TWA occupational exposure level (OEL) that the American conference of Government and Industrial Hygienists (ACGIH) indicates should not be exceeded any time during the work day. Exposures at the STEL should not be longer than 15 min and should not be repeated more than four times per day. There should be at least 60 min between successive exposure at the STEL.

Slope factor: See Cancer potency factor (CPF).

SNUR: Significant new use rule.

Standardized mortality ratio: The number of deaths, either total or cause specific, in a given group expressed as a percentage of the number of deaths that could have been expected if the group has the same age- and sex-specific rates as the general population. Used in epidemiologic studies to adjust mortality rates to a common standard so that comparisons can be made among groups.

STEL: See Short-term exposure limit (STEL).

Surface area scaling factor: The intra- and interspecies scaling factor most commonly used for cancer risk assessment by the US EPA to convert an animal dose to an HED: milligrams per square meter surface area per day. Body surface area is proportional to basal metabolic rate; the ratio of surface area to metabolic rate tends to be constant from one species to another. Since body surface area is approximately proportional to an animal's body weight to the 2/3 power, the scaling factor can be reduced to milligrams per body weight$^{2/3}$.

TC$_{LO}$ (toxic concentration low): The lowest concentration of a substance in air required to cause a toxic effect in some of the exposed population.

TD$_{LO}$ (toxic dose low): The lowest dose of a substance required to cause a toxic effect in some of the exposed population.

Threshold limit value (TLV): The TWA concentration of a substance below which no adverse health effects are expected to occur for workers assuming exposure for 8 h per day, 40 h per week. TLVs are published by the American Conference of Governmental Industrial Hygienists (ACGIH).

TWA: An approach to calculating the average exposure over a specified time period.

UF: One of several, generally 10-fold factors, applied to a NOAEL or a LOAEL to derive an RfD from experimental data. UFs are intended to account for (a) the variation in the sensitivity among the members of the human population, (b) the uncertainty in extrapolating animal data to human, (c) the uncertainty in extrapolating from data obtained in a less-than-lifetime exposure study to chronic exposure, (d) the uncertainty in using an LOAEL rather than an NOAEL for estimating the threshold region, and (e) uncertainty with extrapolation when the database is incomplete.

Unit cancer risk: A measure of the probability of an individual's developing cancer as a result of exposure to a specified unit ambient concentration. For example, an inhalation unit cancer risk of 3.0×10^{-4} near a point source implies that if 10,000 people breathe a given concentration of a carcinogenic agent (e.g., 1 µg/m^3) for 70 years, three of the 10,000 will develop cancer as a result of this exposure. In water, the exposure unit is usually 1 µg/L, whereas in air it is 1 µg/m^3.

Upper bound cancer risk assessment: A qualifying statement indicating that the cancer risk estimate is not a true value in that the dose–response modeling used provides a value that is not likely to be an underestimate of the true value. The true value may be lower than the upper bound cancer risk estimate, and it may even be close to zero. This results from the use of a statistical upper confidence limit and from the use of conservative assumptions in deriving the cancer risk estimate.

Upper 95% confidence limit: Assuming random and normal distribution, this is the range of values below which a value will fall 95% of the time.

Voluntary risk: Risk that an individual has consciously decided to accept.

Safety Pharmacology

Action potential amplitude (APA): The amplitude of the cardiac action potential in millivolts (mV) and is a measure of the total amplitude of the action potential from initial resting level to peak depolarization. APA is determined by the combined activity of voltage-gated sodium and potassium channels.

Action potential duration (APD): The duration of the cardiac action potential in milliseconds, measured from the initial upstroke to the point of return to either 60% (APD_{60}) or 90% (APD_{90}) of the initial resting potential. APD prolongation corresponds to an increase in the electrocardiographic QT interval.

Basic cycle length (BCL): In cardiac electrophysiology, is the BCL of repetitive stimulation and is the inverse of stimulation frequency. BCL = 2, 1, and 0.5 s, respectively, corresponds to bradycardic (30 beats/min), normocardic (60 beats/min), and tachycardic (120 beats/min) heart rates.

hERG: Refers to both the gene that encodes the pore-forming subunit of human cardiac ether-a-go-go-related potassium channel and the protein subunit itself. hERG channels are responsible for the delayed rectifier potassium current (I_{Kr}) that regulates action potential repolarization.

IC$_{50}$: Refers to the concentration or dose of test article that produces 50% inhibitory response in a test system assay.

Purkinje cells: Modified cardiac muscle cells specialized for the conduction of electric excitation from the atrioventricular node through the ventricular septum and throughout the walls of the ventricle. Purkinje cells are organized into fiber bundles that can be readily dissected from the working myocardium and studied *in vitro* using electrophysiological recording methods.

Purkinje fiber stimulation frequency: The repetition rate for the application of brief electric shocks to a Purkinje fiber preparation. A frequency rate of 1 Hz corresponds to 1 shock/s and represents the normal heart rate of 1 beat/s. Frequencies of 0.5 and 2 Hz correspond to bradycardic and tachycardic heart rates.

QT interval: The time interval in the electrocardiogram extending from the start of the QRS complex to the end of the T wave and is the measure of the duration of ventricular depolarization and repolarization. The QRS complex corresponds to the upstroke of the cardiac action potential, and the end of the T wave corresponds to the return of the action potential baseline.

Resting membrane potential (RMP): In cardiac electrophysiology, RMP is the resting membrane potential in mV and is obtained from the membrane voltage measured immediately before the action potential upstroke. RMP is controlled primarily by inwardly rectifying potassium channels. A decrease in RMP may indicate inhibition of the inward rectifier current.

V_{max}: In cardiac electrophysiology, this is the maximum rate of depolarization measured in Volts/seconds (V/s) obtained by taking the first derivative of the rising phase of the action potential. V_{max} amplitude is determined primarily by the activity of cardiac sodium channels. A decrease in V_{max} indicates sodium channel blockade and corresponds to a broadening of the electrographic QRS interval and, potentially, a slowing of conduction velocity in the intact heart.

17

Glossary—Alphabetical

Abbreviated new drug application (ANDA): An ANDA contains data that, when submitted to FDA's Center for Drug Evaluation and Research, Office of Generic Drugs, provides for the review and ultimate approval of a generic drug product. Generic drug applications are called "abbreviated" because they are generally not required to include preclinical (animal) and clinical (human) data to establish safety and effectiveness. Instead, a generic applicant must scientifically demonstrate that its product is bioequivalent (i.e., performs in the same manner as the innovator drug). Once approved, an applicant may manufacture and market the generic drug product to provide a safe, effective, low-cost alternative.

Aberration: In developmental toxicology, a minor structural change. It may be a retardation (a provisional delay in morphogenesis), a variation (external appearance controlled by genetic and extragenetic factors), or a deviation (resulting from altered differentiation).

Ablepharia: Absence or reduction of the eyelid(s).

Abrachius: Without arms, forelimbs.

Absorbed dose: The amount of a substance penetrating across the exchange boundaries of an organism and into body fluids and tissues after exposure.

Acanthosis: In dermatology, hypertrophy of the stratum spinosum and granulosum.

Acardia: Absence of the heart.

Acaudia (anury): Agenesis of the tail.

Acceptable daily intake (ADI): A value used for noncarcinogenic effects that represents a daily dose that is very likely to be safe over an extended period of time. An ADI is similar to a reference dose (RfD) but less strictly defined.

Accessory spleen: An additional spleen.

Acephaly: Congenital absence of the head.

Achondroplasia: A hereditary defect in the formation of epiphysial cartilage, resulting in a form of dwarfism with short limbs, normal trunk, small face, normal vault, etc.

Acrania: Partial or complete absence of the skull.

Action potential amplitude (APA): The amplitude of the cardiac action potential in mV and is a measure of the total amplitude of the action potential from initial resting level to peak depolarization. APA is determined by the combined activity of voltage-gated sodium and potassium channels.

Action potential duration (APD): The duration of the cardiac action potential in milliseconds, measured from the initial upstroke to the point of return to either 60% (APD_{60}) or 90% (APD_{90}) of the initial resting potential. APD prolongation corresponds to an increase in the electrocardiographic QT interval.

Activated partial thromboplastin time: A measure of the relative activity of factors in the intrinsic clotting sequence and the common pathway necessary in normal blood coagulation.

Active ingredient: Any component that provides pharmacological activity or other direct effects in the diagnosis, cure, mitigation, treatment, or prevention of disease or affects the structure or any function of the body of man or animals.

Acystia: Absence of the urinary bladder.

Adactyly: Absence of digits.

ADCC (antibody-dependent cell-mediated cytotoxicity): Cell-mediated immunity in which a specific antibody binds to a target cell, targeting it for cytolytic activity by an effector cell (generally a macrophage or natural killer [NK] cell).

Adduct: The covalent linkage or addition product between an alkylating agent and cellular macromolecules such as protein, RNA, and DNA.

Adjuvant (immunological): A material that enhances an immune response but does not confer immunity by itself. Examples include oil emulsions, aluminum salts, and toll-like receptor agonists. Complete Freund's adjuvant (CFA) was first developed by Jules Freund in the 1940s and contains a water-in-oil emulsion and mycobacterial cell fragments. Incomplete Freund's Adjuvant does not contain the mycobacterial cell fragments.

ADME: Absorption, distribution, metabolism, and excretion.

Administered dose: The amount of a substance given to a human or test animal in determining dose–response relationships, especially through ingestion or inhalation (see Applied dose). Administered dose is actually a measure of exposure, because even though the substance is "inside" the organism, once ingested or inhaled, administered dose does not account for absorption (see Absorbed dose).

Agastria: Absence of the stomach.

Agenesis: Absence of an organ or part of an organ.

Agenesis of the kidney: Absence of the kidney(s).

Agenesis of the lung (lobe): Complete absence of a lobe of the lung.

Aggregate risk: The sum of individual increased risks of an adverse health effect in an exposed population.

Aglossia: Absence of the tongue.

Agnathia: Absence of lower jaw (mandible).

Akinesia: Absence or the loss of power of voluntary motion; immobility.

Alanine amino transferase (ALT): An enzyme, primarily of liver origin, whose blood levels can rise in response to hepatocellular toxicity. Also known as serum glutamic pyruvic transaminase (SGPT).

Albumin: The most abundant blood protein synthesized by the liver.

Alkaline phosphatase: An enzyme whose blood levels can rise in response to hepatobiliary disease or increased osteoblastic (bone cell) activity. Serum alkaline phosphatase activity can decrease in fasted rats because the intestinal isozyme is an important component of serum enzyme activity.

Alkylating agent: A chemical compound that has positively charged (electron-deficient) groups that can form covalent linkages with negatively charged portions of biological molecules such as DNA. The covalent linkage is referred to as an adduct and may have mutagenic or carcinogenic effects on the organism. The alkyl species is the radical that results when an aliphatic hydrocarbon loses one hydrogen atom to become electron deficient. Alkylating agents react primarily with guanine, adding their alkyl group to N7 of the purine ring.

Allogeneic: From a different genetic background. In the context of immunotoxicology, this usually refers to the use of genetically dissimilar cells in *in vitro* assays to elicit a cell-mediated immune reaction.

Altered focus: A histologically identifiable clone of cells within an organ that differs phenotypically from the normal parenchyma. Foci of altered cells usually result from increased cellular proliferation, represent clonal expansions of initiated cells, and are frequently observed in multistage animal models of carcinogenesis. Foci of cellular alteration are most commonly observed in the liver of carcinogen-treated rodents and are believed by some to represent preneoplastic lesions.

Anal atresia: Congenital absence of the anus.

Anemia: Any conditions in which RBC count, hemoglobin concentration, and hematocrit are reduced.

Anencephaly: Congenital absence of the cranial vault with missing or small brain mass.

Aneuploidy: An abnormal number of chromosomes in a cell or organisms that is not an exact multiple of the haploid number.

Anisocytosis: Variations in the size of red blood cells.

Anomaly or abnormality: In developmental toxicology, a morphological or functional deviation from normal limit. It can be a malformation or a variation.

Anophthalmia: Absence of eye(s).

Anorchism: Congenital absence of one or both testes.

Anotia: Absence of the external ear(s).

Anterior chamber: The aqueous-containing cavity of the eye, bounded by the cornea anteriorly, the chamber angle structures peripherally, and the iris and lens posteriorly.

Antibody: Are also referred to as immunoglobulins (Ig). They consist of two basic units: the antigen-binding fragment (Fab), which contains variable regions coding for antigen recognition, and the constant fragment (Fc), which determines the function of the antibody. Based on the Fc region, immunoglobulins are designated IgA, IgD, IgE, IgG, or IgM. The cross-linking of antibody molecules on the surface of a cell leads to the activation of complement, resulting in the destruction of the target by lytic cells or in phagocytosis by macrophages.

Antibody forming cell (AFC)/Plaque forming cell (PFC) assay: The AFC assay measures the ability of animals to produce either IgM or IgG antibodies against a T-dependent or T-independent antigen following *in vivo* sensitization. Due to the involvement of multiple cell populations in mounting an antibody response, the AFC assay actually evaluates several immune parameters simultaneously. It is considered to be one of the most sensitive indicator systems for immunotoxicology studies.

Antigen: A molecule that is the subject of a specific immune reaction. Antigens are recognized in a cognate fashion by either immunoglobulins or the antigen receptor on the surface of T cells. Antigens are usually proteinaceous in nature.

Antibody presenting cell (APC): Cells that are responsible for making antigens accessible to immune effector and regulatory cells. Following internalization and degradation of the antigen (e.g., by phagocytes), a fragment of the antigen molecule is presented on the APC cell surface in association with histocompatibility molecules. The resulting complex is subsequently recognized either by B cells via surface-bound Ig molecules or by T cells via the T cell antigen receptor (TCR). The induction of a specific immune response then proceeds. Representative APC includes macrophages, dendritic cells, and certain B cells.

Aphalangia: Absence of a finger or a toe; corresponding metacarpals not affected.

Aplasia: Lack of development of an organ, frequently used to designate complete suppression or failure of development of a structure from the embryonic primordium.

Aplasia of the lung: The trachea shows rudimentary bronchi, but pulmonary and vascular structures are absent.

Apodia: Absence of one or both feet.

Applied dose: The amount of a substance given to a human or test animal in determining dose–response relationships, especially through dermal contact. Applied dose is actually a measure of exposure, since it does not take absorption into account (see Absorbed dose).

Aproctia: Imperforation or absence of anus.

Arrhinia: Absence of nose.

Arthrogryposis: Persistent flexure or contracture of a joint; flexed paw (bent at wrist) is the most common form of arthrogryposis.

Aspartate aminotransferase (AST): An enzyme whose blood levels can rise in response to hepatotoxicity, muscle damage, or hemolysis. Also known as serum glutamic oxaloacetic transaminase (SGOT).

Aspermia: No ejaculate.

Asthenozoospermia: More than 50% spermatozoa with poor (<2 grade) forward progression.

Astomia: Absence of oral orifice.

Ataxia: Incoordination; the inability to coordinate the muscles in the execution of voluntary movement.

Autoimmunity: Reaction of the immune system against the host organism. In the context of drug development, autoimmunity may take the form of escape from tolerance, as when a drug modifies a host antigen, which is subsequently seen as foreign. Drug-induced autoimmunity may also result from bystander damage to host tissues from a drug-specific immune reaction.

Azoospermia: No spermatozoa in the ejaculate.

Azotemia: An increase in serum urea nitrogen and/or creatinine levels.

Base substitution: The substitution of one or more base(s) for another in the nucleotide sequence.

Basic cycle length (BCL): In cardiac electrophysiology is the BCL of repetitive stimulation and is the inverse of stimulation frequency. BCL = 2, 1, and 0.5 s, respectively, corresponds to bradycardic (30 beats/min), normocardic (60 beats/min), and tachycardic (120 beats/min) heart rates.

B cell/B lymphocyte: Lymphocytes that recognize antigen via surface-bound Ig. B cells that have been exposed to cognate antigen subsequently proliferate and differentiate into plasma cells, which are responsible for producing specific antibody. B cells differentiate in the bone marrow in mammals and in an organ known as the bursa of Fabricius in birds.

Benign: A classification of anticipated biological behavior of neoplasms in which the prognosis for survival is good. Benign neoplasms grow slowly, remain localized, and usually cause little harm to the patient.

Bioaccumulation: Ecotoxicology term describing a process by which chemicals are taken up by aquatic organisms directly from water as well as from exposure through other routes, such as consumption of food and sediment containing chemicals.

Bioaccumulation factor (BAF): The ratio of tissue chemical residue to chemical concentration in an external environmental phase (i.e., water, sediment, or food). BAF is measured as steady state in situations where organisms are exposed from multiple sources (i.e., water, sediment, and food), unless noted otherwise.

Biochemical oxygen demand (BOD): Sometimes called *biological oxygen demand*, a measure of the rate at which molecular oxygen is consumed by microorganisms during oxidation of organic matter. The standard test is the 5-day BOD test, in which the amount of dissolved oxygen required for oxidation over a 5-day period is measured. The results are measured in mg of oxygen/L (mg/L) or parts per million (ppm).

Bioconcentration: A process by which there is a net accumulation of a chemical directly from water into aquatic organisms resulting from simultaneous uptake (e.g., by gill or epithelial tissue) and elimination.

Bioconcentration factor (BCF): A term describing the degree to which a chemical can be concentrated in the tissues of an organism in the aquatic environment as a result of exposure to waterborne chemical. At steady state during the uptake phase of a bioconcentration test, the BCF is a value that is equal to the concentration of a chemical in one or more tissues of the exposed aquatic organisms divided by the average exposure water concentration of the chemical in the test.

Biodegradation: The transformation of a material resulting from the complex enzymatic action of microorganisms (e.g., bacteria, fungi). It usually leads to disappearance of the parent chemical structure and to the formation of smaller chemical species, some of which are used for cell anabolism. Although typically used with reference to microbial activity, it may also refer to general metabolic breakdown of a substance by any living organism.

Biologic license application (BLA): Biological products are approved for marketing under the provisions of the Public Health Service Act (PHS Act). The Act requires a firm who manufactures a biologic for sale in interstate commerce to hold a license for the product. A biologics license application is a submission that contains specific information on the manufacturing processes, chemistry, pharmacology, clinical pharmacology, and the medical effects of the biological product. If the information provided meets FDA requirements, the application is approved, and a license is issued allowing the firm to market the product.

Biological product: A biological product is any virus, serum, toxin, antitoxin, vaccine, blood, blood component or derivative, allergenic product, or analogous product applicable to the prevention, treatment, or cure of diseases or injuries. Biological products are a subset of "drug products" distinguished by their manufacturing processes (biological process vs. chemical process). In general, the term "drugs" includes biological products.

Biological significant effect: A response in an organism or other biological system that is considered to have a substantial or noteworthy effect (positive or negative) on the well-being of the biological system. Used to distinguish from statistically significant effects or changes, which may or may not be meaningful to the general state of health of the system.

Biomaterial: A material that has direct or indirect patient contact. A biomaterial (also termed a *biomedical material*) may be composed of any synthetic or natural rubber or fiber, polymeric or elastomeric formulation, alloy, ceramic, bonding agent, ink, or other nonviable substance, including tissue rendered nonviable, used as a device or any part thereof.

Blanching: In dermal toxicology, to take color from, to bleach. Characterized by a white or pale discoloration of the chemically exposed area of the skin due to decreased blood (ischemia).

Blepharitis: Inflammation of the eyelids.

Blepharospasm: Involuntary spasm of the lids.

Brachydactyly: Shortened digits.

Brachyury (short tail): Tail that is reduced in length.

Bulbous rib: Having a bulge or balloon-like enlargement somewhere along its length.

Cancer potency factor (CPF): The statistical 95% upper confidence limit on the slope of the dose–response relationship at low doses for a carcinogen. Values are in units of lifetime risk per unit dose (mg/kg/day). A plausible upper bound risk is derived by multiplying the extended lifetime average daily dose (LADD) by the CPF.

Cardiomegaly: Hypertrophy (enlargement) of the heart.

Cardiovascular situs inversus: Mirror-image transposition of the heart and vessels to the other side of the body.

Case-control study: A retrospective epidemiologic study in which individuals with the disease under study (cases) are compared with individuals without the disease (controls) in order to contrast the extent of exposure in the diseased group with the extent of exposure in the controls.

Catalepsy: Condition in which there is waxy rigidity of the limbs that may be placed in various positions that will be maintained for a time.

Cataract: An opacity of the lens or its capsule.

CBER: FDA's Center for Biologics Evaluation and Research—the division charged with regulating biological products.

Cluster of differentiation (CD): The CD series in immunology is used to denote cell surface markers (e.g., CD4, CD8). These markers, used experimentally as a means of identifying cell types, also serve physiological roles usually as receptors or ligands. The CD nomenclature was standardized in 1982 and has been expanded since then as new molecules are characterized. The CD nomenclature is especially useful in immunophenotyping immune cell types based on the proteins present on their surface.

CDER: FDA's Center for Drug Evaluation and Research—the division charged with developing and enforcing policy with regard to the safety, effectiveness, and labeling of all drug products for human use.

Ceiling limit: A concentration limit in the work place that should not be exceeded, even for a short time, to protect workers against frank health effects.

Cell-mediated immunity (CMI): Antigen-specific reactivity mediated primarily by T cells. CMI may take the form of immunoregulatory activity (mediated by CD4+ helper T cells) or immune effector activity (mediated by CD8+ killer T cells). Other forms of direct cellular activity in host defense (e.g., NK cells, macrophages) are not antigen specific and are more accurately referred to as innate immunity.

Cephalocele: A protrusion of a part of the cranial contents, not necessarily neural tissue.

CFR: Code of Federal Regulations (United States).

Chemical oxygen demand (COD): COD is measured instead of BOD when organic materials are not easily degraded by microorganisms. Strong oxidizing agents (e.g., potassium permanganate) are used to enhance oxidation. COD values will be larger than BOD values.

Chemokine/chemotactic cytokines: Small chemoattractant proteins important in the stimulation and migration of immune cells. Chemokines are divided into four groups depending on the positions and spacing of the cysteine residues.

Chemosis: In ophthalmology, intense edema of the conjunctiva. The conjunctiva is loose fibrovascular connective tissue that is relatively rich in lymphatics and responds to noxious stimuli by swelling to the point of prolapse between the lids.

Choristoma: A mass of well-differentiated cells from one organ included within another organ, for example, adrenal tissue present in the lung.

Choroid: The vascular middle coat between the retina and sclera of the eye.

Chromosomal aberration: A numerical or structural chromosomal abnormality.

Ciliary body: Portion of the uveal tract between the iris and the choroid of the eye consisting of ciliary processes and the ciliary muscle.

Class testing: The testing of medical device plastics for biological reactivity according to predetermined testing requirements defined by the United States Pharmacopeia (USP).

Clastogen: An agent that produces structural changes of chromosomes.

Clonic convulsion: A convulsion in which the muscles alternately relax and contract.

Clonus: A form of movement characterized by contractions and relaxations of a muscle.

Cocarcinogen: An agent not carcinogenic alone but that potentiates the effect of a known carcinogen.

Cocarcinogenesis: The augmentation of neoplasm formation by simultaneous administration of a genotoxic carcinogen and an additional agent (cocarcinogen) that has no inherent carcinogenic activity by itself.

Cohort study: A study of a group of persons sharing a common experience (e.g., exposure to a substance) within a defined time period; the experiment is used to determine if an increased risk of a health effect (disease) is associated with that exposure.

Combination (medical device) product: A product containing both a drug and a device component that are physically, chemically, or otherwise combined to result in a medical product that is used therapeutically as a single entity. The medical device component must be evaluated for safety according to device requirements, the drug component must be evaluated for safety as per drug requirements, and the safety of the finished combined product must be also be evaluated.

Complement: A group of approximately 20 protein precursor molecules assemble into a complex that intercalates into the membrane of a cell and forms a pore resulting in osmotic lysis of the target cell.

Conceptus: The sum of derivatives of a fertilized ovum at any stage of development from fertilization until birth. Also, term used when the stage of prenatal development at the time of initial insult is not known; also referred to as *embryo–fetus.*

Confidence limit: The confidence interval is a range of values that has a specified probability (e.g., 95%) of containing a given parameter or characteristic. The confidence limit often refers to the upper value of the range (e.g., upper confidence limit).

Conjunctiva: Mucous membrane that lines the posterior aspect of the eyelids (palpebral conjunctiva) and the anterior sclera (bulbar conjunctiva).

Conjunctivitis: Inflammation of the conjunctiva.

Contact dermatitis: A delayed type of induced sensitivity (allergy) of the skin with varying degrees of erythema, edema, and vesiculation, resulting from cutaneous contact with a specific allergen.

Contact urticaria: Wheal-and-flare response elicited with 30–60 min after cutaneous exposure to test substance. May be IgE mediated or nonimmunologically mediated.

Convulsion: A violent spasm of the face, trunk, or extremities.

Cornea: Transparent portion of the outer coat of the eyeball forming the anterior wall of the anterior chamber.

Corpus luteum: The yellow endocrine body formed in the ovary at the site of the ruptured Graafian follicle.

Corrosion: In dermal toxicology, direct chemical action on normal living skin that results in its disintegration and irreversible alteration at the site of contact. Corrosion is manifested by ulceration and necrosis with subsequent scar formation.

Craniorachischisis: Exencephaly and holorrachischisis (fissure of the spinal cord).

Cranioschisis: Abnormal fissure of the cranium; may be associated with meningocele or encephalocele.

Creatine kinase (CK): An enzyme that is concentrated in skeletal muscle, brain, and heart tissue.

Creatinine: The end product of creatine metabolism in muscle. Elevated blood levels can indicate renal (glomerular) injury.

Critical endpoint: A chemical may elicit more than one toxic effect (endpoint), even in one test animal, in tests of the same or different duration (acute, subchronic, and chronic exposure studies). The doses that cause these effects may differ. The critical endpoint used in the dose–response assessment is the one that occurs at the lowest dose. In the event that data from multiple species are available, it is often the most sensitive species that determines the critical endpoint. This term is applied in the derivation of risk reference doses (RfDs).

Cross-sectional study: An epidemiologic study assessing the prevalence of a disease in a population. These studies are most useful for conditions or diseases that are not expected to have a long latent period and do not cause death or withdrawal from the study population. Potential bias in case ascertainment and exposure duration must be addressed when considering cross-sectional studies.

Cryptorchidism (undescended testes, ectopic testes): Failure of the testes to descend into the scrotum (can be unilateral).

Cumulative irritation: In dermal toxicology, primary irritation resulting from repeated exposures to materials that do not in themselves cause acute primary irritation.

Cyclopia: In developmental toxicology, one central orbital fossa with none, one, or two globes.

Cytokine: Small peptides produced primarily by cells of the immune system, particularly helper T cells. Cytokines are grouped into nonexclusive categories including interleukins, tumor necrosis factors, interferons, colony-stimulating factors, and various miscellaneous cytokines. Related molecules include peptide growth factors, transforming growth factors, and chemokines. Cytokines form an interactive network with both hormones and neuropeptides. Cytokines may be referred to in the older literature as lymphokines.

Cytotoxic T-lymphocyte (CTL): A subset of CD8+ T cells able to kill target cells following the induction of a specific immune response. The mechanism of lysis appears to be a combination of direct lysis by extravasation of lytic molecules (such as perforins and granzymes), as well as the induction of apoptosis in the target cell. The measurement of CTL activity is a sensitive indicator of cell-mediated immunity (CMI).

De minimus risk: From the legal maxim, "de minimus non curat lex" or the law is not concerned with trifles. As relates to risk assessment of carcinogens, it is commonly interpreted to mean that a lifetime risk of 1×10^{-6} is a *de minimus* level of cancer risk (i.e., insignificant and therefore acceptable) and is of no public health consequence.

Deflection: In developmental toxicology, a turning, or state of being turned, aside.

Deformity: Distortion of any part or general disfigurement of the body.

Delayed-type hypersensitivity (DTH): A form of cell-mediated immunity in which secondary exposure to an antigen results in an inflammatory reaction mediated by CD4+ T cells.

Dermatitis: Inflammation of the skin.

Desquamation: The shedding of the cuticle in scales or the outer layer of any surface. In dermal toxicology, to shred, peel, or scale off, as the casting off of the epidermis in scales or shred or the shedding of the outer layer of any surface.

Deviation: Variation from the regular standard or course.

Dextrocardia: Location of the heart in the right side of the thorax; a developmental disorder that is associated with total or partial situs inversus (transposition of the great vessels and other thoracoabdominal organs) or occurs as an isolated anomaly.

Dextrogastria: Having the stomach on the right side of the body.

Direct carcinogen: Carcinogens that have the necessary structure to directly interact with cellular constituents and cause neoplasia. Direct acting carcinogens do not require metabolic conversion by the host to be active. They are considered genotoxic because they typically undergo covalent binding to DNA.

Direct contact: When the materials of a medical device are in direct (i.e., intimate) contact with the surface or tissues of the body (e.g., adhesive bandages, pacemaker leads).

Dispersion model: A mathematical model or computer simulation used to predict the movement of airborne or water contaminants. Models take into account a variety of mixing mechanisms that dilute effluents and transport them away from the point of emission.

Displaced rib: Out of normal position.

DNEL (Derived no effect level): A European occupational exposure level of chemical determined for REACH above which humans should not be exposed. A DNEL is typically derived from the No Observed Adverse Effect Level (NOAEL) for a principal toxic effect in an animal study divided by assessment (safety/uncertainty) factors. The risk for a chemical substance is considered acceptable if the exposure levels do not exceed the DNEL. DNELs are communicated in Safety Data Sheets (SDS).

Dose: The amount of substance administered to an animal or human generally expressed as the weight or volume of the substance per unit of body weight (e.g., mg/kg, mL/kg).

Dose–response relationship: A relationship between the dose, often actually based on "administered dose" (i.e., exposure) rather than absorbed dose, and the extent of toxic injury produced by that chemical. Response can be expressed as either the severity of injury or proportion of exposed subjects affected. A dose–response assessment is one of the steps in a risk assessment.

Drug: A substance recognized by an official pharmacopoeia or formulary. A substance intended for use in the diagnosis, cure, mitigation, treatment, or prevention of disease. A substance (other than food) intended to affect the structure or any function of the body. A substance intended for use as a component of a medicine but not a device or a component, part, or accessory of a device. Biological products are included within this definition and are generally covered by the same laws and regulations, but differences exist regarding their manufacturing processes (chemical process vs. biological process.)

Drug label: The FDA-approved label is the official description of a drug product that includes indication (what the drug is used for); who should take it; adverse events (side effects); instructions for uses in pregnancy, children, and other populations; and safety information for the patient. Labels are often found inside drug product packaging.

Duration of exposure: Generally referred to in toxicology as acute (one time), subacute (repeated over several weeks), subchronic (repeated for a fraction of a lifetime), and chronic (repeated for nearly a lifetime).

Dust: The airborne state of a chemical that is solid at room temperature but is dispensed into a particulate atmosphere.

Dysarthria: In neurotoxicology, disturbance of articulation due to emotional stress or to paralysis, incoordination, or spasticity of muscles used in vocalizing.

Dysgenesis: Defective development; malformation.

Dyskinesia: Difficulty in performing voluntary movements; a movement disorder characterized by insuppressible, stereotyped, and automatic movements.

Dysmelia: Absence of a portion of one or several limbs.

Dysplasia: Disordered tissue formation characterized by changes in size shape and orientational relationships of adult types of cells. Primarily seen in epithelial cells.

Dystocia: Abnormal labor.

Dystonia: Abnormal tonicity (hyper or hypo) in any tissues.

EC_{50} (median effective concentration): The concentration of chemical in water to which test organisms are exposed that is estimated to be effective in producing some sublethal response in 50% of the test organisms. The EC_{50} is usually expressed as a time-dependent value (e.g., 24 h or 96 h EC_{50}). The sublethal response elicited from the test organisms as a result of exposure to the chemical must be clearly defined (e.g., test organisms may be immobilized, lose equilibrium, or undergo physiological or behavioral changes).

Ectocardia: Displacement of the heart inside or outside the thorax.

Ectopic: Out of the normal place.

Ectopic esophagus: Displacement of the esophagus (description of position should be included).

Ectopic pinna: Displaced external ear.

Ectrodactyly: Absence of all or of only a part of digit (partial ectrodactyly).

Ectromelia: Aplasia or hypoplasia of one or more bones of one or more limbs (this term includes amelia, hemimelia, and phocomelia).

Eczema: Inflammatory condition in which the skin becomes red and small vesicles, crusts, and scales develop.

Edema: An excessive accumulation of serious fluid or water in cells, tissues, or serous cavities.

ELISA (Enzyme-linked immunosorbent assay): A type of immunoassay in which specific antibodies are used to capture and detect molecules of interest from a fluid matrix. The most common form is a "sandwich" ELISA in which the antibodies are bound to a substrate such as a plastic culture plate, and a second labeled antibody is used to detect the bound molecules.

Encephalocele: A partial protrusion of brain through an abnormal cranial opening; not as severe as exencephaly.

Endemic: Present in a community or among a group of people; said of a disease prevailing continually in a region.

Environmental fate: The destiny of a chemical or biological pollutant after release into the environment. Environmental fate involves temporal and spatial considerations of transport, transfer, storage, and transformation.

Epigenetic: Change in phenotype without a change in DNA structure. One of the two main mechanisms of carcinogen action, epigenetic carcinogens are nongenotoxic, that is, they do not form reactive intermediates that interact with genetic material in the process of producing or enhancing neoplasm formation.

Erythema: An inflammatory redness of the skin, as caused by chemical poisoning or sunburn, usually a result of congestion of the capillaries.

Eschar: A dry scab, thick coagulated crust or slough formed on the skin as a result of a thermal burn or by the action of a corrosive or caustic substance.

Estrus: Phase of the sexual cycle of female mammals characterized by willingness to mate.

Exencephaly: Brain outside of the skull as a result of a large cranial defect.

Exfoliation: To remove in flakes or scales, peel. To cast off in scales, flakes, or the like. To come off or separate, as scales, flakes, sheets, or layers. In dermal toxicology, detachment and shedding of superficial cells of an epithelium or from any tissue surface. Scaling or desquamation of the horny layer of epidermis, which varies in amount from minute quantities to shedding the entire integument.

Exomphalos: Congenital herniation of abdominal viscera into umbilical cord.

Exophthalmos: Abnormal protrusion of the eyeball.

Exophthalmos: Protrusion of the eyeball ("pop" eye).

Exposure: Contact of an organism with a chemical, physical, or biological agent. Exposure is quantified as the amount of the agent available at the exchange boundaries of the organism (e.g., skin, lungs, digestive tract) and available for absorption.

Exposure assessment: The determination of the extent of human exposure.

Exposure frequency: The number of times an exposure occurs in a given period. The exposure(s) may be continuous, discontinuous but regular (e.g., once daily), or intermittent.

Extract: A solution produced by the incubation of a material/medical device in an appropriate vehicle. After incubation, the vehicle contains the soluble chemicals (or leachables) that have dissolved out of, or off, the material/medical device.

Extrapolation: An estimate of response or quantity at a point outside the range of the experimental data. Also refers to the estimation of a measured response in a different species or by a different route than that used in the experimental study of interest (i.e., species to species, route to route, acute to chronic, high to low).

Fasciculations: Involuntary contractions, or twitching, of groups of muscle fibers.

Fate: Disposition of a material in various environmental compartments (e.g., soil or sediment, water, air, biota) as a result of transport, transformation, and degradation.

Fecundity: Ability to produce offspring rapidly and in large numbers.

Fence line concentration: Modeled or measured concentrations of pollutants found at the boundaries of a property on which a pollution source is located. Usually assumed to be the nearest location at which an exposure of the general population could occur.

Feticide: The destruction of the fetus in the uterus.

Fibrinogen: A glycoprotein that is involved in the formation of fibrin.

Fluorescein (fluorescein sodium): A fluorescent dye, the simplest of the fluorane dyes and the mother substance of eosin, which is commonly used intravenously in ophthalmology to determine the state of adequacy of circulation in the retina and a lesser degree the choroid and iris. Another important use is to detect epithelial lesions of the cornea and conjunctiva. Peak excitation occurs with light at a wavelength between 485 and 500 nm, and peak emission occurs between 520 and 530 nm.

Fovea: Depression in the macula of the eye adapted for most acute vision.

Frameshift mutation: A mutation in the genetic code in which one base or two adjacent bases are inserted or deleted to the nucleotide sequence of a gene.

Frank effect level (FEL): Related to biological responses to chemical exposures (compare with NOAEL and LOEL); the exposure level that produces an unmistakable adverse health effect (such as inflammation, severe convulsions, or death).

Fume: The airborne state of a chemical that is liquid or solid at room temperature and pressure but is heated and allowed to condense into a particulate atmosphere.

Fundus: The posterior portion of the eye visible through an ophthalmoscope.

Gamete: A male (spermatozoa) or female (ovum) reproductive cell.

Gamma glutamyltransferase (γGT): An enzyme of liver origin, whose blood concentration can be elevated in hepatobiliary disease.

Gas: The airborne state of a chemical that boils at or below room temperature and pressure.

Gastroschisis: Fissure of the abdominal wall (median line) not involving the umbilicus, usually accompanied by protrusion of the small part of the large intestine, not covered by membranous sac.

Gene mutation: A detectable permanent change (point mutation, insertion, or deletion) within a single gene or its regulating sequences.

Genotoxic carcinogen: An agent that interacts with cellular DNA either directly in its parent form (direct carcinogen) or after metabolic biotransformation.

Geometric standard deviation (GSD): In inhalation toxicology, the relative dispersion of the MMAD such that a value approaching 1 indicates a monodisperse atmosphere.

Globulin: A group of blood proteins synthesized by lymphatic tissue in the liver.

Hapten: Low molecular weight molecules that are not antigenic by themselves but that are recognized as antigens when bound to larger molecules, usually proteins.

Hazard: The inherent ability of a substance to cause an adverse effect under defined conditions of exposure.

Hazard identification: The determination of whether a particular chemical is or is not causally linked to particular health effects.

Hazard index: The ratio of the maximum daily dose (MDD) to the acceptable daily intake (ADI) used to evaluate the risk of noncarcinogens. A value of less than 1 indicates that the risk from the exposure is likely insignificant; a value greater than 1 indicates a potentially significant risk.

Head-only exposure: A system for exposing test animals via inhalation in which they are restrained in a tube in such a way that only their heads are exposed directly to the test material.

Hemivertebra: Presence of only one-half of a vertebral body.

Hemolysis: The destruction of red blood cells resulting in the liberation of hemoglobin into plasma.

Hepatic lobe agenesis: Absence of a lobe of the liver.

Hepatomegaly: Abnormal enlargement of the liver.

hERG: Refers to both the gene that encodes the pore-forming subunit of human cardiac ether-a-go-go-related potassium channel and the protein subunit itself. hERG channels are responsible for the delayed rectifier potassium current (I_{Kr}) that regulates action potential repolarization.

Host defense: The ability of an organism to protect itself against disease associated with exposure to infectious organisms, foreign tissue, or neoplasia. Host defense assays measure protection from infectious or neoplastic disease and are mediated by an immunological effect involving all components of the immune system—innate, cell-mediated, and humoral-mediated immunity. Host resistance assays are the only way to demonstrate immunological reserve.

Human equivalent dose (HED): A dose in humans anticipated to provide the same degree of effect as that observed in animals at a given dose. In drug development, the term HED is usually used to refer to the HED of the NOAEL.

Humoral-mediated immunity (HMI): Specific immune responses that are mediated primarily by humoral factors including antibodies and complement. The induction of HMI generally, although not exclusively, requires the cooperation of cellular immune mechanisms.

Hybridoma: A genetically engineered cell clone that produces antibodies of a single type (i.e., monoclonal antibodies). Monoclonal antibodies are highly specific for their cognate antigen and make highly useful tools for immunotoxicological studies.

Hydrocephaly: Enlargement of the head caused by abnormal accumulation of cerebrospinal fluid in subarachnoid cavity (external hydrocephaly) or ventricular system (internal hydrocephaly).

Hydronephrosis: Dilatation of the renal pelvis usually combined with destruction of renal parenchyma and often with dilation of the ureters (bilateral, unilateral). Note: This is a pathology term and should have histological confirmation.

Hyperemia: Excess of blood in a part due to local or general relaxation of the arterioles. Blood vessels become congested and give the area involved a reddish or red-blue color.

Hyperkeratosis: Hypertrophy and thickening of the stratum corneum.

Hyperkinesia: Excessive muscular activity.

Hyperplasia: A numerical increase in the number of phenotypically normal cells within a tissue or organ.

Hypersensitivity: A vigorous and often inappropriate immune response to seemingly innocuous antigens. Hypersensitivity is classified into subtypes depending on the mechanisms of action and the target cells or tissues.

Hypertrophy: Increase in the size of an organelle, cell, tissue, or organ within a living organism. To be distinguished from hyperplasia, hypertrophy refers to an increase in size rather than an increase in number. Excessive hyperplasia in a tissue may produce hypertrophy of the organ in which that tissue occurs.

Hypoplasia of the lung: Bronchial tree is poorly developed, and pulmonary tissue shows an abnormal histological picture (total or partial); incomplete development, smaller.

Hypospadias: Urethra opening on the underside of the penis or on the perineum (males) or into the vagina (females).

IC$_{50}$: Refers to the concentration or dose of test article that produces 50% inhibitory response in a test system assay.

Icteric: Relating to a jaundiced condition, typically as a result of elevated serum bilirubin levels.

ICH: The International Council for Harmonization of Technical Requirements for Pharmaceuticals for Human Use (ICH) is a project that brings together the regulatory authorities of Europe, Japan, and the United States and experts from the pharmaceutical industry in the three regions to discuss scientific and technical aspects of pharmaceutical product registration. The ICH develops scientific guidelines that are harmonized between the three regions.

Identification threshold: A limit (%) above which a drug impurity or degradation product should be identified.

Immediately dangerous to life and health (IDLH): A concentration representing the maximum level of a pollutant from which an individual could escape within 30 min without escape-impairing symptoms or irreversible health effects.

Immunological reserve: The concept that the immune response exhibits multiple immunological functions acting in an orchestrated manner, such that a decrease in one function is compensated by other immune functions. Immunological reserve prevents infectious or neoplastic disease due to acute reductions in one or two immune functions. This reserve would theoretically prevent a severe reduction in host resistance following a temporary immunosuppression of selected immune functions. This concept is important in the interpretation of immunotoxicology data.

Immunostimulation: Enhancement of immune function above an accepted baseline (control) level. Immunostimulation may be beneficial, for example, therapeutics designed to restore a suboptimal immune response. It may conversely be detrimental, as would be the case with autoimmunity or hypersensitivity.

Immunosuppression: Depression/reduction of immune function below an accepted baseline (control) level. Immunosuppression may result from inadvertent exposure to drugs or other chemical or physical agents, intentional modification for therapeutic reasons (e.g., organ transplantation), or following exposure to certain infectious agents (e.g., HIV). An important consideration in immunotoxicology is the ability to determine the amount of immunosuppression necessary to alter host defense. Immunosuppression may result in a state of immunodeficiency.

Imperforate: Not open; abnormally closed.

In situ carcinoma: A localized intraepithelial form of epithelial cell malignancy. The cells possess morphological criteria of malignancy but have not yet gone beyond the limiting basement membrane.

Incidence: The number of new cases of a disease within a specified time period. It is frequently presented as the number of new cases per 1,000, 10,000, or 100,000. The incidence rate is a direct estimate of the probability or risk of developing a disease during a specified time period.

Incomplete ossification (delayed, retarded): The extent of ossification is less than what would be expected for that developmental age, not necessarily associated with reduced fetal or pup weight.

IND: Investigational New Drug Application—a request to initiate a clinical study of a new drug product.

Indirect contact: When materials of a medical device do not contact the surface or tissues of the body, the materials of the device may influence the body. In this case, a solution or other materials that contact the device may become contaminated with leachables from the device that in turn contacts tissues of the body (e.g., intravenous infusion bag).

Initiation: The first step in carcinogenesis whereby limited exposure to a carcinogenic agent produces a latent but heritable alteration in a cell, permitting its subsequent proliferation and development into a neoplasm after exposure to a promoter.

Initiator: A chemical, physical, or biological agent that is capable of irreversibly altering the genetic component (DNA) of the cell. While initiators are generally considered to be carcinogens, they are typically used at low noncarcinogenic doses in two-stage initiation–promotion animal model systems. Frequently referred to as a "tumor initiator."

Injection: In ophthalmology, congestion of blood vessels.

Innate (natural or nonspecific) immunity: Host defense mechanisms that do not require prior exposure to antigen; often are antigen nonspecific in nature. Nonspecific immunity is mediated by NK cells, macrophages, neutrophils, γδT cells, and complement.

Intratracheal dosing: A method of delivering, via a syringe and blunt needle, test material directly into the trachea of a test animal.

Involuntary risk: A risk that impinges on an individual without their awareness or consent.

Iris: The circular pigmented membrane behind the cornea and immediately in front of the lens; the most anterior portion of the vascular tunic of the eye. It is composed of the dilator and sphincter muscles and the two-layered posterior epithelium and mesodermal components that form the iris stroma.

Iritis: Inflammation of the iris, manifested by vascular congestion (hyperemia). An outpouring of serum proteins into the aqueous (flare) may accompany the inflammatory reaction.

Irritant: A substance that causes inflammation and other evidences of irritation, particularly of the skin, on first contact or exposure; a reaction of irritation not dependent on a mechanism of sensitization.

Irritation: A local reversible inflammatory response of normal living skin to direct injury caused by a single application of a toxic substance, without the involvement of an immunological mechanism.

ISO: International Standards Organization.

IUCLID: An acronym for International Uniform Chemical Information Database software application used to capture, store, maintain, and exchange data on intrinsic and hazard properties of chemical substances. Co-developed by the European Chemicals Agency (ECHA) and the OECD, ECHA maintains and makes the software available for the submission of chemical information under REACH.

Keratitis: Inflammation of the cornea.

Lactate dehydrogenase: An enzyme found in several organs, including liver, kidney, heart, and skeletal muscle.

Latency: The period of time between exposure to an injurious agent and the manifestation of a response.

LC_{50} (median lethal concentration): The concentration of a chemical required to cause death in 50% of the exposed population when exposed for a specified time period and observed for a specified period of time after exposure. Refers to inhalation exposure concentration in the context of air toxics and may refer to water concentration for tests of aquatic organisms. In aquatic toxicology, the LC_{50} is often expressed as a time-dependent value (e.g., 24 h or 96 h LC_{50}).

LC_{LO} (lethal concentration low): The lowest concentration of a chemical required to cause death in some of the population after exposure for a specified period of time and observed for a specified period of time after exposure. Refers to inhalation time exposure in the context of air toxics (may refer to water concentration for tests of aquatic organisms).

LD_{50} (median lethal dose): The dose of a chemical required to cause death in 50% of the exposed population after noninhalation exposure (e.g., injection, ingestion) for a specified observation period after exposure.

LD_{LO} (lethal dose low): The lowest dose of a chemical required to cause death in some of the population after noninhalation exposure (e.g., injection, ingestion) for a specified observation period after exposure.

Lens: A transparent biconvex structure suspended in the eyeball between the aqueous and the vitreous. Its function is to bring rays of light to a focus on retina. Accommodation is produced by variations in the magnitude of this effect.

Levocardia: Displacement of the heart in the extreme left hemithorax.

Lifetime average daily dose (LADD): The total dose received over a lifetime multiplied by the fraction of a lifetime during which exposure occurs, expressed in mg/kg body weight/day.

Lifetime risk: A risk that results from lifetime exposure.

Liquid aerosol: The airborne state of a chemical that is liquid at room temperature and pressure but is nebulized into a particulate atmosphere.

Lordosis: Anterior concavity in the curvature of the cervical and lumbar spine as viewed from the side.

Lowest-observed adverse-effect level (LOAEL): The lowest dose or exposure level of a chemical in a study at which there is a statistically or biologically significant increase in the frequency or severity of an *adverse* effect in the exposed population as compared with an appropriate, unexposed control group.

Lowest-observed effect level (LOEL): In a study, the lowest does or exposure level of a chemical at which a statistically or biologically significant effect is observed in the exposed population compared with an appropriate unexposed control group. The effect is generally considered not to have an adverse effect on the health and survival of the animal. This term is occasionally misused in place of a LOAEL.

Lung compliance: A measure of the ease in which the lung is extended usually expressed as mL/cmH_2O.

Lung resistance: The total resistance to the movement of air during the expiratory phase of the respiratory cycle expressed as $cmH_2O/mL/s$.

Macrobrachia: Abnormal size or length of the arm.

Macrodactylia: Excessive size of one or more digits.

Macroglossia: Enlarged tongue, usually protruding.

Macrophage: A bone marrow-derived cell present in the peripheral tissue that serves a wide variety of host defense functions, acting as both nonspecific phagocytes and killer cells, as well as regulators of specific immune reactions. Macrophages have different designations depending on the tissue in which they are located, such as Kupffer cells in the liver and veiled cells in the lymphatic system.

Macrophthalmia: Enlarged eye(s).

Major histocompatibility complex (MHC): Cell surface molecules that determine tissue compatibility and regulate self-recognition and tolerance. Two major classes are recognized Class I (present on all nucleated cells) and Class II (present on B cells, T cells, and macrophages). MHC molecules direct the course of immune reactivity and are presented in association with antigens by antibody presenting cells (APCs). In humans, MHCs are specifically referred to as human leukocyte antigen (HLA).

Malformation: Defective or abnormal formation; deformity. A permanent structural deviation that generally is incompatible with, or severely detrimental to, normal survival or development.

Malignant: A classification of anticipated biological behavior of neoplasms in which the prognosis for survival is poor. Malignant neoplasms grow rapidly, invade, destroy, and are usually fatal.

Margin of exposure (MOE): The ratio of the NOAEL to the estimated human exposure. The MOE was formerly referred to as the margin of safety (MOS).

Mass median aerodynamic diameter (MMAD): In inhalation toxicology, the median-sized particle based on mass measurement relative to a unit density sphere.

Maximal acceptable toxicant concentration (MATC): In aquatic toxicology, the hypothetical toxic threshold concentration lying in a range bounded at the lower end by the highest tested concentration having no observed effect (NOEC) and at the higher end by the lowest concentration having a statistically significant toxic effect (LOEC) in a life cycle (full chronic) or a partial life cycle (partial chronic) test. This can be represented by NOEC < MATC < LOEC.

Maximum contaminant level (MCL): The maximum level of a contaminant permissible in water as defined by regulations promulgated under the Safe Drinking Water Act.

Maximum daily dose (MDD): Maximum dose received on any given day during a period of exposure generally expressed in mg/kg body weight/day.

Maximum recommended starting dose (MRSD): The highest dose recommended as the initial dose of a drug in a clinical trial. The MRSD is predicted to cause no adverse reactions.

Maximum tolerated dose (MTD): The highest dose of a toxicant that causes toxic effects without causing death during a chronic exposure and that does not decrease the body weight by more than 10%.

Mean corpuscular hemoglobin (MCH): The average amount of hemoglobin per red blood cell.

Mean corpuscular hemoglobin concentration (MCHC): The average hemoglobin concentration per red blood cell.

Mean corpuscular volume (MCV): The average size of the red blood cell.

Medical device: Any instrument, apparatus, appliance, material, or other articles, including software, whether used alone or in combination, intended by the manufacturer for use by human beings solely or principally for the purpose of diagnosis, prevention, monitoring, or treatment; alleviation of disease, injury, or handicap; investigation, replacement, or modification of the anatomy or of a physiological process; control of conception; and that which does not achieve its principal intended action of the body by pharmaceutical, immunological, or metabolic means but may be assisted in its function by such means.

Meiosis: Cell division occurring in maturation of the sex cell (gametes) by means of which each daughter nucleus receives half the number of chromosomes characteristic of the somatic cells of the species.

Metaplasia: The substitution in a given area of one type of fully differentiated cell for the fully differentiated cell type normally present in that area, for example, squamous epithelium replacing ciliated epithelium in the respiratory airways.

Metastasis: The dissemination of cells from a primary neoplasm to a noncontiguous site and their growth therein. Metastases arise by dissemination of cells from the primary neoplasm via the vascular or lymphatic system and are an unequivocal hallmark of malignancy.

Methemoglobin: Oxidized hemoglobin incapable of carrying oxygen.

Microcephaly: Small head.

Micrognathia: Shortened lower jaw (mandible); tongue may protrude.

Micronucleus: A microscopically detectable particle in a cell that contains nuclear DNA, usually 1/20th to 1/5th the size of the main nucleus. It may be composed of a broken centric or acentric part of a chromosome or a whole chromosome.

Microphthalmia: Small eyes.

Microstomia: Small mouth opening.

Microtia: Small external ear.

Miotic: A drug causing pupillary constriction.

Mitogen: Mitogens are molecules capable of inducing cellular activation and may include sugars or peptides. The ability of a cell to respond to stimulation with a mitogen (generally assessed by cellular proliferation) is thought to give an indication of the cell's immune responsiveness. Mitogens most commonly employed in immunotoxicology assays include the T cell mitogens concanavalin A (ConA) and phytohemagglutinin (PHA). Mitogens routinely used for assessing B cell proliferation include pokeweed mitogen (PWM) and *Escherichia coli* lipopolysaccharide (LPS).

Mitogenesis: The generation of cell division or cell proliferation.

Mitotic index: The ratio of the number of cells in a population in various stages of mitosis to the number of cell in the population not in mitosis.

Mixed lymphocyte response/reaction (MLR): An *in vitro* assay that measures the ability of lymphocytes to respond to the presence of allogeneic cells. This proliferation represents the initial stage of the acquisition of cytotoxic T-lymphocyte (CTL) function by $CD8^+$ T cells, and thus serves as a measure of cell-mediated immunity (CMI). The MLR is a form of lymphoproliferation. Also referred to as mixed lymphocyte culture (MLC).

Modifying factor (MF): A factor that is greater than 0 and less than or equal to 10; used in the operational derivation of an RfD. Its magnitude depends upon an assessment of the scientific uncertainties of the toxicological database not explicitly treated with standard UF (e.g., the completeness of the overall database). The default value for the MF is 1. The use of an MF was discontinued by the EPA in 2004.

Monocardium: Possessing a heart with only one atrium and one ventricle.

Mononuclear phagocyte system: Previously known as the reticuloendothelial system (RES), this system is composed of all phagocytic cells of the body, including monocytes/macrophages and polymorphonuclear cells (i.e., neutrophils).

Mucosa-associated lymphoid tissue (MALT): Previously known as gut-associated lymphoid tissue (GALT). Lymphoid cells and tissues lining the mucosa which serve as the first point of contact with antigen encountered via this route. MALT comprises Peyer's patches, the appendix, tonsils, and lymphoid cells in the lamina propria of the gut.

Multigravida: A female pregnant for the second (or more) time.

Multistage model: A mathematical function used to extrapolate the probability of incidence of disease from a bioassay in animals using high doses to that expected to be observed at the low doses that are likely to be found in chronic human exposure. This model is commonly used in quantitative carcinogenic risk assessments where the chemical agent is assumed to be a complete carcinogen and the risk is assumed to be proportional to the dose in the low region.

Mutation: A structural alteration of DNA that is hereditary and gives rise to an abnormal phenotype. A mutation is always a change in the DNA base sequence and includes substitutions, additions, rearrangements, or deletions of one or more nucleotide bases.

Mydriatic: A drug causing pupillary dilatation.

Myoclonus: Brief, involuntary twitching of a muscle or a group of muscles.

Myotonia: Delayed relaxation of a muscle after an initial contraction.

Naris (nostril) atresia: Absence or closure of nares.

Nasal agenesis: Absence of the nasal cavity and external nose.

Natural (innate or nonspecific) immunity: Host defense mechanisms that do not require prior exposure to antigen; often are antigen-nonspecific in nature. Nonspecific immunity is mediated by NK cells, macrophages, neutrophils, $\gamma\delta$T cells, and complement.

Natural killer (NK) cells: A population of lymphocytes distinct from T and B cells, also referred to as large granular lymphocytes (LGLs) because of their microscopic appearance. NK cells exhibit cytotoxicity against virally infected cells and certain tumor cells. The assessment of NK cell function provides a good measure of innate immunity.

Necrosis: Pathological death of one or more cells or of a portion of tissue or organ, resulting from irreversible damage.

New drug application (NDA): When the sponsor of a new drug believes that enough evidence on the drug's safety and effectiveness has been obtained to meet FDA's requirements for marketing approval, the sponsor submits to FDA an NDA. The application must contain data from specific technical viewpoints for review, including chemistry, pharmacology, medical, biopharmaceutics, and statistics. If the NDA is approved, the product may be marketed in the United States. For internal tracking purposes, all NDAs are assigned an NDA number.

New molecular entity (NME): A novel molecule under development for pharmaceutical purposes. The term encompasses both NCEs, such as small molecule drugs, and new biological entities (NBEs).

Nonocclusive: The site of dermal application of a test substance is open to the air.

Nonthreshold toxicant: An agent considered to produce a toxic effect from any dose; any level of exposure is deemed to involve some risk. Usually only in regard to carcinogenesis.

No-observed adverse-effect level (NOAEL): The highest experimental dose at which there is not statistically or biologically significant increases in frequency or severity of *adverse* health effects, as seen in the exposed population compared with an appropriate, unexposed population. Effects may be produced at this level, but they are not considered to be adverse.

No-observed effect level (NOEL): The highest experimental dose at which there is not statistically or biologically significant increase in the frequency or severity of effects seen in the exposed compared with an appropriate unexposed population.

Normozoospermia: Normal semen sample.

Nose-only exposure: A system for exposing test animals via inhalation in which they are restrained in a tube in such a way that only their nose or snout is exposed directly to the test material.

Nulliparous: A female that never has born viable offspring.

Nystagmus: An involuntary, rapid movement of the eyeball that may be horizontal, vertical, rotatory, or mixed.

Occlusive covering: A bandage or dressing that covers the skin and excludes it from air. Prevents loss of a test substance by evaporation and by increasing tissue penetration.

Occupational exposure limit (OEL): A generic term denoting a variety of values and standards, generally time-weighted average (TWA) concentrations of airborne substances to which a worker can be safely exposed during defined work periods.

Octanol–water partition coefficient (K_{ow}): The ratio of the solubility of a chemical in n-octanol and water at steady state; also expressed as P. The logarithm of P or K_{ow} (i.e., log P or K_{ow}) is used as an indication of the propensity of a chemical for bioconcentration by aquatic organisms.

Oligodactyly: Fewer than normal number of digits.

Oligohydramnios (oligoamnios): Reduction in the amount of amniotic fluid.

Oligozoospermia: A subnormal sperm concentration in ejaculate.

Omphalocele: Midline defect in the abdominal wall at the umbilicus, through which the intestines and often other viscera (stomach, spleen, and portions of the liver) protrude. These are always covered by a membranous sac. As a rule, the umbilical cord emerges from the top of the sac.

Oncogene: The activated form of a proto-oncogene. Oncogenes are associated with development of neoplasia.

Opisthotonus: A form of muscle spasm of the back and rear neck muscles producing abnormal posturing in which the head, neck, and spine are arched backwards so that the heels approximate to the head.

Optic disk: Ophthalmoscopically visible portion of the optic nerve.

Pachynsis: Abnormal thickening.

Packed cell volume: The percent of blood that contains RBC components; synonymous with hematocrit.

Palpebral: Pertaining to the eyelid.

Pannus: Vascularization and connective-tissue deposition beneath the epithelium of the cornea.

Paresthesia: An abnormal sensation, such as burning, prickling, tickling or tingling.

Patent ductus arteriosus (ductus botalli): An open channel of communication between the main pulmonary artery and the aorta may occur as an isolated abnormality or in combination with other heart defects.

Peripheral blood mononuclear cells (PBMCs)/Peripheral blood mononuclear leukocytes (PBMLs): Leukocytes derived from the peripheral circulation. Because of their accessibility, these cells are often used in *ex vivo* immune function assessment.

Permissible exposure limit (PEL): Similar to an occupational exposure limit (OEL).

Pharmacologically active dose (PAD): The lowest dose tested in an animal species with the intended pharmacological activity.

Photoallergy: An increased reactivity of the skin to UV and/or visible radiation produced by a chemical agent on an immunological basis. Previous allergy sensitized by exposure to the chemical agent and appropriate radiation is necessary. The main role of light in photoallergy appears to be in the conversion of the hapten to a complete allergen.

Photoirritation: Irritation resulting from light-induced molecular changes in the structure of chemicals applied to the skin.

Photosensitization: Sensitization of the skin to UV light, usually due to the action of certain drugs, plants, or other substances; may occur shortly after administration of the substance or may occur only after latent period of days to months. The processes whereby foreign substances, either absorbed locally into the skin or systemically, may be subjected to photochemical reactions within the skin, either leading to chemically induced photosensitivity reactions or altering the "normal" pathologic effects of light. UV-A is usually responsible for most photosensitivity reactions.

PLA: Product License Application for a biologic.

Plasmid: An autonomously replicating DNA molecule distinct from the normal genome. A plasmid may insert into the host chromosome or form an extra chromosomal element.

PNEC (Predicted no effect concentration): A European exposure concentration of a chemical determined for REACH, which marks the limit at which below no adverse effects in an ecosystem are expected to occur during short-term or long-term exposure. PNECs are calculated by dividing endpoints of toxicity such as mortality (LC50), or the No Observed Effect Concentration (NOAEC) for growth and reproduction by assessment (safety/uncertainty) factors.

Poikilocytosis: Variations in the shape of red blood cells.

Point mutation: Change in the genetic code, usually confined to a single base pair.

Polychromasia: Increased basophilic staining of erythrocytes.

Polycythemia: An increase in the number of red blood cells.

Polydactyly: Extra digits.

Polysomia: A fetal malformation consisting of two or more imperfect and partially fused bodies.

Posterior chamber: Space filled with aqueous anterior to the lens and posterior to the iris of the eye.

Potency: A comparative expression of chemical or drug activity measured in terms of the relationship between the incidence or intensity of a particular effect and the associated dose of a chemical, to a given or implied standard of reference. Can be used for ranking the toxicity of chemicals.

ppb: Parts per billion.

ppm: Parts per million.

Predicate device: A previously marketed medical device that is substantially equivalent to a proposed device. The predicate device is used as a comparison to the proposed device to establish safety and efficacy.

Preneoplastic lesion: A lesion usually indicative that the organism has been exposed to a carcinogen. The presence of preneoplastic lesions indicates that there is enhanced probability for the development of neoplasia in the affected organ. Preneoplastic lesions are believed to have a high propensity to progress to neoplasia.

Prevalence: The percentage of a population that is affected with a particular disease at a given time.

Procarcinogen: An agent that requires bioactivation in order to give rise to a direct acting carcinogen. Without metabolic activation, these agents are not carcinogenic.

Processing aid: A material that contacts a medical device product during the manufacturing process and, therefore, has a potential for affecting product quality and/or may elicit a biological response following the use of a medical device. Solvents, cleaning products, lubricants, and mold-release agents are examples of processing aids.

Progression: Processes associated with the development of an initiated cell to a biologically malignant neoplasm. Sometimes used in a more limited sense to describe the process whereby a neoplasm develops from a benign to a malignant proliferation or from a low-grade to a high-grade malignancy. Progression is that stage of neoplastic development characterized by demonstrable changes associated with increased growth rate, increased invasiveness, metastases, and alterations in biochemical and morphological characteristics of a neoplasm.

Promoter: *Use in multistage carcinogenesis*—an agent that is not carcinogenic itself but when administered after an initiator of carcinogenesis stimulates the clonal expansion of the initiated cell to produce a neoplasm. *Use in molecular biology*—a DNA sequence that initiates the process of transcription and is located near the beginning of the first exon of a structural gene.

Promotion: The enhancement of neoplasm formation by the administration of a carcinogen followed by an additional agent (promoter) that has no intrinsic carcinogenic activity by itself.

Prothrombin time: A measure of the relative activity of factors in the extrinsic clotting sequence and the common pathway necessary in normal blood coagulation.

Protooncogene: A normal cellular structural gene that, when activated by mutations, amplifications, rearrangements, or viral transduction, functions as an oncogene and is associated with development of neoplasia. Proto-oncogenes regulate functions related to normal growth and differentiation of tissues.

Pseudopregnancy: (a) False pregnancy: condition occurring in animals in which anatomical and physiological changes occur similar to those of pregnancy; (b) the premenstrual stage of the endometrium: so called because it resembles the endometrium just before implantation of the blastocyst.

Ptosis: Drooping of the upper eyelid.

Pupil: The round opening at the center of the iris that allows transmission of light to the posterior of the eyeball.

Purkinje cells: Modified cardiac muscle cells specialized for conduction of electric excitation from the atrioventricular node through the ventricular septum and throughout the walls of the ventricle. Purkinje cells are organized into fiber bundles that can be readily dissected from the working myocardium and studied *in vitro* using electrophysiological recording methods.

Purkinje fiber stimulation frequency: The repetition rate for application of brief electric shocks to a Purkinje fiber preparation. A frequency rate of 1 Hz corresponds to 1 shock/s and represents the normal heart rate of 1 beat/s. Frequencies of 0.5 and 2 Hz correspond to bradycardic and tachycardic heart rates.

q1*: The symbol used to denote the 95% upper bound estimate of the linearized slope of the dose–response curve in the low-dose region as determined by the multistage model.

QT interval: The time interval in the electrocardiogram extending from the start of the QRS complex to the end of the T wave and is the measure of the duration of ventricular depolarization and repolarization. The QRS complex corresponds to the upstroke of the cardiac action potential, and the end of the T wave corresponds to the return of the action potential baseline.

Qualification: For pharmaceuticals, the process of acquiring and evaluating data that establishes the biological safety of an individual impurity or degradation product or a given impurity or degradation profile at the level(s) specified.

Qualification threshold: A limit (%) above which a drug impurity or degradation product should be qualified.

Rachischisis: Absence of vertebral arches in limited area (partial rachischisis) or entirely (rachischisis totalis).

REACH: Regulation (EC) No. 1907/2006 of the European Parliament and of the Council concerning the Registration, Evaluation, Authorization and Restriction of Chemicals, known in short as the "REACH," requires manufacturers and importers of chemicals to gather information on the properties of their chemical substances, which will allow their safe handling, and to register the information in a central database in the European Chemicals Agency (ECHA) in Helsinki. The Agency is the central point in the REACH system: it manages the databases necessary to operate the system, coordinates the in-depth evaluation of suspicious chemicals and it builds a public database for chemical hazard information.

Reference dose (RfD): An estimate (with uncertainty spanning perhaps an order of magnitude or more) of the daily exposure to the human population (including sensitive subpopulations) that is likely to be without deleterious effects during a lifetime. The RfD is reported in units of mg of substance/kg body weight/day for oral exposures or mg/substance/m^3 of air breathed for inhalation exposures (RfC).

Regulatory gene: A gene that controls the activity of a structural gene or another regulatory gene. Regulatory genes usually do not undergo transcription into messenger RNA.

Renal hypoplasia: Incomplete development of the kidney.

Reporting threshold: A limit (%) above which a drug impurity or degradation product should be reported.

Reticuloendothelial system (RES): The system composed of all phagocytic cells of the body, including monocytes and tissue macrophages. This system is now more commonly known as the mononuclear phagocyte system.

Resorption: A conceptus that, having implanted in the uterus, subsequently died and is being or has been resorbed.

Resting membrane potential (RMP): In cardiac electrophysiology, RMP is the resting membrane potential in mV and is obtained from the membrane voltage measured immediately before the action potential upstroke. RMP is controlled primarily by inwardly rectifying potassium channels. A decrease in RMP may indicate inhibition of the inward rectifier current.

Reticulocyte: An immature (polychromatic) erythrocyte.

Reticulocytosis: Increased numbers of reticulocytes in the circulation, typically seen in response to regenerative anemia.

Retina: The innermost or nervous tunic of the eye that is derived from the optic cup (the outer layer develops into the complex sensory layer).

Rhinocephaly: A developmental anomaly characterized by the presence of a proboscis-like nose above the eyes, partially or completely fused into one.

Risk: The probability that an adverse effect will occur under a particular condition of exposure.

Risk assessment: The scientific activity of evaluating the toxic properties of a chemical and the conditions of human exposure to it in order both to ascertain the likelihood that exposed humans will be adversely affected and to characterize the nature of the effects they may experience. May contain some or all of the following four steps:

Hazard identification: The determination of whether a particular chemical is or is not causally linked to particular health effect(s)

Dose–response assessment: The determination of the relation between the magnitude of exposure and the probability of occurrence of the health effects in question

Exposure assessment: The determination of the extent of human exposure

Risk characterization: The description of the nature and often the magnitude of human risk, including attendant uncertainty

Risk management: The decision-making process that uses the results of risk assessment to produce a decision about environmental action. Risk management includes the consideration of technical, scientific, social, economic, and political information.

Rudimentary rib: Imperfectly developed riblike structure.

Safety pharmacology: The investigation of the pharmacological effects of a drug other than the desired primary therapeutic effect. Safety pharmacology studies are focused on identifying adverse effects on physiological functions.

Schistoglossia: Cleft tongue.

Sclera: The white tough covering of the eye that, with the cornea, forms the external protective coat of the eye.

Seminiferous epithelium: The normal cellular components within the seminiferous tubule consisting of Sertoli cells, spermatogonia, primary spermatocytes, secondary spermatocytes, and spermatids.

Semiocclusive: The site of dermal application of test substance is covered; however, the movement of air through covering is not restricted.

Sensitization (allergic contact dermatitis): An immunologically mediated cutaneous reaction to a substance.

Septal agenesis: Absence of nasal septum.

Sertoli cells: Cells in the testicular tubules providing support, protection, and nutrition for the spermatids.

Short-term exposure limit (STEL): A time-weighted occupational exposure limit that the ACGIH indicates should not be exceeded any time during the work day. Exposures at the STEL should not be longer than 15 min and should not be repeated more than four times per day. There should be at least 60 min between successive exposures at the STEL.

Sister chromatid exchange: The morphological reflection of an interchange between DNA molecules at homologous loci within a replicating chromosome.

Skin immune system (SIS): Cells associated with the skin that participate in immunity. Includes Langerhans cells, dendritic cells, and keratinocytes. Alternatively known as skin-associated lymphoid tissue (SALT).

Slope factor: See Cancer potency factor (CPF).

Smoke: The airborne state of a chemical that is combusted and allowed to condense into a particulate matter.

SNUR: Significant new use rule.

Somatic cell: A normal diploid cell of an organism as opposed to a germ cell, which is haploid. Most neoplasms are believed to begin when a somatic cell is mutated.

Sorbitol dehydrogenase (SDH): An enzyme of liver origin, whose blood concentration rises in response to hepato-cellular injury.

Spermatocytogenesis: The first stage of spermatogenesis in which spermatogonia develop into spermatocytes and then into spermatids.

Spermiation: The second stage of spermatogenesis in which the spermatids transform into spermatozoa.

Spina bifida: Defect in closure of bony spinal cavity.

Standardized mortality ratio: The number of deaths, either total or cause specific, in a given group expressed as a percentage of the number of deaths that could have been expected if the group has the same age- and sex-specific rates as the general population. Used in epidemiologic studies to adjust mortality rates to a common standard so that comparisons can be made among groups.

STEL: See *Short-term exposure limit (STEL)*.

Stereotypy: The constant repetition of gestures or movements that appear to be excessive or purposeless.

Superficial sloughing: Characterized by dead tissue separated from a living structure. Any outer layer or covering that is shed. Necrosed tissue separated from the living structure.

Surface area scaling factor: The intra- and interspecies scaling factor most commonly used for cancer risk assessment by the US EPA to convert an animal dose to an HED: milligrams per square meter surface area per day. Body surface area is proportional to basal metabolic rate; the ratio of surface area to metabolic rate tends to be constant from one species to another. Since body surface area is approximately proportional to an animal's body weight to the 2/3 power, the scaling factor can be reduced to milligrams per body weight$^{2/3}$.

Sympodia: Fusion of the lower extremities.

Syndactyly: Partially or entirely fused digits.

T cell/T lymphocyte: Lymphocytes that recognize specific antigens via a complex of molecules known collectively as the T cell antigen receptor (TCR). T cells are primarily responsible for the induction and maintenance of cell-mediated immunity (CMI), although they also regulate humoral-mediated immunity (HMI) and some nonimmune effector mechanisms. A variety of T cell subpopulations exist, including helper T cells, cytotoxic T cells, inducer T cells, and regulatory T cells. T cells mature in the thymus.

TC$_{LO}$ (toxic concentration low): The lowest concentration of a substance in air required to cause a toxic effect in some of the exposed population.

TD$_{LO}$ (toxic dose low): The lowest dose of a substance required to cause a toxic effect in some of the exposed population.

Tetralogy of Fallot: An abnormality of the heart that includes pulmonary stenosis, ventricular septal defect, dextroposition of the aorta overriding the ventricular septum and receiving blood from both ventricles, and right ventricular hypertrophy.

Teratozoospermia: Fewer than 30% spermatozoa with normal morphology.

Thoracogastroschisis: Midline fissure in the thorax and abdomen.

Threshold limit value (TLV): The TWA concentration of a substance below which no adverse health effects are expected to occur for workers assuming exposure for 8 h per day, 40 h per week. TLVs are published by the American Conference of Governmental Industrial Hygienists (ACGIH).

Time-weighted average (TWA): An approach to calculating the average exposure over a specified time period.

TLm or TL$_{50}$ (median tolerance limit): In aquatic toxicology, the concentration of material in water at which 50% of the test organisms survive after a specified time of exposure. The TL$_m$ (or TL$_{50}$) is usually expressed as a time-dependent value (e.g., 24 h or 96 h TL$_{50}$).

Tonic convulsion: A convulsion in which muscle contraction is sustained.

Totalis or partialis: Total or partial transposition of viscera (due to incomplete rotation) to the other side of the body; heart most commonly affected (dextrocardia).

Tracheal stenosis: Constriction or narrowing of the tracheal lumen.

Transformation: Typically refers to tissue culture systems where there is conversion of normal cells into cells with altered phenotypes and growth properties. If such cells are shown to produce invasive neoplasms in animals, malignant transformation is considered to have occurred.

Triglycerides: Synthesized primarily in the liver and intestine; the major form of lipid storage.

Ulceration: The development of an inflammatory, often suppurating lesion, on the skin or an internal mucous surface of the body caused by superficial loss of tissue, resulting in necrosis of the tissue.

Ultimate carcinogen: That form of the carcinogen that actually interacts with cellular constituents to cause the neoplastic transformation. The final product of metabolism of the procarcinogen.

Uncertainty factor (UF): One of several, generally 10-fold factors, applied to a NOAEL or a LOAEL to derive a RfD from experimental data. UFs are intended to account for (a) the variation in the sensitivity among the members of the human population, (b) the uncertainty in extrapolating animal data to human, (c) the uncertainty in extrapolating from data obtained in a less-than-lifetime exposure study to chronic exposure, (d) the uncertainty in using a LOAEL rather than a NOAEL for estimating the threshold region, and (e) uncertainty with extrapolation when the database is incomplete.

Unilobular lung: In the rat fetus, a condition in which the right lung consists of one lobe instead of four separate lobes.

Unit cancer risk: A measure of the probability of an individual's developing cancer as a result of exposure to a specified unit ambient concentration. For example, an inhalation unit cancer risk of 3.0×10^{-4} near a point source implies that if 10,000 people breathe a given concentration of a carcinogenic agent (e.g., $1 \mu g/m^3$) for 70 years, 3 of the 10,000 will develop cancer as a result of this exposure. In water, the exposure unit is usually $1 \mu g/L$, whereas in air it is $1 \mu g/m^3$.

UDS: Unscheduled DNA synthesis that occurs at some stage in the cell cycle other than S-phase in response to DNA damage and is usually associated with DNA excision repair.

Upper 95% confidence limit: Assuming random and normal distribution, this is the range of values below which a value will fall 95% of the time.

Upper bound cancer risk assessment: A qualifying statement indicating that the cancer risk estimate is not a true value in that the dose–response modeling used provides a value that is not likely to be an underestimate of the true value. The true value may be lower than the upper bound cancer risk estimate, and it may even be close to zero. This results from the use of a statistical upper confidence limit and from the use of conservative assumptions in deriving the cancer risk estimate.

Urea nitrogen (BUN): The end product of protein catabolism. Blood levels can rise after renal (glomerular) injury.

USP negative control plastic RS: A standardized plastic produced by the USP for use as a control material in some biocompatibility assays.

Vaginal plug: A mass of coagulated semen that forms in the vagina of animals after coitus; also called copulation plug or bouchon vaginal.

Vapor: The airborne state of a chemical that is liquid at room temperature and pressure but is volatile.

Variation: In developmental toxicology, an alteration that represents a retardation in development, a transitory alteration, or a permanent alteration not believed to adversely affect survival, growth, development, or functional competence.

Vitreous: Transparent, colorless, and mass of soft, gelatinous material filling the space in the eyeball posterior to the lens and anterior to the retina.

V_{max}: In cardiac electrophysiology, this is the maximum rate of depolarization measured in Volts/seconds (V/s) obtained by taking the first derivative of the rising phase of the action potential. V_{max} amplitude is determined primarily by the activity of cardiac sodium channels. A decrease in V_{max} indicates sodium channel blockade and corresponds to a broadening of the electrographic QRS interval and, potentially, a slowing of conduction velocity in the intact heart.

Voluntary risk: Risk that an individual has consciously decided to accept.

Whole-body exposure: A system for exposing test animals via inhalation in which they are placed in a chamber in such a way that their entire bodies are exposed directly to the test material.

Xenobiotic: Any substance that is foreign to the immune system; in the context of immunotoxicology, the term generally refers to nonbiological chemicals or drugs.

References

Balls, M., Blaauboer, B., Brusick, D., Frazier, J., Lamb, D., Pemberton, M., Reinhart, C., Roberfroid, M., Rosenkrantz, H., Schmid, B., Spielmann, H., Stammati, A.-L., and Walum, E. (1990) *Report and Recommendations of the CAA/ERGATT Workshop on the Validation of Toxicity Test Procedures*, ALTA 18,313.

Cronin, E. (1980) *Contact Dermatitis*, Churchill Livingstone, New York, Chapters 1–17.

Environ Corporation (1980) *Risk Assessment Guidance Manual*, AlliedSignal, Inc., Morristown, NJ.

Hallenbeck, W.G. and Cunningham, K.M., eds. (1986) *Quantitative Risk Assessment for Environmental and Occupational Health*, Lewis Publishers, Chelsea, MI, Appendix 2.

Klaassen, C.D., Amdur, M.O., and Doull, J., eds. (1991) *Casarett and Doull's Toxicology, The Basic Science of Poisons*, 4th edn., Pergamon Press, New York.

Maronpot, R.R. (1991) *Handbook of Toxicologic Pathology*, Academic Press, San Diego, CA, pp. 127–129.

Marzulli, F.N. and Maibach, H.I., eds. (1977) *Dermatotoxicology*, 2nd edn., Hemisphere Publishing Corporation, Washington, DC.

Middle Atlantic Reproduction and Teratology Association (1989) *A Compilation of Terms Used in Developmental Toxicity Evaluations*.

Morris, W., ed. (1978) *The American Heritage Dictionary of the English Language*, New College edition, Houghton Miffin Company, Boston, MA.

Rand, G.M., ed. (1995) *Fundamentals of Aquatic Toxicology*, 2nd edn., Taylor & Francis Group, Washington, DC.

Stedman's Medical Dictionary, 25th edn., Williams & Wilkins, Baltimore, MD, 1990.

United States Environmental Protection Agency (1984) *Federal Insecticide, Fungicide, Rodenticide Act, Pesticide Assessment Guidelines*, Hazard Evaluation Division, Guidance for Evaluation of Dermal Sensitization.

United States Environmental Protection Agency (1989) *Glossary of Terms Related to Health, Exposure and Risk Assessment*, Air Risk Information Support Center, EPA No.450/3-88/016.

United States Food and Drug Administration (2007) www.fda.gov

Notes